# Beyond Lamarckism

Over the past 20 years, the role of phenotypic plasticity in Darwinian evolution has become a hotly debated topic among biologists and philosophers of science. For instance, in the Extended Evolutionary Synthesis, a new form of evolutionary theory that aims to include processes not taken into account by standard theory (the Modern Synthesis), the question of the remarkable plasticity of living beings is central.

*Beyond Lamarckism: Plasticity in Darwinian Evolution, 1890–1970* shows that the evolutionary impact of plasticity was in fact debated long before the emergence of the current debate on the limits of the Modern Synthesis. The question of how the plasticity of organisms could play a causal role in Darwinian evolution was raised on two separate occasions: first, around 1900, with the emergence of the theory of "organic selection" and, second, during the formation of the Modern Synthesis itself, in the mid-20th century. Out of these reflections came a very large number of concepts, models, and many different terms ("organic selection", "stabilizing selection", "genetic assimilation", "Baldwin effect", etc.), which were often developed independently in various research traditions and empirical contexts. This book also looks at the reasons why these conceptions have been downplayed in the standard understanding of adaptive evolution.

Showing the extraordinary complexity of this history, *Beyond Lamarckism* is aimed at readers interested in evolutionary theory, whether philosophers, biologists, or historians.

**Laurent Loison** is a CNRS researcher in history and philosophy of biology (Paris, France). A major part of his work focuses on the history of the various forms of Lamarckism in biology over the past two centuries.

# History and Philosophy of Biology
Series Editor: Rasmus Grønfeldt Winther is Professor of Humanities at the University of California, Santa Cruz (UCSC).

This series explores significant developments in the life sciences from historical and philosophical perspectives. Historical episodes include Aristotelian biology, Greek and Islamic biology and medicine, Renaissance biology, natural history, Darwinian evolution, Nineteenth-century physiology and cell theory, Twentieth-century genetics, ecology, and systematics, and the biological theories and practices of non-Western perspectives. Philosophical topics include individuality, reductionism and holism, fitness, levels of selection, mechanism and teleology, and the nature-nurture debates, as well as explanation, confirmation, inference, experiment, scientific practice, and models and theories vis-à-vis the biological sciences.

Authors are also invited to inquire into the "and" of this series. How has, does, and will the history of biology impact philosophical understandings of life? How can philosophy help us analyze the historical contingency of, and structural constraints on, scientific knowledge about biological processes and systems? In probing the interweaving of history and philosophy of biology, scholarly investigation could usefully turn to values, power, and potential future uses and abuses of biological knowledge.

The scientific scope of the series includes evolutionary theory, environmental sciences, genomics, molecular biology, systems biology, biotechnology, biomedicine, race and ethnicity, and sex and gender. These areas of the biological sciences are not silos, and tracking their impact on other sciences such as psychology, economics, and sociology, and the behavioral and human sciences more generally, is also within the purview of this series.

**Beyond Lamarckism**
Plasticity in Darwinian Evolution, 1890–1970
*Laurent Loison*

## The Riddle of Organismal Agency
New Historical and Philosophical Reflections
*Edited by Alejandro Fábregas-Tejeda, Jan Baedke, Guido I. Prieto and Gregory Radick*

For more information about this series, please visit: www.routledge.com/History-and-Philosophy-of-Biology/book-series/HAPB

# Beyond Lamarckism

Plasticity in Darwinian Evolution, 1890–1970

**Laurent Loison**

Routledge
Taylor & Francis Group

LONDON AND NEW YORK

First published 2025
by Routledge
4 Park Square, Milton Park, Abingdon, Oxon OX14 4RN

and by Routledge
605 Third Avenue, New York, NY 10158

*Routledge is an imprint of the Taylor & Francis Group, an informa business*

*British Library Cataloguing-in-Publication Data*
A catalogue record for this book is available from the British Library

*Library of Congress Cataloging-in-Publication Data*
Names: Loison, Laurent, author.
Title: Beyond Lamarckism : plasticity in Darwinian evolution, 1890–1970 / Laurent Loison.
Description: Abingdon, Oxon ; New York, NY : Routledge, 2025. | Series: History and philosophy of biology | Includes bibliographical references and index.
Identifiers: LCCN 2024015582 (print) | LCCN 2024015583 (ebook) | ISBN 9781032719689 (hardback) | ISBN 9781032729114 (paperback) | ISBN 9781003422990 (ebook)
Subjects: LCSH: Evolution (Biology)—History. | Phenotypic plasticity.
Classification: LCC QH361 .L74 2025 (print) | LCC QH361 (ebook) | DDC 576.8—dc23/eng/20240605
LC record available at https://lccn.loc.gov/2024015582
LC ebook record available at https://lccn.loc.gov/2024015583

ISBN: 978-1-032-71968-9 (hbk)
ISBN: 978-1-032-72911-4 (pbk)
ISBN: 978-1-003-42299-0 (ebk)

DOI: 10.4324/9781003422990

Typeset in Sabon
by Apex CoVantage, LLC

«Travailler un concept c'est en faire varier l'extension et la compréhension, le généraliser par l'incorporation de traits d'exception, l'exporter hors de sa région d'origine, le prendre comme modèle ou inversement lui chercher un modèle, bref, lui conférer progressivement, par des transformations réglées, la fonction d'une forme.»

—Georges Canguilhem, 1963

# Contents

**PART III**
**(Dis?)Integration into the Modern Synthesis: thinking**
**plasticity genetically**                                              173

# Figures

# Acknowledgments

The work that led to this book began in 2017, and I am indebted to many people, not all of whom I can name here. My greatest debt goes to the biologists with whom I interacted most regularly in exploring some of these issues: Guillaume Achaz, Amaury Lambert, and Arnaud Le Rouzic. I hope these collaborations will lead to others. I have also learned a lot from my exchanges with the biologists in the PlashPhen group (CNRS), and I would like to thank Patricia Gibert, Vincent Debat, Luis-Miguel Chevin, and Tom Van Dooren in particular. On several occasions over the years, I have benefited from Greg Radick's invaluable advice, particularly on the little-known role played by William Thorpe in this history. I would also like to thank Adam Wilkins for his encouragement in bringing this project to fruition. In exploring the work of the Russian school, the help of Sergey Shalimov has been invaluable, and I thank Aleksandra Traykova for translating two important texts, including Schmalhausen's 1941 article on stabilizing selection. Chapter 8 on quantitative genetics has been carefully re-read by my colleague Jean-Baptiste Grodwohl, who has made a number of well-informed criticisms. I would also like to thank Stanislas Leibler for inviting me to a very stimulating seminar on certain aspects of Waddington's experimental work at a time when he, Archishman Raju, and BingKan Xue were grappling with the question of the theoretical significance of this work. A preliminary version of this book was defended in June 2023 as part of my *Habilitation à Diriger des Recherches* (Université Paris 1 Panthéon-Sorbonne). Many thanks to the colleagues who sat on the jury, in addition to Patricia and Arnaud, Anouk Barberousse, Denis Forest, Philippe Huneman, and Stéphane Tirard. In its final form, the book owes a great deal to Emily Herring, who expertly corrected the language. Finally, my warmest thanks to Rasmus Winther for his enthusiasm in welcoming this book into the *History and Philosophy of Biology* collection he directs at Routledge.

# Abbreviations

| | |
|---|---|
| FA | Fluctuating Asymmetry |
| GA1 | Genetic Assimilation Sense 1 |
| GA2 | Genetic Assimilation Sense 2 |
| MS | Modern Synthesis |
| NCT | Niche Construction Theory |
| OS1 | Organic Selection Sense 1 |
| OS1' | Organic Selection Sense 1' |
| OS2 | Organic Selection Sense 2 |

# Introduction

The concept of adaptation is notoriously ambiguous in biology. It refers either to a state (being adapted) or to a process. When we consider the question of adaptation as a process, different phenomena can be envisaged depending on the time scale considered. We speak of individual adaptation when, physiologically, the phenotype accommodates new environmental constraints. For example, the ability to tan in the sun is an adaptation in this sense, that is, a plastic accommodation. But we can also refer to an evolutionary process, taking place over a great number of generations, which results in a fruitful match between the characteristics of a population and the requirements of its environment. From the 1870s to the 1920s,[1] during the golden age of Lamarckian theories, the link between these two adaptation processes was conceived as direct: evolutionary adaptation was seen as the long-term result of the physiological accommodations made by individuals. To this end, many biologists endorsed the idea of the "inheritance of acquired characters": all or part of the results of the individual process of accommodation were supposed to be transferred to the next generation by heredity. Somehow, there had to be a physiological mechanism capable of enabling such cumulative action.

In the 1880s, from August Weismann onwards, neo-Darwinism emerged largely in opposition to neo-Lamarckism. In a strict Darwinian framework, plasticity – the ability to transform one's traits over the course of a lifetime – was no longer seen as the driving force behind evolution but solely as the result of natural selection. In other words, while plasticity was on the side of evolutionary causes for neo-Lamarckian biologists, it became a product of evolution by natural selection for neo-Darwinian biologists (and thereby a secondary topic). While the opposition between Lamarckians and Darwinians constituted one of the major divides in the history of evolutionary theory from Weismann to the consolidation of the Modern Synthesis (MS) in the 1950s and 1960s, there was still room for a lesser-known median perspective, a "third way" between Lamarckism (where plasticity is the main causal factor in evolution) and Darwinism (where plasticity is not a causal factor but only a product of evolution). Indeed, during the period 1890–1970,

DOI: 10.4324/9781003422990-1

several biologists, and sometimes leading biologists, tried to understand how, within a rigorously Darwinian framework (i.e., excluding the possibility of a physiological process of inheritance of acquired characters), plasticity could play a causal role in the evolution of species.

This book is a conceptual history of the various causal relationships that, *Beyond Lamarckism*, were postulated between the adaptive plasticity of organisms and the evolution of species from the late 19th century until the early 1970s. Like any history, it describes a change, in this case, a conceptual change. "Conceptual history" represents one way of conducting studies in the history of science among others – for example, sociological or anthropological approaches (*Science Studies*), cultural history, or material history. While the history discussed in this book focuses primarily on scientific content, it is not an abstract, disembodied, or decontextualized history. Scientists are systematically situated in their intellectual environment, experimental work is detailed in all the complexity of its multiple material dimensions, and the circulation of ideas is highlighted by the close attention given to the different modes of knowledge dissemination throughout countries and languages. However, none of these aspects are studied for themselves, but insofar as they help us to understand the object under investigation: *what happened*, that is to say how biologists thought about the causal relationship. In so doing, the methodology of this work is situated at the intersection of the history and philosophy of science and seeks to follow an already long-standing tradition which, going back to the 19th century and Ernst Mach,[2] still finds a number of representatives today, foremost among whom is Hasok Chang.[3] This is not, therefore, a triumphalist history that seeks to ensure the legitimacy of the present but one that is attentive to moments of opposition, abandoned paths, and ideas that today no longer necessarily have a clearly identified place within biological theory. In short, a critical history of concepts that is aware of the present state of biology but is not standardized by it.

Why focus on a topic which, at first sight, may seem rather narrow and technical in the grand scheme of biological knowledge? The main reason for the present work is that it is precisely the question of the causal relationships between the adaptive plasticity of organisms and their evolution, which is at the core today of the debate on the possibility and the merits of an "Extended Evolutionary Synthesis". For the past 20 years or so, the question of the theoretical framework of evolutionary biology has indeed been debated again by some biologists and philosophers.[4] At the center of this debate was the question of whether the standard framework, that is, that of the MS, should be extended by including a whole set of phenomena which were supposedly ignored when this framework was put in place in the 1940s and 1950s. These include, in particular, niche construction, phenotypic plasticity, epigenetic inheritance, and developmental variation by and large (Evo-Devo). All these processes, and all the mechanisms postulated to account for them, are in fact more or less specific forms of causal relationships between what an organism does during its individual life to conform to the requirements of its environment and the evolutionary posterity of such plastic accommodations. Thus, in the face of the apparent heterogeneity of formulations, it is basically a single question that is being asked in the current debate, the one that is the focus of the present investigation.

## A discontinuous history of concepts

Unlike other histories, this is not one of continuous and progressive deepening of a theoretical furrow. On the contrary, it is an extremely erratic history, which has met numerous dead ends, multiple reconfigurations, and periods of almost complete eclipse. It is, in a word, a fundamentally discontinuous history in which the concepts at stake have been reinvented more than they have been taken up or inherited, hence the extreme profusion of terms to qualify these causal links between adaptability and evolution: from 1896 and Baldwin's "organic selection" to the "coincident selection" of the Russian school; from Hovasse's "postadaptation" to Waddington's "genetic assimilation"; from Schmalhausen's "stabilizing selection" to what Simpson called the "Baldwin effect" in 1953. These are among the dozen terms invented during the first half of the 20th century.

There were in fact two pivotal moments in this tortuous history. The first lasted only about ten years, from some incidental speculations by August Weismann (1894) to a book – so important at the time but so quickly forgotten afterward – by John T. Gulick, enigmatically titled *Evolution: Racial and Habitudinal* (1905). It was during this decade that James M. Baldwin theorized organic selection, in a dialogue that is difficult to reconstruct with the often very similar ideas elaborated at the same time by Conwy Lloyd Morgan and Henry F. Osborn. This decade was the founding moment of what can be called the theory of organic selection. We shall see that the main theoretical questions linked to this evolutionary mechanism were indeed posed at this time, and the debate was from the outset of remarkable complexity.

This momentum came to a halt after the publication of Gulick's book. During the period 1910–1935, it went through a veritable eclipse. The relation between adaptability and adaptation became a blind spot in evolutionary theory, and all the concepts in the vicinity of organic selection were mentioned only occasionally and almost exclusively in textbooks. For almost 25 years, there were no works on this question which had interested the most central figures in evolutionary biology, such as Alfred Russel Wallace.

We will see that it was in the context of Russian-speaking biology that things began to change around 1935–1936. For reasons that remain to be clarified, a number of zoologists took up this question again, mostly in almost total ignorance of the founding debates of the late 1890s and early 1900s. However, this was not a mere reinvention of the previous concepts. There are many reasons for the gap between these two moments, to which I will return in due course. One of them was that genetics now occupied a central position in the life sciences. Another, perhaps even more important, was that this second wave of work no longer consisted solely of a series of speculations: these questions were now incorporated into rigorous experimental and quantitative work. Therefore, there was, at the end of the 1930s, a form of experimental turn, a characteristic that we will also find in the work of the British entomologist William H. Thorpe at the same time.

The two major actors of this period, Ukrainian zoologist Ivan I. Schmalhausen and English embryologist Conrad Waddington, are still well known. Without question, they are the biologists who produced the most developed concepts to account for the causal links between plasticity and evolution. Indeed, Waddington's term "genetic assimilation" continues to be used by evolutionary biologists today. However, the precise position of these thinkers within this history is difficult to place, for although they arrived in the wake of the previous debates, they were at first not fully aware of them and often very explicitly displayed their skepticism toward the mechanism of organic selection. Thus, with Schmalhausen's stabilizing selection and Waddington's genetic assimilation, this history undergoes a new bifurcation which this book aims to map.

To understand what happened after Waddington and Schmalhausen, in the late 1950s and 1960s, we will study the reception of their speculations within the then-dominant theoretical framework, that is, the MS.[5] As the MS was itself in the process of stabilizing, if not "hardening", one could have thought that it was more or less impervious to this kind of speculation. This was the case with some concepts put forth at the time. It was to undermine the importance of organic selection in adaptive evolution that Simpson schematized the broad lines of what was henceforth called the "Baldwin effect".[6] It was also to undermine Waddington's contribution that Mayr preferred to speak of "threshold selection" rather than "genetic assimilation".[7] However, beyond this apparent non-reception, we will see that other conceptual dimensions had a genuine history within the MS and evolutionary genetics.

### Disentangling concepts from the same family

After this quick overview, the complexity of this history is already apparent. This complexity was at least threefold: terminological (how should we name the phenomena?), conceptual (how should we understand their underlying mechanisms?), and experimental (do we really have experimental data attesting to their reality?). None of these questions have been fully answered yet, so their history is particularly uncertain in its outcome, which, in my opinion, while making it more difficult to tell, also gives it a scope that exceeds historical scholarship.

From this point of view, the first ambition of this book is to characterize, as completely as possible, all the concepts brought into play over the 80 years under examination. My main hypothesis is that there was a conceptual dimension to this great diversity of terminology. There have been *different* ways of conceiving the phenomena that start with adaptive plasticity and end with genetic evolution, which are not without overlap, but which are not reducible to each other. In other words, we are neither in a Kuhnian configuration of incommensurability nor in a logicist scheme of possible inter-theoretical reduction. As far as these two classical alternatives can find effective cases in the life sciences, this was not the case here.

What we observe, rather, are two main divergent perspectives on the same phenomenon, which have progressively become individualized in the course of history but never in a clear and total manner, always maintaining areas of overlap. Simply put, these two perspectives were, on the one hand, that of ecology and, on the other, that of embryology. The possible evolutionary impact of adaptive plasticity was first considered from the point of view of its interaction with natural selection. It was in this type of context that the set of concepts that, for convenience, can be grouped under the term "organic selection" was formed. If acquired characteristics are not in themselves inheritable (Lamarckism), they can nevertheless transform the selective context of a population facing altered environmental conditions. Such a perspective was indeed the basis of the seminal debate between Baldwin, Lloyd Morgan, and Osborn. It is particularly evident in Gulick's post-1896 work. It was also the basis of the experimental research carried out in USSR and then in England at the end of the 1930s and the beginning of the 1940s. For all these biologists, it was a question of understanding how plasticity interacts with selection pressures to which a population is subjected and how, under certain conditions, this interaction can lead to the genetic stabilization of an evolutionary adaptation and/or to speciation and phyletic divergence.

This was not the perspective favored by Schmalhausen and Waddington. For them, the important question was not that of linking plastic variation to its ecological consequences but that of the reconfiguration of embryonic development in the wake of what developmental plasticity allowed. In short (but this will of course require much more detailed explanations in the chapters that follow), they substituted the question of the ecological "why?" for that of the embryological "how?" With the gain in fitness, which was the central question of organic selection, now taken for granted (this is particularly evident when examining the experimental designs of Waddington's best-known work on genetic assimilation), they focused on the way natural ("stabilizing") selection is able to reshape the so-called epigenetic landscape (i.e., the developmental phase-space) to stabilize genetically what was initially only a conditional response.

As a result, I believe that there have been two main ways of posing the question of the causal links between adaptive plasticity and evolution. (1) An ecological way, where the causality sought is outside the organism. The organism is then conceived as part of a population interacting with its environment. This was the way of organic selection. (2) An embryological way, where the causality sought is situated inside the organism, as it is itself a totality. This way was that of genetic assimilation. These two major perspectives have each generated a variety of concepts, the contours of which will be gradually outlined in the chapters to come. We shall see that, overall, four major conceptual schemes can be identified; two of them related to organic selection and the other two to genetic assimilation.

The stake of this history is thus unquestionably that of a conceptual clarification, and in this sense the history of science largely joins the philosophy of science. If there is no doubt that concepts are the main point of intersection between science, philosophy, and history, each does not work on them in exactly the same way. For the scientist, concepts are tools whose operativity, that is, their capacity to produce experimental research, is the only thing that counts. For the philosopher, concepts are first of all puzzles to be solved: it is a question of making distinctions that can clarify their content. What is the historian's position vis-à-vis the scientist or the philosopher? There are obviously large areas of overlap, which makes possible and even often fruitful collaborations between science, philosophy, and history. But the historian's gaze is also at odds with that of the scientist and the philosopher because what interests him or her in the first place is to unfold the meanings that, like strata, have accumulated (and sometimes amalgamated) over the course of the history of science. The history of science thus should consider concepts as archives, as collections of meanings nested within each other, which must be reactivated in their specificity. This book can therefore be read as a patient attempt to unfold these multiple layers of meanings. My primary objective is to gain historical depth. That conceptual depth might also be gained is, I hope, equally likely.

### Where to start?

The narrative I propose begins with some speculations by German zoologist August Weismann in 1894,[8] a few months before the seminal papers by Baldwin, Lloyd Morgan, and Osborn (1896) were published. There are two main reasons for this choice: on the one hand, Weismann had clearly envisaged certain conceptual aspects of what Baldwin called "organic selection"; on the other hand, Weismann did indeed serve as a resource (if not a starting point) for certain actors of this founding moment, foremost among them Lloyd Morgan. His presence is therefore justified for both conceptual and historical reasons.

Could it have been otherwise? Would it have been possible to start this history before Weismann? This question comes back to conditions of the logical possibility of a concept. As soon as the said conditions are met, it seems perfectly plausible that the same concept could have been formed several times and independently, even if, of course, slight variations might be observed as a result of the specificities of local intellectual contexts. As we shall see, this is what has happened repeatedly throughout the history of concepts accounting for the causal links between adaptability and evolution. What are the conditions of possibility at stake here? Without going into detail, we can distinguish two main ones. First, it must be conceivable that an organism could transform itself adaptively during its individual life. Such a conception of the properties of living organisms largely predates the debate that was launched in the late1890s, since it was already very explicitly formulated by Lamarck.[9] Second, since acquired characteristics were no longer thought of as hereditary, it was necessary to have at one's disposal the concept of natural selection to account for the process that begins with adaptability and ends with adaptation. These two conditions of possibility being made explicit open a rather significant window: from a logical point of view, "organic selection" (or something close to it) had been thinkable since 1859 and the publication of *The Origin of Species*.

It is therefore quite possible that before Baldwin (1896), and even before Weismann (1894), other naturalists, psychologists, or zoologists may have formulated hypothetical explanations for the origin of certain adaptations that closely resemble organic selection, even if these speculations were to be quickly forgotten and played no role in the way the question was dealt with at the end of the 19th century. Some works in ethology and animal psychology by British ethologist Douglas Alexander Spalding (1841–1877) seem to fit in this category. Spalding died prematurely at the age of 36 and remained little known to his contemporaries and totally absent from the debate on organic selection until the early 1950s.[10] It was in fact J.B.S. Haldane, in a context that remains to be clarified, who exhumed part of his work and republished one of his articles in 1954 in the *British Journal of Animal Behaviour*.[11] The article in question, originally published in 1873, is entitled "Instinct, with Original Observations on Young Animals". As early as 1954, Haldane saw clear similarities between some of Spalding's ideas and the experimental work that Waddington had just done in Edinburgh.[12] After Haldane,[13] it was not until the question of the causal role of plasticity in evolution came back to the forefront in the early 2000s that Spalding's name reappeared on a few occasions,[14] without the question of the precise content of his ideas being pursued. What should we make of this?

The article in question, ten pages long, distinguished itself by the richness of its conceptual analysis of the nature and origin of instincts and by Spalding's willingness to anchor his hypotheses in experimental work, conducted on a few dozen individuals (chickens and turkeys). Spalding was probably the first to highlight what, in the 20th century, would be known as the phenomenon of "imprinting". In this very dense text, it also appears that Spalding adhered to the possibility of the inheritance of acquired characters, whether they be anatomical, morphological, or behavioral.[15] One might therefore think that his Lamarckism rendered unnecessary any recourse to a mechanism of the organic selection type. In fact, it is only on the very last page of his text that Spalding addresses this issue.

Here again, the richness of his conceptualization, consisting of only two paragraphs, is striking. What is of particular interest is that Spalding amalgamates the two perspectives that will progressively tend to individualize themselves over time: the embryological perspective and the ecological perspective. In the first instance, and it is this passage that particularly caught Haldane's attention in 1954, Spalding proposes a thought experiment which, indeed, in its broad outlines at least, is very close to the actual experiments performed by Waddington in the 1950s:

> Suppose a Robinson Crusoe to take, soon after his landing, a couple of parrots, and to teach them to say in very good English, "How do you do, sir?" – that the young of these birds are also taught by Mr. Crusoe and their parents to say, "How do you do, sir?" – and that Mr. Crusoe, having little else to do, sets to work to prove the doctrine of Inherited Association by direct experiment. He continues his teaching, and every year breeds from the birds of the last and previous years that say "How do you do, sir?" most frequently and with the best accent. After a sufficient number of generations his young parrots, continually hearing their parents and a hundred other birds saying "How do you do, sir?" begin to repeat these words so soon that an experiment is needed to decide whether it is by instinct or imitation; and perhaps it is part of both. Eventually, however, the instinct is established.[16]

Most likely, in 1873 Spalding already envisioned three key ingredients of the concept of genetic assimilation: (1) the capacity to produce developmental variation (here in the form of learning); (2) the selection of individuals capable of the most marked developmental variation; (3) the progressive stabilization of the phenotype once a certain undefined "threshold" is reached ("After a sufficient number of generations"). As will be the case with Waddington and Schmalhausen, what is at stake here are not the selective reasons that guide the process (in other words, the gain in fitness linked to the ability to pronounce the phrase "How do you do, sir?" is not immediately considered[17]) but, rather, *such a gain being taken for granted*, the description (here very schematic) of the progressive consequences of the selective regime on embryonic development. But, remarkably, Spalding does not stop at this first thought experiment and immediately considers a second one, this time involving not parrots but turkeys:

> Again, turkeys have an instinctive art of catching flies, which, it is manifest, the creatures in their present shape may have acquired by experience. But suppose the circumstances of their life to change; flies steadily become more abundant, and other kinds of food scarcer: the best fly-catchers are now the fittest to live, and each generation they are naturally selected. This process goes on, experience probably adding to the instinct in ways that we need to attempt to conceive, until a variety or species is produced that feeds on flies alone. To look at, this new bird will differ considerably from its turkey ancestors; for change in food and in habits of life will have affected its physical conformation,

and every useful modification of structure will have been preserved by natural selection.[18]

This time, it is the issue of the *ecological consequences* that interests Spalding: what would happen if environmental conditions were to change drastically? Then, the organisms most able to plastically accommodate this change could survive, which would be the starting point for the selective fixation of the new phenotype. Very explicitly, what Spalding envisages here is the fact that plasticity makes it possible to avoid the extinction of a population ("the best fly-catchers are now the fittest to *live*"), which is precisely the most central aspect in Baldwin's own account and in the long history of the concept of organic selection.

Thus, nearly a quarter of a century before Baldwin, Lloyd Morgan, and Osborn's first publications and more than three-quarters of a century before Waddington's experimental work, we find in the space of a single page in a single article some of the most essential ideas of the concepts of organic selection and genetic assimilation, albeit in a necessarily underdeveloped form. Spalding did consider *both* possible perspectives, that of ecology and that of embryology, and did not distinguish them. His case is therefore all the more remarkable because he was one of the few who did so. Most of the biologists and psychologists interested in organic selection during the period 1890–1950 focused on the question of the links between plasticity and fitness, whereas Waddington and Schmalhausen neglected this issue to focus on the embryonic effects of stabilizing selection.

## Where to end?

Ideally, any history of science should be able to trace its chronological course back to the present to produce the most satisfactory answer possible to the founding question of the historical undertaking: how did we get here? As we shall see, knowledge of the present has been a precious help in problematizing the different episodes marking out a history of almost a century. Nevertheless, our investigation essentially stops at the beginning of the 1970s, that is, around the time of Waddington's death. There are several reasons for this choice.

The first is that the following period, from the early 1970s to the late 1980s, consisted, once again, of an almost total eclipse of the concepts discussed here. In the form Simpson had given it, the "Baldwin effect" was largely marginalized. At the same time, the rise of the concept of the genetic program of development invisibilized the Waddingtonian perspective articulated around the concept of developmental canalization. Research on either of these aspects was therefore extremely rare during this period.

The second reason is the extremely erratic way in which interest has picked up since the late 1980s, but especially since the second half of the

1990s. Indeed, if there is renewed interest in canalization and genetic assimilation (but mostly, for the latter concept, in a strictly descriptive sense often far from Waddington's ideas), it has been from very different perspectives depending on the research contexts. There is no such thing as a homogeneous movement of return to these questions but rather multiple attempts at revival/reformulation/re-elaboration, which may have had very little in common. Thus, simply mapping what has happened over the past 20 years would require a considerable effort that would have made this book far too voluminous to be useful. I hope to be able to come back to this in future works, based on the knowledge of the history reactivated in the next nine chapters.

## Notes

1  Bowler (1983); Loison (2010); Gliboff (2011).
2  Mach (1893).
3  Chang (2012, 2021).
4  For a general treatment, see Pigliucci & Müller (2010).
5  Mayr & Provine (1980).
6  Simpson (1953).
7  See especially Mayr (1963).
8  He developed his speculations in the *Romanes Lectures* he gave in 1894 (Weismann, 1894).
9  Lamarck (1802, 1809).
10  Even today he remains poorly known, but see Gray (1967).
11  It is this version of his text that I use here. The original version appeared in *MacMillan's Magazine* (1873, 27, pp. 282–293).
12  "While parts of Spalding's discussion refer to controversies current eighty years ago, other parts are completely up-to-date. Thus the technique of Crusoe's suggested experiment is quite analogous to that by which Waddington (1953, *Evolution*, 7, 118–126) has evoked a new morphological character in Drosophila melanogaster" (Haldane, 1954, p. 1).
13  Waddington, through Haldane, was himself aware of this article by Spalding. He referred to it directly in his 1957 *The Strategy of the Genes*. What he criticizes then is that Spalding's reasoning is incomplete: according to Waddington, it lacks the concept of canalization to give a satisfactory account of the way in which natural selection ends up producing the phenotype in a constitutive way (Waddington, 1957a, pp. 162–164).
14  Avital & Jablonka (2000, p. 321), Price et al. (2003, p. 1434); Dennett (2003); Pigliucci et al. (2006, p. 2362).
15  Spalding (1873, p. 9).
16  Spalding (1873, p. 11).
17  Nevertheless, it should be noted that in the following sentences Spalding comes very quickly to this question and even exposes a perfectly assumed causal hypothesis capable of accounting for the evolutionary reasons for the *consolidation* of this adaptation: "And if the parrots themselves have acquired a taste for good English the best speakers will be sexually selected, and the instinct will certainly endure to astonish and perplex mankind, though in truth we may as well wonder at the crowing of the cock or the song of the skylark" (Spalding, 1873, p. 11).
18  Spalding (1873, p. 11).

## Part I

# The rise of organic selection

Thinking plasticity ecologically

# 1 Framing the issue from the viewpoint of natural selection

## The confusing birth of organic selection in the pre-Mendelian era

Four years before the so-called rediscovery of Mendel's laws in 1900, "organic selection",[1] as it would come to be called, was also independently elaborated at the same time by three different scholars: American psychologist and philosopher James Mark Baldwin (1861–1934), British ethologist and psychologist Conwy Lloyd Morgan (1852–1936), and American paleontologist Henry Fairfield Osborn (1857–1935). Yet, in contrast to what happened for genetics, it remains unclear what exactly was "discovered" at the beginning of 1896.[2] Though reconstructions of the paths that led these biologists to formulate organic selection are already available,[3] the different issue of *what concepts* were exactly developed around 1900 remains far less investigated. Here, historical work faces at least three cumulative difficulties: (a) the conceptual issue at stake is intrinsically difficult; (b) the theorization occurred before the stabilization of standard Mendelian vocabulary; and (c) Baldwin was especially concerned with making his case for priority, which led him to develop a highly confusing system of partial and biased self-quotations and rewriting of the history that culminated in his puzzling 1902 volume *Development and Evolution*, which is an inextricable mixture of pieces of previously published texts edited and altered into a "new" book.

For the most part, Robert Richards offers a convincing account regarding (c). In chapter 10 of his *Darwin and the Emergence of Evolutionary Theories of Mind and Behavior*, he carefully deconstructs Baldwin's strategy and rhetoric to make clear that it was not until January 1896 that Baldwin was able to conceive organic selection in the sense that partly matches the later "Baldwin effect".[4] Even if the term "organic selection" was already present in the 1895 first edition of Baldwin's *Mental Development in the Child and the Race*, and despite Baldwin's numerous self-quotations of this book from 1896 onward, organic selection did not, in this text, refer to a specific evolutionary mechanism but to a process of intra-individual form of selection supposedly at work during ontogenesis. Given that this aspect has already been clarified by Richards, I will not develop this point further in the present chapter and instead refer the reader to his 1987 book for a more detailed treatment.

Point (b) concerns not only Baldwin but was common to all the scientists involved: the vocabulary fluctuated a great deal, and each of them, at

DOI: 10.4324/9781003422990-3

first, described the phenomenon in his own terms (many neologisms were coined at the time, as we shall see with John T. Gulick's example). Note that only Baldwin proposed a generic name for this new evolutionary mechanism, which was of course in close connection with his priority claim. Special attention is thus needed not to interpret this period's wording anachronistically. In most cases, authors wrote "congenital variations" when they were referring to hereditary variations, an idea rephrased "genetic mutations" in the interwar period. For individual and strictly phenotypic variation, they used different terms like "modifications" or "accommodations" ("plasticity" was not rare). Lloyd Morgan's "modifications" eventually became the most common wording.

It therefore seems that (b) and (c) are difficulties that can be overcome to get a better characterization of (a), that is, how the idea of organic selection was conceptually framed during these years. From the publication of the first articles by Baldwin, Lloyd Morgan, and Osborn (1896) until Gulicks' extensive argument in *Evolution, Racial and Habitudinal* (1905), the definition, causality, scope, and limits of organic selection were a real concern for several major evolutionary biologists, for instance, Alfred Russel Wallace, Edward B. Poulton, or Yves Delage. In contrast to what is sometimes thought and written,[5] organic selection was a significant dimension of the evolutionary debate in Western science around 1900.[6] Why was this so? Why did organic selection suddenly become such an important issue in the debate about the mechanisms of species evolution?

It can be hypothesized that at least four elements played a major role in creating the theoretical context that made this new evolutionary mechanism thinkable and debatable. The first, and, most general, was the impossibility of experimentally demonstrating the inheritance of acquired characters. Indeed, while many Lamarckian biologists were confident of this possibility in the 1870s and 1880s,[7] the accumulation of negative results (or at least the virtual absence of clearly positive results) made the hypothesis more fragile with each passing year. So, if "acquired characters" (i.e., plasticity) were to play an evolutionary role, it could not be the simple, direct one conferred to it by the various Lamarckian theories: in the mid-1890s, it was thought necessary by some scientists to go beyond standard Lamarckism. In a recent work, David Ceccarelli has shown how Baldwin's constant opposition to Lamarckian heredity and Cope's ideas played a decisive role in the genesis of his concept of organic selection.[8]

The second significant piece of context, aptly highlighted by David Depew,[9] is the understanding of natural selection that was dominant at the time. Indeed, in the midst of the "eclipse of Darwinism", natural selection was rarely seen as a creative force in evolution. For many scientists, it consisted only of a negative cleaver eliminating the unfit. This was precisely the understanding of both Baldwin and Lloyd Morgan. For this reason, plasticity was first understood as a means of *counteracting* the eliminatory effect of natural selection, which would otherwise lead to extinction. Plasticity and natural selection were thus seen as opposing causalities, playing against each other, as will be detailed in the second section of this chapter.

Third, the rapid development of ethology at the time brought the concept of plasticity itself to the fore. Indeed, the study of animal behavior offered abundant examples of plastic accommodations to environmental demands. It should be remembered that ethology and psychology were Baldwin and Lloyd Morgan's main disciplines, which is of course not unrelated to the way they developed the model of organic selection. For them, adaptive plasticity was often equated with behavioral learning. It means that they were especially interested in labile characters that can be transformed more than once in an individual lifetime.[10] Nonetheless, I will not restrict their ideas only to this form of plasticity and examine their views as instances of a more general conception regarding the relations between plasticity, natural selection, and hereditary adaptation. I will do so first because they themselves already extended, on several occasions, organic selection far beyond specific cases in which animal behavior might take the leading role.[11] Second, right after the founding years, this extended view quickly became what was at stake and adaptive plasticity was never limited only to learning (see Chapters 2 and 3).

The fourth and last contextual element to consider is probably the most important and yet the one that has so far been ignored in the historiography. This is the long-running debate about the usefulness of the first stages of traits which only become efficient adaptations to a specific function (e.g., wings for flying) when fully developed. This is a foundational debate of Darwinism itself,[12] since Darwin had to respond to such objections (notably as formulated by Mivart[13]) in later editions of *The Origin of Species*.[14] Darwin's answers did not, on the whole, convince biologists active in the period 1860–1900, and the preservation of an adaptation in the early stages of its making thus remained an open and controversial issue. It was precisely this question that Weismann (1894) and Baldwin (1896) sought to answer. For both Weismann and Baldwin, plasticity in a way "secured"[15] (in their own terminology) the smallest congenital variations, which could then be accumulated until they crossed the threshold of definitive usefulness. Even though this line of reasoning suffered from numerous weaknesses (some of which were already highlighted at the time[16]), it is essential, historically speaking, to stress that the genesis of the original concept of organic selection largely took place within the framework of this debate, which accompanied the reception of Darwin's seminal book for decades.

## 1.1   Plasticity, development, and evolution: where Weismann left the issue

German zoologist and cytologist August Weismann (1834–1914) was (and sometimes still is, as I have experienced during informal discussions with colleagues) credited for having formulated in print the very first instance of the new evolutionary mechanism that Baldwin would term "organic selection" in 1896. This idea was already endorsed by Lloyd Morgan and Osborn and, unsurprisingly, strongly contested by Baldwin. The aim of this section is to clarify Weismann's specific contribution to the genesis of the concept of organic selection.

In the 1890s, Weismann continued the fight against Lamarckism begun in the previous decade. His opposition to Lamarckian heredity took him to several fronts, often intimately intertwined. On the one hand, as is well known,[17] Weismann and Spencer clashed over the evolutionary causes leading to the phylogenetic regression of functionless organs. It was on this occasion that, to maintain the "allmacht" of natural selection, Weismann elaborated his theory of germinal selection. On the other hand – and this is where Weismann's theoretical trajectory intersects with the path of organic selection – because of his longstanding opposition to Lamarckian biologists, Weismann was also concerned with the problem of (the origin of) adaptive plasticity. Nineteenth-century Lamarckians, most usually, took for granted that organisms were able to react adaptively to environmental variations: the inheritance of acquired characters was equated with the inheritance of required characters. For instance, the powerful French school elaborated a physiological neo-Lamarckism inspired by Claude Bernard where physiological laws were supposed to explain individual adaptive variations.[18] Adaptive plasticity was thus rarely envisioned in an evolutionary perspective. Weismann could not be satisfied with such conceptions, and, to him, adaptability was not just about physiology but, as a purposeful property, necessarily about evolution and natural selection.

This was his main motivation in giving his 1894 *Romanes Lecture* titled "The Effect of External Influences Upon Development".[19] The whole 69-page essay is a reflection on the evolutionary meaning of adaptive plasticity for organisms, both plants and animals. In short, Weismann thought of two non-exclusive processes that could explain adaptability during ontogenesis. The first was "intra-individual selection" (or "intra-selection"), that is, a selective sorting of entities during embryogenesis, especially cells and "biophors" (the ultimate atoms of living matter in Weismann's framework). This hypothesis was directly linked to his model of germinal selection that would be published the following year, in 1895.[20] Yet, he thought that intra-individual selection could not be the only mechanism at work and that adaptive plasticity needed to also be conceived as an evolved and selected property of living things.[21] In his (highly speculative) view, the two mechanisms, intra-selection and "personal selection" (standard natural selection acting on plasticity itself understood as an evolved adaptation), reinforced each other.[22]

This dual solution to the problem of the origin of adaptive plasticity deserves to be emphasized because around 1900 it remained a rather confused issue in the debate regarding organic selection. Indeed, it is sometimes very difficult to pinpoint whether a biologist understood plasticity as something akin to selection in *present*-day ontological processes (Weismann's "intra-selection") or as a true evolutionary adaptation, that is, a physiological property selected in the *past*. While Weismann was especially clear about the fact that these mechanisms were different solutions, things were often confused after him, with the two explanations being conflated, for instance, in most of Baldwin's writings.

Weismann is sometimes credited for having anticipated, if not properly conceived, organic selection because of a lengthy passage from his *Romanes Lecture*. In his 1896 article entitled "On Modification and Variation",[23] where he distinguished for the first time between non-hereditary accommodations ("modifications") and congenital variations ("variations"), Lloyd Morgan almost entirely based his argument on Weismann's 1894 lecture and more specifically on this long passage that deserves to be integrally cited:

> Let us take the well-known instance of the gradual increase in development of the deer's antlers, in consequence of which the head, in the course of generations, has become more and more heavily loaded. The question has been asked as to how it is possible for the parts of the body which have to support and move this weight to vary simultaneously and harmoniously if there is no such thing as the transmission of the effects of use or disuse, and if the changes have resulted from processes of selection only? This is the question put by Herbert Spencer as to "*co-adaptation*," and the answer is to be found in connection with the process of intra-selection. It is by no means necessary that all the parts concerned – skull, muscles and ligaments of the neck, cervical vertebrae, bones of the fore-limbs, etc. – should simultaneously adapt themselves *by variation of the germ* to the increase in size of the antlers; for in each separate individual the necessary adaptation will be temporarily accomplished by intra-selection – by the struggle of parts – under the trophic influence of functional stimulus.
>
> The improvement of the parts in question, when so acquired, will certainly not be transmitted, but yet the primary variation is not lost. Thus when an advantageous increase in the size of the antlers has taken place, it does not lead to the destruction of the animal in consequence of the other parts being unable to suit themselves to it. All parts of the organism are in a certain degree variable and capable of being determined by the strength and nature of the influences that affect them; and this capacity to respond conformably to functional stimulus must be regarded as the means which make possible the maintenance of a harmonious co-adaptation of parts in the course of the phyletic metamorphosis of a species. Herbert Spencer has given it as his opinion that in the harmonious working together of parts a cogent reason is to be found for accepting the doctrine of the transmission of acquired characters: but in so doing he has overlooked the fact that there is a never-resting principle at work which is uninterruptedly concerned with the production of harmony, alike in respect of size and functional activity, among parts that co-operate: I mean the principle of intra-selection.

Naturally the degree of discord among the parts may sometimes be such that intra-selection is not able to produce harmony; for there must be definite limits to the scope of adaptation, and we well know that the exercise of a function for too long a time or too violently ceases to produce strengthening of the organ, and causes weakening instead. But as the primary variations in the phyletic metamorphosis occurred little by little, the secondary adaptations would probably as a rule be gained till, in the course of generations, by constant selection of those germs the primary constituents of which are best suited to one another, the greatest possible degree of harmony may be reached, and consequently a definitive metamorphosis of the species involving all the parts of the individual may occur.[24]

First, it must be stated that Weismann himself never claimed he should be acknowledged as the discoverer of organic selection. This deserves mentioning but in no way determines whether he did or not in fact come up with organic selection before 1896. It only offers us an interesting distinction between Weismann's intention, his own appreciation of what he was doing, and what he really did and published.

On a strictly conceptual level, it remains that, in the previous detailed passage, though expressed in a scattered manner, Weismann formulated most of the ideas that can be read in Baldwin's, Lloyd Morgan's, and Osborn's seminal papers, and crucially that individual accommodations, while not heritable by themselves, secure germinal variations for subsequent phyletic evolution. This specific notion was the cornerstone of the concept of organic selection as it was first conceived around 1900 (see Section 1.2) and has remained central ever since: plasticity is above all about intergenerational survival in a changing selective context, a phenomenon thought to buy time for congenital variations to arise and then be selected.

Baldwin, and to a lesser extent Richards, contested Weismann's priority in the discovery of organic selection on the grounds that, in Weismann's framework, "acquired variations would follow congenital variations instead of leading the way for them".[25] This argument is highly interesting because it points to a difficulty in defining organic selection (and later the Baldwin effect). In Weismann's conception, the driving change – the "advantageous increase in the size of the antlers" in his example – is not environmental but comes from the inside: it is a fortuitous congenital modification of the germ plasm that is initially accommodated during embryogenesis until natural selection catches up and hereditarily stabilizes and optimizes the new conformation.[26] Such a conception can also be found in Lloyd Morgan's formulation, directly inspired by Weismann's ideas:

According to [Weismann's] conception, variations of germinal origin occur from time to time. By its innate plasticity the several parts of an organism implicated by their association with the varying part are

modified in individual life in such a way that their modifications coop-
erate with the germinal variation in producing an adaptation of double origin, partly congenital, partly acquired. The organism then waits,
so to speak, for a further congenital variation, when a like process of
adaptation again occurs; and thus race-progress is effected by a series
of successive variational steps, assisted by a series of cooperating individual modifications.

If now it would be shown that, although on selectionist principles
there is no transmission of modification due to individual plasticity,
yet these modifications afford the conditions under which variations
of like nature are afforded an opportunity of occurring and of making
themselves felt in race-progress, a further step would be taken towards
a reconciliation of opposing views. Such it appears to me, may well be
the case.[27]

Thereby, from the start, there was a substantial disagreement (or at least
a plurality of viewpoints) about what drives developmental and evolutionary change: for Baldwin and also for Osborn,[28] the impetus comes from an
environmental perturbation, whereas for Weismann and Lloyd Morgan, the
initial perturbation is, or at least might be, an internal modification of the
hereditary substance. This divergence has never been fully resolved. Indeed,
although the standard formulation of the Baldwin effect puts forward a
causality that begins with an environmental perturbation, other theorizations have promoted a complete symmetry between the "outside" and the
"inside", between plastically accommodating an environmental variation or
a hereditary variation. This is the case, for example, of the model of "genetic
accommodation" recently devised by Mary-Jane West Eberhard where the
novel input could be either environmental or mutational.[29]

While Weismann did not propose that such an evolutionary process
could start with an environmental variation, he did identify – although
allusively and with very little detail – the role that plasticity could play in
the conservation of certain incipient congenital variations: the variations
in question are initially preserved in the population because the individuals who carry them, thanks to their developmental plasticity, are able to
survive. In this way, Weismann, in an admittedly rather vague terminology,
nonetheless, formulated a central aspect of the mechanism of organic selection: *plasticity as a means of withdrawing certain hereditary variations
from the blade of natural selection, until such time as they have reached a
certain threshold.* In this sense, the path he mapped out in his 1894 essay
is indeed congruent with the fundamental logic of organic selection, which
will be detailed in the next section. It therefore seems well-founded, both
conceptually (the content of Weismann's speculative theses) and historically (the way these theses were understood by Lloyd Morgan), to begin
this history with his *Romanes Lecture* "The Effect of External Influences
upon Development".

## 1.2   Blocking natural selection: avoiding extinction until evolutionary adaptation

Baldwin's highly developed two-part article published in *The American Naturalist* in June 1896, a few months after the first presentations of the mechanism of "organic selection", is often considered to set out the first consistent formulation of this "new factor in evolution" (the title given by Baldwin to his article). Yet this long text (29 pages) is a very difficult read, and it is by no means easy to find a non-ambiguous formulation of anything resembling the Baldwin effect. For example, it seems that Baldwin continued to use the expression "organic selection" in an extremely ambiguous sense, not necessarily linked to an evolutionary mechanism based on phenotypic plasticity.[30]

So, to give a true picture of what Baldwin, Lloyd Morgan, Osborn, and others conceived as a new evolutionary mechanism, we need to focus on the overall content of their publications at the time, without expecting complete coherence: here and there, we will necessarily find statements that are partly contradictory. Nonetheless, a conception emerges in which plasticity was fundamentally understood as a property enabling the suspension of natural selection following a significant environmental change. *To block the action of natural selection* (Organic Selection Sense 1, thereafter OS1) was the first and most explicitly acknowledged role ascribed to individual "modifications", and it remained the most central for decades, as we shall see in the next two chapters. Plasticity was supposed to avoid extinction in altered environmental conditions. When one compares most of the various formulations of organic selection that were given in the late 19th century by Lloyd Morgan, Osborn, and Baldwin, this is their more significant commonality. For instance, in his first article on the subject, Lloyd Morgan writes: "This individual plasticity is undoubtedly of great advantage in race progress. The adapted individual will *escape elimination* in the life-struggle."[31]

This aspect was especially central to Baldwin, who always strongly emphasized that plastic organisms can "stand the 'storm and stress' of the physical influences of the environment" and thus are "kept alive",[32] which "prevents the incidence of natural selection".[33] As I have already stressed in the introduction of the present chapter, this emphasis on the fact that plasticity is key in blocking natural selection was motivated by the then prevalent understanding of natural selection: after Darwin and before the rise of the Modern Synthesis, natural selection was for Baldwin, Lloyd Morgan, and many others during the so-called eclipse of Darwinism an exclusively negative force, able to destroy maladaptive variants but unable of any creative action of its own. Baldwin explicitly expanded on this aspect in most of his writings devoted to the organic selection, representing natural selection as a "negative law"[34] or a "negative principle".[35]

Since, for Baldwin, adaptive plasticity amounts to buying time,[36] species have "all the time necessary to get the variations required for the full instinctive performance of the function".[37] This is why the causal characterization of the Baldwin effect recently proposed by Godfrey-Smith as a "breathing space" is historically grounded and meaningful:[38] all the actors involved around 1900 explicitly supported the idea that plasticity is first about preventing natural selection to operate in new environmental conditions and then to avoid extinction (OS1).

But if plasticity accommodates environmental change, what drives evolution? Why, then, does the process not stop there? Why does plastic variation gradually give way to congenital variation? This is in fact one of the major challenges faced by the concept of organic selection (and by the Baldwin effect) throughout the 20th century right up to the present day. It is striking to note that this theoretical problem was addressed from the very beginning. In France (see Chapter 2), Sorbonne professor, zoologist, and prominent figure of the period, Yves Delage, was the founder of the journal *L'Année biologique*, which compiled most of the significant work published in Western science. In 1899, in collaboration with Georges Poirault, the journal's editorial assistant, he published a short review that expressed skepticism about the scope and significance of organic selection precisely because plastic accommodations, in suppressing any fitness differences between individuals, make natural selection ineffective.[39] Baldwin was aware of these objections. In *Development and Evolution*, he rephrased the theoretical problem as follows:

> Such a criticism takes the form of the question as to the further utility of congenital variations, especially those of the coincident tendency, when by the use of accommodations, the individuals can already cope with the environment. Put generally, this criticism would read: does not the theory of organic selection, by showing that accommodation does supplement imperfect organs and functions, make it unnecessary that variation and natural selection should be further operative? This leads, it would appear, to the extreme position of the organicists, as is illustrated by the quotation made from Pfeffer on an earlier page. It seems to be also the opinion of Delage (See the *Année Biologique*, III, 1899, p. 512).

Baldwin was by no means comfortable with Delage's criticism, which points to a genuine theoretical difficulty within the framework of organic selection. He mostly remained unclear about the conditions that would favor the replacement of a modifiable phenotype by an inflexible phenotype and addressed the issue only incidentally on a few occasions (as Robinson and Dukas proposed,[40] Baldwin's difficult positioning might also explain why he sometimes suggested that organic selection could lead to the maintenance or even the increase of developmental plasticity). To Delage's challenge, Baldwin answered that, at least in some cases, congenital variations could be the basis of a much more refined adaptation, that is, that individual accommodation through plasticity was too limited a process (except, maybe, for learning). In the end, natural selection "would give the gradual shifting of the congenital mean toward the *full* endowment".[41] In modern terms, the fitness peak could not be reached by plastic variations alone, and, therefore, the standard cumulative effect of natural selection could take place to fix the final (and more refined) adaptive trait.[42]

The non-attainment of the adaptive optimum by purely plastic means, only briefly and marginally sketched in Baldwin's writings, would be more explicitly expanded and assumed by others in later times. As one could expect, this was the case decades later, when organic selection became a topic of renewed interest within evolutionary theory during the rise of the Modern Synthesis (Chapter 3). In a different theoretical context, it was also the main aspect of Waddington's complex notion of the "tuning" of canalization, as we shall see in Chapter 6. But already in the early 1900s, that is, just after the publication of the seminal papers that we examined, Gulick, a neglected figure in the historiography of organic selection, made a remarkable (but still overlooked) contribution to the explicit characterization of the causal steps that lead to the final hereditary fixation of plastic accommodations (see Section 1.4).

### 1.3    Polarizing selection: creating new selective pressures

For several contributors of the 2003 edited volume *Evolution and Learning, The Baldwin Effect Reconsidered*, for instance, Godfrey-Smith or Dennett, there is a conceptual distinction to be made between what remains the standard account of organic selection and what the volume presents as a different, updated understanding rooted in the modern conception of Niche Construction Theory. Godfrey-Smith opposes "Baldwin's mechanism", that is, the breathing space conception (OS1), to "Deacon's mechanism",[43] the latter being characterized as the transformation of the "social ecology of the population"[44] that will selectively favor specific genetic variants. In other words, *the plastic transformations of a population facing new environmental challenges significantly alter the selective pressures to such an extent that it defines a new optimum in a new adaptive landscape* (Organic Selection Sense 2, thereafter OS2). For Godfrey-Smith, the main causal connection between adaptive plasticity and adaptation has to be looked for in such an active transformation of the selective regime according to a niche construction perspective.[45]

Such a distinction between blocking and polarizing natural selection represents the main investigative pathway of the first part of the present book. What I disagree is the chronology of its inception within biology: do we really have to wait until the 1980s and 1990s for the second account (OS2) to be conceived? In my reading, the idea that plastic accommodations transform selection pressures actually predated this period by several decades and can even be traced back to some of Lloyd Morgan's and Baldwin's writings, before John T. Gulick gave it a considerable refinement already in 1905. Other evident instances of OS2 were also common in the second phase of the debate when organic selection once again became a topic of interest in evolutionary theory from the late 1930s onward. In the next chapter, I will detail what happened in France in the 1930s and 1940s and will pay special attention to Raymond Hovasse's case. In his main book on the subject, he did not hesitate to postulate that the plastic response "polarized natural selection".[46]

At first sight, it might appear doubtful to find such clear-cut formulations half a century earlier. Yet, as documented in the following section, in a neglected book from 1905, Gulick went very far in conceptualizing OS2, much more indeed than Lloyd Morgan, Osborn, or even Baldwin. Among the three "co-discoverers" of organic selection, Osborn, who was the least interested in this evolutionary explanation, seems to not have conceived anything close to the idea that plasticity polarizes natural selection into a new adaptive niche. In Lloyd Morgan's writings, and as early as 1896, there are some sentences that could be interpreted as going in that direction but that remain equivocal and underdeveloped. For example, at the end of "On Modification and Variation", he wrote:

> It is here suggested that the modification as such is not inherited, but is *the condition under which congenital variations are favored* and given time to get a hold on the organism, and are thus enabled by degrees to reach the fully adaptive level.[47]

For Baldwin, the case is, unsurprisingly, more complex. Most probably, Baldwin already understood the dual causal role of the plastic response in relation to natural selection, even if, again, his formulations are often equivocal. Baldwin's writings are indeed filled with ambiguous statements about this issue. On the one hand, he argued that (behavioral) plasticity might provide organisms with a buffer against natural selection (OS1), but, on the other, he also emphasized that the acquired modifications pave the way for natural selection (OS2). He often repeated that accommodations "determine"[48] the direction of evolution. In what sense precisely did he understand the term "determine"? One thing he made clear was that congenital variations remain "fortuitous in the strict sense",[49] thus "determine" does not imply a Lamarckian understanding, like, for instance, in Osborn's account. If some paragraphs remain difficult to read, others, less frequent, are not. For example, as soon as April 1897, he phrased the issue in remarkably modern statistical terms:

> It follows also that the likelihood of the occurrence of coincident variations will be greatly increased with each generation, under this "screening" influence of modification [plasticity]; for the mean of the congenital variations will be shifted in the direction of the adaptive modification [plastic response], seeing that under the operation of natural selection upon each preceding generation variations which are not coincident tend to be eliminated.[50]

For Baldwin, plastic variation was not conceived as the first stage (in an embryological sense) of hereditary variation (as would be the case with Waddington, for example). Therefore, for him, the term "determine" most likely meant that there was an ecological or functional (and not structural) causal connection between provisoire adaptability and final adaptation: accommodations "determine" the direction of evolution inasmuch they increase the likelihood of new congenital variation to be selected because of the setting up of new ecology. This is how the previous quotation can be interpreted. The idea that plasticity actively alters the selective regime is also the most well-founded interpretation of the following quote, in which Baldwin compares organic selection to a special kind of artificial "self-selection":[51]

> But any influence, such as the individual's own accommodation to his environment, which is important enough to keep him and his like alive [OS1], while others go under in the struggle for existence, may be considered with reason as a real cause in producing just such effects. Thus by the processes of accommodation, a weapon *analogous to artificial selection* is put into the hands of the organism itself, and the species profits by it [OS2].[52]

Other arguments support such a reading. Godfrey-Smith argues that "Deacon's mechanism requires that the facultatively acquired behaviors become *entrenched in the social life of the population*".[53] This is exactly the reason why Baldwin always puts emphasis on his concept of "social heredity", a component of organic selection that is missing in Osborn's account and less developed in Morgan's characterization. Paul Griffiths, another contributor to the 2003 edited volume, acknowledges Baldwin's originality on this point and even proposes that "the influence of social heredity on natural selection [operates] via the process we call niche construction".[54] According to Griffiths, "Baldwin's account of social heredity was a theory of what would today be called 'niche-construction'".[55] My only disagreement here regards the relationship between organic selection and social heredity. Griffiths seems to suggest that, for Baldwin, the mechanism of organic selection was in a sense alien to social heredity,[56] whereas it seems to me that Baldwin conceived them as self-reinforcing processes.[57]

For Baldwin, organic selection was one component of a larger theory he called his "theory of orthoplasy", which also encompassed social heredity. Baldwin opposed two forms of heredity, "physical" and "social". Physical heredity was standard biological heredity, that is, intergenerational similarities based on the transmission of a material substance, whatever it may be (Weismann's germ plasm, Mendelian genes, etc.). Yet, as a psychologist, Baldwin also paid special attention to imitative learning that allows the social transmission of behavioral characters. This form of social transmission among higher animals, and especially humans, is what he called social heredity. In a new environment, it is behavior that first accommodates change.

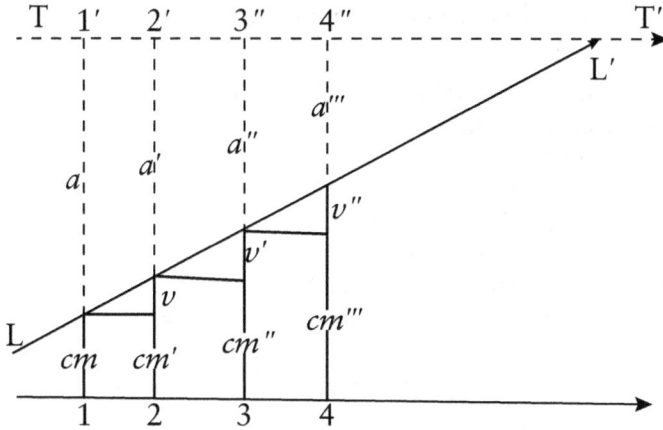

Theory of Orthoplasy. LL', line of evolution; 1, 2, etc., successive generations
by physical heredity; TT', line of tradition (social transmission); *cm*, *cm'*, etc.,
congenital mean; *a, a'*, etc., accommodations (and modifications) supplementing
or screening *cm*, etc. *v, v'*, etc., (congenital) variations added to *cm*, etc.,
by natural selection. The species is kept alive by *a, a'*, etc., and TT', during
the evolution of *cm*. The line TT', considered as 'tradition,' is of varying
importance according to the character in question and to the grade of the organism
in the scale of life; but it signify any utility for which the accommodations
are necessary, it is always present, and may be called the 'line of utility.'

*Figure 1.1 Baldwin diagrammatic representation of his "Theory of Orthoplasy".* As
a general evolutionary mechanism, this diagram shows that organic selection was supposed to lead to a complete hereditary fixation of the initial
plastic response (see Baldwin's explanations below the diagram). Reproduced from: J.M. Baldwin, *Development and Evolution*, New York, The
MacMillan Company, 1902.

If this new behavioral pattern is transmitted from one generation to the next
by "tradition", then it allows time for "congenital variations" that follow
the same direction to progressively reinforce the adaptive behavior, until it
becomes completely instinctive and fixed. In Figure 1.1,[58] individual plasticity
for the trait of interest decreases from the left to the right, until stage *L'* when
the evolutionary process is completed.

It seems, therefore, that the conceptual tension, so difficult to resolve,
between two kinds of relationship between adaptive plasticity and natural
selection (plasticity as a force *opposing* natural selection vs. plasticity as a force
*guiding* natural selection) was already one of the major elements in Baldwin's
thinking, much more so than what can be found at the same time in Lloyd
Morgan's or Osborn's articles and books. This higher level of complexity probably also explains why Baldwin's writings are often less clear than those of his
contemporaries: he was probably the first to foresee this conceptual difficulty,
which would continue to frame many debates throughout the 20th century.

## 1.4   More on polarizing selection: John T. Gulick's neglected contribution

American naturalist and missionary John Thomas Gulick (1832–1923) remains an intriguing figure in the early history of evolutionary theory. His contribution to the theoretical understanding of speciation was hotly debated in the period 1870–1910 by the most central actors of the time, including Darwin, Wallace, and Romanes. His work was then marginalized during the decades leading up to the consolidation of the Modern Synthesis, which is when part of his thinking re-emerged. Mayr and Wright especially saw in Gulick one of the very first to design concepts close to the founder effect and genetic drift, concepts that acknowledge the role of chance in speciation and evolution,[59] even though they were formulated in a pre-Mendelian terminology.

Despite Mayr's endorsement,[60] and despite Romanes openly aligning himself with Gulick's views on speciation in the 1880s and 1890s,[61] Gulick was for a long time almost entirely ignored in the historiography. Indeed, it was not until 1975 that the first comprehensive study of his ideas about "segregation", reproductive isolation, and speciation was published.[62] And then, it was only in 2006, that a two-part article on Gulick's theory was published by Brian K. Hall.[63] In comparison to Lesch's 1975 work, Hall's is not only about "cumulative segregation and geographical isolation" (Part I) but also about "coincident or ontogenetic selection" (Part II). According to Hall, "in 1872, [. . .] 24 years before Baldwin, Morgan and Osborn proposed their theories of organic selection, Gulick developed a theory of *coincident or ontogenetic selection*".[64] If this were indeed the case, Gulick would be a genuine forerunner of the concept of organic selection, given that Baldwin, Morgan, and Osborn, as well as standard historiography on the subject, simply ignored his work.[65] Was Gulick the Mendel of organic selection?

I do not share this reading. Even though, by 1905, Gulick did indeed go much further than his contemporaries in conceptualizing OS2, that is, in emphasizing organisms' agency in creating a new selective regime, he did so because he benefited from what had been done in the period 1896–1902. In other words, in my view, it is too much to credit Gulick for conceptualizing organic selection as soon as 1872, or even in the 1880s or 1890s, as Hall does. No doubt, in the late 19th century, Gulick had many tools at hand to do so, and he even came close on a couple of occasions, as we shall see. However, he did not connect plasticity, natural selection, and hereditary variation in the way required within the causal hypothesis of organic selection. What he did do was to conceive a general framework within which speciation could perfectly occur without the need for natural selection to act as a diversifying factor in distinct environments. He might have been one of the first to assume, in a strictly Darwinian perspective, that what was evolutionary relevant was not the extrinsic environment per se but more precisely the way an organism interacts with it. Such an idea was already present in his first detailed synthesis on speciation, in 1888.[66] Nonetheless, it was only in 1905 that Gulick connected organism's agency with organic selection.

During his entire life, even if his scientific work was regularly interrupted or at least delayed because of his involvement as a missionary in Japan and in China,[67] Gulick focused on one single problem: why is evolution divergent, that is, why does it produce several distinct lines from a common origin? The answer to this puzzle, for Gulick, could not be found in the standard Darwinian account. To put it briefly, in his view (shared by others at the time, including Romanes[68]), natural selection is an explanation of organic adaptation, not of diversification and speciation. To him, one major principle was at the basis of divergent evolution: the principle of "segregate breeding" or "segregation".[69] His lifetime endeavor was to elucidate the various causal factors able to produce segregation and, in the end, reproductive isolation. It was exactly the same motive that made Romanes introduce his own concept of "physiological selection". According to Romanes too, Darwin did not say much about the origin of species as such, because the way isolation is produced was simply ignored in Darwin's framework. Yet, in contrast to Romanes, Gulick was able to ground his thinking on an impressive empirical basis.

Gulick was born and raised in Hawaii, where very early he became highly interested in natural history in general, and in collecting land snails in particular.[70] Despite health issues, he spent a considerable amount of time collecting snails on various islands of the archipelago. It has been estimated that Gulick's collection counted around no less than 44,500 snail shells[71] and that his intense collecting activity might have participated in the extinction of several endemic species. He was especially interested in the tree-living snails of the family Achatinellidae that lived in the forests of island Oahu, which were among the most brilliantly and variably colored achatinellids. Gulick identified approximately 800 varieties in the several valleys of the island. These land snails are exceptional in that an individual usually lives its entire life on a single tree: *Achatinellidae* populations have a very low power of migration.[72] Thus, in studying varieties and species from different valleys, Gulick studied a remarkable form of diversifying evolution in natural conditions. The most stimulating aspect to him was the fact that each valley offered more or less the same environment to snail populations: the same temperatures, predators (birds), trees, amount of sunlight, and so on (Figure 1.2). And, despite uniform environmental conditions, he observed a stunning evolutionary radiation. As early as 1872, he doubted that natural selection could be responsible for such "divergent evolution".[73]

On this empirical basis, he elaborated a complex and evolving framework which aimed to explain what he termed "polytypic evolution"[74] (as opposed to "monotypic evolution", later labeled anagenesis). I will not enter his entire sophisticated framework here. Gulick developed it over decades and coined dozens of terms that mostly did not live beyond him (e.g., "intensive segregation", "filio-parental election", "impregnational method", "environal partition", and "dominational selection"). All of these concepts made his explanatory system especially complex to understand, as has already been pointed out.[75] Note that this must have been the case already in the late 19th

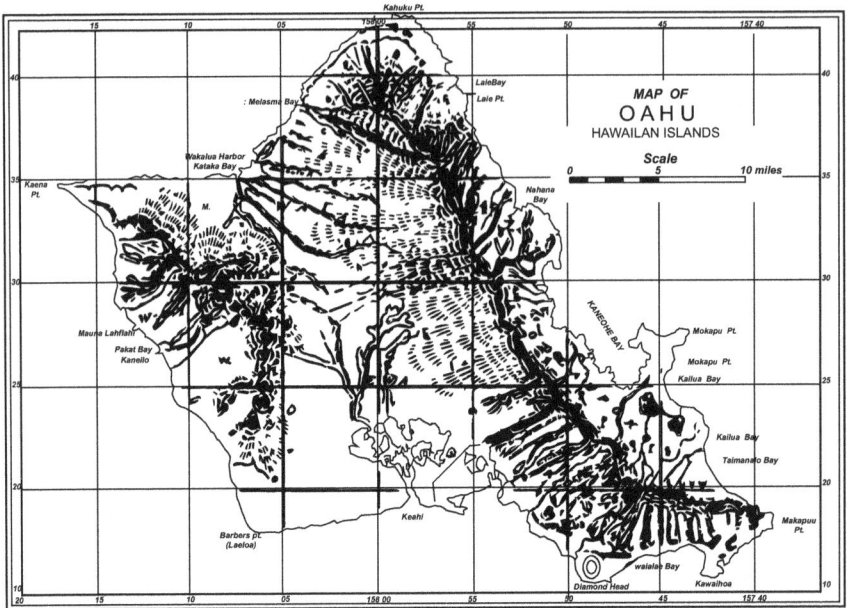

*Figure 1.2 A map of Oahu published by Gulick.* The island consists of several valleys, with most of them sharing the same orientation. Reproduced from: J.T. Gulick, *Evolution, Racial and Habitudinal,* Washington D.C., Carnegie Institution of Washington, 1905.

century, which also might explain why Gulick was progressively forgotten during the first decades of the 20th century. Overall, Gulick's conception of divergent evolution emphasized the causal role of what would later be termed allopatric or geographic isolation.[76] As Rundell puts it, Gulick "was among the first to describe such patterns of allopatric ecologically similar species, and he used them to assert a potential role for geographical isolation in speciation".[77]

Here, I shall focus on only one aspect of Gulick's theory of segregation, namely, his idea that the way an organism interacts with its environment is key in polarizing the selective pressures that will direct its subsequent evolution. This was only a limited part of Gulick's repertoire to explain divergent evolution. Other factors were considered, that, to some extent, predated the founding effect and genetic drift.[78] Hall's interpretation that Gulick also anticipated organic selection for decades rests on the fact that, very early on, Gulick emphasized this neglected theoretical dimension in the working of natural selection. To avoid misinterpretation, it is necessary to emphasize how Gulick's ideas on the link between plasticity and evolution changed throughout his career.

First, in organic selection, whether OS1 or OS2, what paves the way for evolutionary change is phenotypic plasticity. In one way or another, plasticity accommodates (OS1) or produces (OS2) a new relationship between the organism and its environment. In short, phenotypic change *starts* genetic evolution. In Gulick's understanding before 1905, that is, before he published his only book on the evolutionary process, the role he ascribed to plasticity was both rather limited and ambiguous. In his 1872–1873 papers, the issue is simply completely ignored.[79] At that time, Gulick only focused on geographic isolation and accidental events to explain divergent segregation and subsequent isolation and speciation. In a second series of works, in the late 1880s and early 1890s, he again put the emphasis on the need to prevent "free crossing" for speciation to occur.[80] In his 1888 synthesis "Divergent Evolution through Cumulative Segregation", there are passages that indeed bring organic selection to mind. Yet, a significant difficulty is that it is hard to decide whether, in his view, the various resources of a single environment were differently used because of plasticity (as in organic selection) or because of some initial hereditary heterogeneity within a single population. Most of the time at least, it seems that the second reading must be favored:

> For the production of Industrial Segregation it is necessary that there should be, in the same environment, a diversity of fully and of approximately available resources more or less separated from each other, and in the organism some diversity of adaptation to these resources, accompanied by powers of search and of discrimination, by which it is able to find the resources for which it is best fitted and to adhere to the same when found.[81]

Here, nascent adaptive radiation might be the direct consequence of standard natural selection acting on pre-existing variants, and it seems problematic to credit Gulick for anticipation of organic selection on such a tenuous argument. In 1891, in another important paper titled "Intensive Segregation, or Divergence through Independent Transformation", Gulick made another step toward OS2 but the same ambiguity prevented him from clearly ascribing a causal role to phenotypic plasticity. He then contrasted two forms of selection, "passive" and "active".[82] Passive natural selection is the standard Darwinian formulation in which the selective regime is acted upon organisms, whereas active natural selection is the consequence of the ecology actively produced by organisms. What remains equivocal is that he did not see the "diversity in the uses to which different sections of one species put their powers, when appropriating resources from the same environment" as originating in plastic accommodations but rather in "some variation in the powers of the organism", which rather seems to designate, in Gulick's wording, some kind of innate differences.[83]

In comparison to his ambiguous statements of the 1880s and 1890s, his 1905 formulations are crystal clear and endorse a complete and explicit appropriation of the mechanism of organic selection. Gulick went indeed much further than Baldwin and Lloyd Morgan in emphasizing the causal role of organismal agency in evolution, and it makes perfect sense to consider that, in the early 20th century, it was Gulick who elaborated the most refined concept of organic selection. At least four aspects of Gulick's account are worth noting.

(1) First, unsurprisingly, Gulick's conceptualization was mostly about OS2. Causal initiative was on the side of plastic organisms, able to create a new way to interact with the environment, and as such to produce new selective pressures. He termed this mode of selection, in which the selective regime is decided by the way of life "chosen" by an organism,[84] "endonomic selection" or "active selection" (see the previous text). Gulick goes so far in that direction that on some occasions his wording is almost equivalent to our modern Niche Construction Theory:

> *By means of accommodation birds and beasts determine their relations to the environment, and so determine the kind of selection to which they and their descendants are thereafter subjected.* The choice of conditions that have brought pleasant experiences and the avoidance of those that have brought unpleasant experiences is a degree of intelligence enjoyed by many species of animals and is a form of accommodation that may determine the mode of life and so determine the forms of selection.[85]

(2) Second, Gulick was perhaps the only one at that time to fully understand the dual causal role played by plasticity. Plastic accommodations could be about blocking natural selection (OS1) or polarizing natural selection (OS2): "*not only are previous forms of environal selection brought to an end by accommodation, but certain forms of reflexive selection may be thus made to cease, and perhaps other forms introduced*".[86] In the first case, the new selective pressures are extrinsic constraints accommodated by plastic abilities to avoid extinction, whereas, in the second case, it is plasticity itself that generates the new ecology. "Passive" natural selection concerned the first case, when "accommodational powers of the individuals, preserving the organism from extinction under the stress of great and sudden change [give] time for the production and accumulation of variations that coincide with the accommodation in adapting the organism to the new conditions".[87] In contrast, "active" (or endonomic) selection rests upon "alternative methods of adjustments of the same environment *till the organism has adopted a particular method of suiting itself to its conditions*".[88]

(3) Third, Gulick, like Baldwin (but again much more comprehensively), understood that organic selection is fully effective to the extent that the adaptive peak is not attainable through plastic accommodation. To him, the latter case might happen only with highly evolved organisms, especially humans, where cultural accommodation is so powerful that it prevents selection from taking place:

> This power is so great that man may suddenly enter a new territory, furnishing not a single product that he can eat, and by his art of agriculture so completely meet his needs that he becomes a permanent settler without even

subjecting himself to any new form of selection. *In other words, accommodation is in some cases so complete that coincident selection is prevented. This suggests that in some spheres of activity coincident (that is, organic) selection, is less liable to occur in man than in animals whose accommodation is less complete.*[89]

(4) Fourth, like Baldwin, and, to a lesser degree Lloyd Morgan (and on this point Gulick might have been especially influenced by them), Gulick ended up stressing the complex and reciprocal relationships between biological and cultural evolution, especially in human populations. This was why he titled his 1905 book *Evolution, Racial and Habitudinal.* "Habitudinal" referred to learned "traditions", which could reshape the selective environment and then intervene in "racial" evolution. Gulick contrasted two main forms of selection (and, for each of them, many sub-forms). When characters were determined by biological heredity, he used the term "selection". When he considered learned behaviors, he preferred the word "election". Selection and election were always interacting, to such an extent that he spoke of both "coincident selection" and "coincident election" in his book. He even tried to ground his theorization on empirical cases, for instance, when he focused on a population of cats that had learned to live near water.[90]

(5) Fifth, probably for the first time, the concept of organic selection was not brought into play primarily to account for the genesis of adaptations but first and foremost to shed light on diversification and speciation. Baldwin, Lloyd Morgan, and Osborn, as well as the other naturalists and biologists who until then had taken up the issue, had all wished to show that the adaptations of organisms could result from a mechanism other than variation/selection or the inheritance of acquired characters. For Gulick, the question of adaptation was secondary, what counted earlier was the explanation of divergence, and OS2 was elaborated precisely in this perspective. Therefore, he anticipated in a remarkable way the debates that would resume in the 1940s concerning the links between plasticity and sympatric speciation, notably in the context of the work of ethologist William Thorpe (see Section 3.2.)

All in all, as Gulick made clear in his 1905 preface, coincident selection became a major component of his final understanding of divergent evolution.[91] But this had not been the case beforehand, when he had been unaware of Baldwin, Lloyd Morgan, and Osborn's work. If Gulick should be praised for having produced the most refined reflection on organic selection around 1900 – one that would remain unsurpassed for decades – he cannot be credited for being the "founding father"[92] of this evolutionary mechanism. Because he was prepared to give a leading role to the organisms' agency in creating a new ecology, Gulick went very far in the direction of Niche Construction Theory. Indeed, no one went as far as he did in the early years of the

20th century, and, even in the 1940s and 1950s, only a few biologists (e.g., William Thorpe or Alister Hardy) could stand comparison, as we shall see.

## 1.5    Chapter's conclusion

Organic selection was framed to overcome the shortcomings of the simplistic opposition between (neo)Lamarckism and (neo)Darwinism. For Lamarckian biologists, the link between plasticity and adaptation was simple and straightforward: adaptation was the outcome of the cumulative effect of the physiological process of inheritance of acquired characters. On the contrary, for the few Darwinians of the period, adaptive plasticity was at first not really an issue, because adaptation was built by natural selection acting on chance variations. Weismann stands out by his interest in adaptability, and we have seen how far he went in the direction of conceiving organic selection. For Baldwin, Lloyd Morgan, Gulick, and others, adaptive plasticity was evolutionary significant *because* of its relation to natural selection. Plasticity was supposed to impact natural selection either by blocking natural selection and avoiding extinction (OS1) or by (also) altering selective pressures (OS2). This means that neither the developmental pathway underlying the plastic response nor that which ultimately underlies hereditary adaptation was considered relevant dimensions of the explanation to be produced. In the original organic selection framework, there is no need for the final hereditary ("congenital") adaptation to be homologous to the initial plastic accommodation ("modifications"). Organic selection is blind to *structure* and based solely on *function* since it impacts survival (i.e., fitness). It is in this precise sense that the relationship between plasticity and adaptation can be described as "ecological": insofar as, whatever the structures involved, plastic variation transforms the ecological scene of competitive interactions between living organisms.

To support such a reading of this history, Baldwin's writings are again very useful. First, it must be noted that he always phrased the mechanism of organic selection in terms of "lines of function"[93] and "utility", that is, in functional rather than structural terms. Second, in the wake of Lloyd Morgan, the term "coincident" variations quickly spread during the first years of the debate. "Coincident" points to the idea that, whatever the material basis at hand, only the final outcome in the struggle for life is at stake. Baldwin defined coincident variations as "variations which coincide with or are similar in *direction* to modifications".[94] Third, and moreover, in a few places, Baldwin explicitly acknowledged that the way plasticity works (at the developmental level) is irrelevant for organic selection to be efficient. For example:

> The truth of organic selection is quite distinct, of course, from the truth of any particular doctrine as to how the accommodations in the life of the individual are effected; it may be that there as many ways of doing this as the usual language of daily life implies, i.e., mechanical, nervous, intelligent, etc.[95]

Thus, from the very beginning, organic selection was framed in an ecological perspective primarily focused on the competitive interactions among organisms. This initial focus will not vanish, and we will find it at work again in different theoretical and empirical contexts, as will be documented in the two following chapters. From the outset and until today, organic selection is first and foremost a fitness-based argument.

Another significant aspect pertaining to this first debate needs to be stressed: the fundamentally typological dimension of most of these early speculations. Almost systematically, what was at stake was the survival of *one* organism in which, after a sufficient period of time, *the* right hereditary variation would eventually occur (e.g., in Lloyd Morgan's writings: "*the* adapted *individual* will escape elimination in the life-struggle"[96]). Here again, this early debate on organic selection bears the imprint of its time, when what Mayr called "population thinking" was still virtually non-existent. From this point of view, most of the formulations put forward were to some extent simplifications, a characteristic which would later be used by biologists such as Mayr and Waddington to challenge the validity of such a mechanism.

Finally, it must be highlighted that the complexity and numerous ambiguities of this founding episode are also a consequence of the fact that the debate was restricted to its theoretical dimension: no research program was developed at that time, not even an experimental take. Baldwin, Lloyd Morgan, and Osborn exchanged only abstract arguments, more or less in connection with empirical data that happened to be collected for other reasons. Even Gulick's views lacked empirical support. His large collections of land snails opposed the idea that standard natural selection can be a factor in divergent evolution, but they do not directly support his final synthesis when he eventually gave a leading role to organic selection (most of the shells' characters were indeed non-functional). This too will be a deeply entrenched characteristic of the history of organic selection: the difficulty to engage with experimentation, and the even greater difficulty to obtain non-ambiguous results. In the next two chapters, we shall see that it was only in the late 1920s that the empirical side started (very timidly) to be worked out and that the first experimental programs were designed and launched even later, in the late 1930s and early 1940s.

## Notes

1 "Organic selection" was Baldwin's wording. By 1900, no clear consensus had been reached on terminology, which remained fluctuant, although "organic selection" was the most commonly used label. For the sake of simplicity, I stick (most of the time) to Baldwin's term.

2 Remarkably, it was during the same session of the New York Academy of Sciences, at the end of January 1896, that Lloyd Morgan and Baldwin, each for the first time, developed their ideas on the links between plasticity and adaptation. A few months later, Osborn independently proposed his own version of the evolutionary mechanism (Richards, 1987, p. 399).

3 Robert Richards offers a documented restitution of some important aspects of the debate in chapters 8 and 10 of his 1987 *Darwin and the Emergence of*

*Evolutionary Theories of Mind and Behavior* (Richards, 1987). See also Depew (2003) and Ceccarelli (2019).

4  David Depew and Paul Griffiths came to the same conclusions (Depew, 2003; Griffiths, 2003).
5  Scheiner (2014).
6  Richards (1987, pp. 402–403).
7  Bowler (1983); Loison (2012).
8  Ceccarelli (2019).
9  Depew (2003, p. 10).
10  As far as I know, Breen W. Robinson and Reuven Dukas were the first to emphasize this distinction (Robinson & Dukas, 1999, p. 583).
11  Baldwin was especially concerned in making organic selection a general adaptive mechanism: "Organic selection becomes, accordingly, a *universal principle*, provided, and in so far as, as *accommodation is universal*" (Baldwin, 1902, p. 38, my emphasis).
12  Surprisingly, while this debate played a significant role after 1859, it has received very little attention to date. For example, Peter Bowler does not even mention it in his classic book on the eclipse of Darwinism, when he lists the six main arguments that were formulated against natural selection (Bowler, 1983, pp. 23–26). While Gould mentions this opposition in his book *The Structure of Evolutionary Theory*, he limits himself to a few lines in a work of over 1,300 pages (Gould, 2002, p. 203). To my knowledge, one of the very few articles dealing specifically with this question (with the explicit aim of better understanding Bergson's position) is Tahar (2022).
13  Mivart (1871).
14  Darwin (1872), see especially chapter VI, "Difficulties of the Theory".
15  For example, Baldwin (1902, p. 183).
16  For instance by French zoologist Yves Delage: "Finally, a last objection to the theory [of organic selection]. If innate variation is too weak at the beginning to offer any advantage, and if ontogenetic adaptation plays the greatest role in the final constitution of the animal, this adaptation occurs just as much in individuals with the innate variation in question as in those without. So will the contribution made by germline variation be enough to ensure the survival of some at the expense of others? More than likely not, because if it were otherwise, this variation alone would have been sufficient" (Delage & Goldsmith, 1909, pp. 289–290, my translation).
17  Gould (2002, pp. 203–208).
18  Loison (2010).
19  Weismann gave his lecture on May 2, 1894, in Oxford (Sheldonian Theatre).
20  Weismann (1895).
21  Weismann (1894, p. 15).
22  Weismann (1894, pp. 16–17).
23  Lloyd Morgan (1896a). This paper was a prepublication excerpt from his book *Habit and Instinct* (Morgan, 1896b). The article is equivalent in content and wording of chapter XIV of this book.
24  Weismann (1894, pp. 18–19). Emphasis in the original.
25  Richards (1987, p. 401).
26  In such a discontinuous history that lasted more than a century, it is fascinating to note that a very close conception was independently re-elaborated by anthropologist Gregory Bateson (1904–1980) almost 70 years later. He published in 1963 an article entitled "The Role of Somatic Change in Evolution" where cybernetics holds the place of the main theoretical framework and where he discussed at length the way developmental adaptability might be able to cope with new genetic mutations to produce viable phenotypes (Bateson, 1963).

27 Lloyd Morgan (1896a, p. 737).
28 In 1896–1897, it is indeed in Osborn's writings that one can find the most explicit statements: "The hypothesis, as it appears to myself is, briefly, that ontogenic adaptation is of a very profound character, it enables animals and plants to survive *very critical changes in their environment*" (Osborn, 1897b, p. 946, my emphasis. See also Osborn, 1897a).
29 West-Eberhard (2003).
30 Certain other possible and actual functions and structures decay from disuse. Whatever the method of doing this may be, we may simply, at this point, claim the law of use and disuse, as applicable in ontogenetic development, and apply the phrase, "Organic Selection", to the organism's behavior in acquiring new modes or modifications of adaptive function with its influence of structure" (Baldwin, 1896b, p. 444).
31 Morgan (1896a, pp. 736–737). My emphasis.
32 Baldwin (1896b, p. 445).
33 Baldwin (1902, pp. 64, 99).
34 Baldwin (1896b, p. 550).
35 Baldwin (1902, p. 32).
36 For example: "The individual modification acts, in short, as a *screen* to perpetuate and develop congenital variations and correlated groups of these. *Time* is thus given to the species to develop by coincident variation characters indistinguishable from whose which were due to acquired modification" (Baldwin, 1902, p. 150, my emphasis).
37 Baldwin (1902, p. 99).
38 Godfrey-Smith (2003).
39 Delage, Poirault (1897, pp. 512–513). My translation.
40 Robinson & Dukas (1999, p. 583).
41 Baldwin (1902, p. 210). My emphasis.
42 It must be also noticed that during the earlier years of the 20th century, the idea of a cost of plasticity seems not to have been considered yet, which made the issue at stake even harder to disentangle.
43 After Terence Deacon's (1997) book *The Symbolic Species: The Co-Evolution of Language and the Brain.*
44 Godfrey-Smith (2003, p. 56).
45 Godfrey-Smith (2003, p. 66).
46 The original French formulation is as follows (Hovasse, 1950, p. 128): "D'autre part, l'accommodat, en tant qu'il polarise la Sélection naturelle, guide et tempère les mutations".
47 Morgan (1896a, p. 740). My emphasis.
48 Baldwin (1896c, p. 725).
49 Baldwin (1897b, p. 771).
50 Baldwin (1897a, p. 635). See also Baldwin, 1902, p. 150.
51 At least one other zoologist at that time drew the comparison between organic selection and artificial selection to stress the causal role of individual plasticity in guiding evolutionary change. In a striking anticipation of the logic of Niche Construction Theory (see the conclusion of Part I), English naturalist Frederick Web Headley (1856–1919), building on Baldwin's and Lloyd Morgan's early work on organic selection, put it as follows as soon as 1900 (Headley, 1900, p. 128):

But obviously the quickening up of evolution is not all. The individual gains in importance. He improves his powers, is able to face a change of environment that otherwise would have been fatal [OS1]. He makes an environment for his young in which intelligence can be developed: he chooses the environment which they shall have when out of the nursery and so decides to some extent what qualities shall be the winning qualities in life. In fact he is beginning to take the helm and steer the species. Or we may put it in this way: when the individuals of one generation decide the environment in which the next shall grow up, selection ceases to be purely natural, it is in part artificial [OS2].

52  Baldwin (1902, p. 175). My emphasis.
53  Godfrey-Smith (2003, p. 63). My emphasis.
54  Griffiths (2003, p. 195).
55  Griffiths (2003, pp. 195, 199).
56  "Baldwin anticipated [niche construction], in a primitive way, in his discussion of how social heredity could "set the direction of phylogenetic progress" without the operation of the Baldwin effect" (Griffiths, 2003, p. 204).
57  This might also be the interpretation recently supported by Ceccarelli (Ceccarelli, 2019).
58  Baldwin (1902, p. 188).
59  "The first modern author to attribute population differences to accident seems to have been Gulick (1873, 1894, 1905) in his endeavor to explain the characters of *Achatinella* snails in different valleys on Oahu, Hawaiian Islands" (Mayr, 1963, p. 204).
60  Mayr (1963, 1970, 1982).
61  Lesch (1975).
62  Lesch (1975).
63  Hall (2006a, 2006b).
64  Hall (2006b, p. 493).
65  For instance, Gulick's name appeared only once in the 2003 edited volume by Weber and Depew, in a footnote, in Hall's chapter devoted to genetic assimilation.
66  Gulick (1888).
67  Hall (2006a).
68  Lesch (1975).
69  Gulick (1905, p. 7).
70  For biographical details, see especially Addison Gulick (1924) and Hall (2006a).
71  Rundell (2011, p. 152).
72  Rundell (2011).
73  Gulick (1872, pp. 223–224).
74  Gulick (1888).
75  Rundell (2011, p. 149).
76  Gulick (1888, p. 231; 1905, pp. 85, 125).
77  Rundell (2011, p. 146).
78  See, for instance, Gulick (1905, pp. 16, 140).
79  Gulick (1872, 1873).
80  Gulick (1888, 1891).
81  Gulick (1888, p. 222).
82  Gulick (1891, p. 326).
83  Gulick (1891, p. 326).
84  "These forms of selection are imposed on itself by the forms of accommodation the species has assumed, for the conditions are not so determined by the environment that no alternative remains to the organism" (Gulick, 1905, p. 64).
85  Gulick (1905, p. 63). Emphasis in the original.
86  Gulick (1905, p. 64). Emphasis in the original.
87  Gulick (1905, p. 65).
88  Gulick (1905, p. 65). Emphasis in the original.
89  Gulick (1905, p. 63). Emphasis in the original.

90 "According to the New Orleans Times-Democrat there are on the shores of Loui-
siana, near the mouth of the Mississippi river, two types of cats. The great majority are like the cats of other regions, but a small tribe on Tarpon Island have
apparently lost all aversion to being in the water. Their separation from other
families of cats has allowed of their establishing their habits of feeding on entirely
new lines of tradition, for they all wade freely in the shallow waters of the beach
hunting for small fish; and three or four of the bolder ones swim off to oyster
boats lying at anchor nearby. This is an example of partition allowing an innovation to be established through election as a permanent habitude; and as Captain
Bosco, who owns these cats, says it is many years since they began to go into the
water, we have reason to believe that coincident selection has begun to operate in
producing a breed whose innate instincts are better adapted to this mode of life
than were those of the original stock from which they sprang; or may it not be
possible that the direct inheritance of acquired characters has removed the instinctive aversion to water that belongs so universally to cats?" (Gulick, 1905, p. 68).
91 Gulick (1905, pp. iv–vi).
92 Hall (2006b, p. 490).
93 Baldwin (1896a, p. 441).
94 Baldwin (1902, p. 151). My emphasis.
95 Baldwin (1902, p. 143).
96 Lloyd Morgan (1896a, pp. 736–737). My emphasis.

# 2 The eclipse of organic selection

## The case study of French-speaking post-Lamarckian biology

After a period of debate regarding organic selection (roughly 1896–1905), the interest in this "new factor" in evolution quickly diminished at the end of the first decade of the 20th century. At least in the English-speaking world, published references to such an evolutionary mechanism became very rare and incidental and mostly concerned textbooks on evolution rather than articles engaged with the issue of evolutionary causality.[1] On an international scale, this eclipse of organic selection lasted over 30 years before it became a topic of interest again in the late 1930s, first in the USSR (between Kharkiv, Kiev, and Moscow) and then in other places (see Chapter 3).

What happened in French-speaking biology also fits this discontinuous pattern. After a quick and substantial introduction during the period 1897–1909, organic selection was almost completely forgotten during the following three decades. The French lack of interest is especially telling for at least two reasons. First, the seminal arguments and discussions had rapidly become accessible to French readers, even those who were unable to read English. For example, Baldwin's *Mental Development* was translated into French as early as 1897, and it was indeed in this French version that Baldwin added for the first time a rather detailed discussion of both organic selection and social heredity (see the first section of the present chapter). Second, as is well known, for decades French biologists opposed the idea that natural selection was an important evolutionary factor. During the eclipse of Darwinism, French naturalists and zoologists favored a (neo)Lamarckian understanding of evolution. But, around 1900, it became more and more difficult to support a strictly Lamarckian account of the evolutionary origin of adaptations: various research programs failed to demonstrate the efficiency of possible physiological mechanisms of the inheritance of acquired characters. This should have led French biologists to pay more attention to explanations which offer a plausible alternative to the neo-Darwinian chance variation/selection account. Organic selection was precisely one of those, and yet, as we shall see, this mechanism was barely mentioned between the early 1910s to the late 1930s.

The aim of this chapter is thereby twofold. In the first section, I detail the reception and subsequent disappearance of organic selection in an explicitly

DOI: 10.4324/9781003422990-4

Lamarckian context. Adaptation and teleology were the crux of the matter for most French biologists during the period 1880–1930, whatever their precise theoretical positioning. We will see how organic selection became known and discussed by prominent figures such as zoologist Yves Delage. The later writings of zoologist Lucien Cuénot perfectly illustrate the quick disappearance of this topic within French disputes about adaptive evolution. In the second section, I put forward more evidence in support of this reading, drawing on the complex case of Swiss French-speaking psychologist Jean Piaget. I show that though there were several good reasons for him to be aware of Baldwin's work on organic selection, he was not and failed to find the third way between Darwinism and Lamarckism that he was looking for in the 1920s. It was not until the late 1960s that he retrospectively reinterpreted his early zoological work using both Baldwin and Waddington's terminology (but, as we shall see, in a rather distinct and personal understanding).

These two sections will allow us to better understand Raymond Hovasse's evolutionary synthesis. Sixty years or so before Mary-Jane West-Eberhard, he tried, in two consecutive books (1943 and 1950), to elaborate a synthesis of his own, primarily based on the evolutionary role of phenotypic plasticity. Despite only being published in French, these books were considered important enough by George Gaylord Simpson to be discussed in his seminal article that definitively stabilized the terminology of the Baldwin effect (1953). In resurrecting Hovasse's forgotten contribution, without ignoring his many weaknesses, I aim to show how the ecological dimension of organic selection was then assumed. Even though Hovasse never tried to develop a quantitative or experimental approach to the topic, his writings contain several (albeit underdeveloped) passages about significant theoretical issues: for instance, the crucial idea that the purpose of plasticity was also to alter selective pressures in the building of evolutionary adaptations.

All in all, the French case offers a perfect case study to document the remarkable eclipse and subsequent resurrection of organic selection. It also offers an informative standard to draw a comparison with other national/local traditions that will be instrumental in the experimental turn that started in the USSR and in England in the late 1930s. As we shall see, Hovasse's involvement, even if theoretically highly ambitious, remained only verbal.

## 2.1 The reception of organic selection in a Lamarckian context

From the late 1870s onward French biologists promoted a neo-Lamarckian view of evolution based on the evolutionary efficiency of the inheritance of acquired characters.[2] Natural selection was at best tolerated and was deprived of any creative role. Most French biologists were looking at the evolutionary process from a physiological perspective: given that organisms were physiologically able to adapt themselves to their surroundings and given that species are largely well adapted to their environment, there must exist a causal connection between these two indisputable facts. Plasticity was supposed to lead to adaptation in a straightforward manner, because of some sort of Lamarckian inheritance. This seemed so obvious that, for years these biologists focused primarily on individual plasticity, confident in the fact that the inheritance of acquired characters would soon be experimentally ascertained. However, at the beginning of the 20th century, a clear-cut demonstration was still lacking, and this situation, in connection with the emergence of new alternatives – for instance, Hugo de Vries' mutation theory – started to weaken this theoretical framework. It was in these years, when this first form of French neo-Lamarckism began to slow down, that organic selection started to attract some attention.

On July 16, 1897, just over a year after Baldwin published his famous "A New Factor in Evolution" in *The American Naturalist*, psychologist Léon Marillier wrote the preface to the French translation of Baldwin's 1895 book *Mental Development in the Child and the Race*. He emphasized two new developments[3] which had not been present in the original edition: Baldwin had added a few paragraphs devoted to social heredity and organic selection. Marillier, in a wording very close to Baldwin's,[4] summarized organic selection as follows:

> "Acquired characters, individual modifications or adaptations are not directly inherited, but they act on heredity and evolution by indirectly determining their course".[5] Accommodations of this order, by occurring in certain living animals, remove the variations they constitute from the destructive action of natural selection. Variations in the same direction will thus be able to develop in successive generations, while variations in opposite or different directions will not be fixed and will be lost. The species will thus progress in the directions which have first been indicated by these acquired modifications, and the individual acquisitions will gradually become congenital variations. The difficulties raised by the question of the heredity of acquired characters could thus, according to Mr. Baldwin, be set aside.[6]

This is probably the first definition of organic selection printed in French, and, unsurprisingly, in complete accordance with Baldwin's own account, it emphasizes the idea of "breathing space": plasticity is mostly about preventing organisms "from the destructive action of natural selection".[7] Marillier was faithful to Baldwin, who, in this 1897 French edition, repeatedly argued that (behavioral) plasticity is able to "free" "ontogenetic acquisitions" from the "yoke of natural selection".[8] Therefore, as soon as 1897, a summary of Baldwin's main concepts (organic selection, social heredity, and also orthoplasy) was already accessible to a French-speaking audience.

As already mentioned in the previous chapter, the second introductory channel of the organic selection debate was Yves Delage's journal *L'Année biologique.* In 1895, Delage decided to create a new journal devoted to recording and discussing not only new facts and empirical results but new ideas regarding what was called "General Biology", that is, hypotheses about heredity, protoplasmic functioning, evolution, and so on. But, for practical reasons, the publication was delayed during the first years and the very first 1895 issue was only published in 1897. Similarly, the 1897 volume that contains the first discussion of organic selection was only printed in 1899.

In this volume, organic selection was mentioned and discussed several times, with each of the seminal texts of Baldwin, Morgan, and Osborn explicitly cited. The first short discussion is a single paragraph written by Delage (in collaboration with Georges Poirault, the journal's editorial assistant). This 20-line text is historically remarkable, as already emphasized (Chapter 1) because in it Delage might have been the very first to formulate the idea that plastic responses to environmental challenges prevent the progressive replacement of "accommodations" by "congenital variations" because they mask hereditary diversity. In the same volume, other reviews made accessible the main ideas supported by Baldwin, Morgan, and Osborn about organic selection. This continued in the following years. For instance, in 1902, a three-page review was specifically devoted to the content of Baldwin's new book *Development and Evolution.*[9] The author, L. Defrance, acknowledged the "originality" of this new evolutionary mechanism (in comparison to standard Darwinism and Lamarckism) and explicitly deplored that it was "still not well enough known".[10]

A third significant way of introduction was Delage and Goldsmith's bestseller *Les Théories de l'Evolution,* first published in French in 1909 and then translated into an English edition, which was widely read and known at the time.[11] Delage and Goldsmith devoted a whole chapter to "Organic selection" (chapter 17). This eight-page text offers a very clear account of the conceptual and terminological issues at stake and is of historical interest for at least two specific reasons. The first is that the authors were not convinced by Baldwin's argument regarding the problem of the causality of the replacement of accommodations by congenital adaptations, a problem Delage was one of the first to have stated.[12] The second reason is that this chapter contains a passage in which Delage and Goldsmith seem to have understood that plastic accommodations are evolutionary relevant because they can change the ecology of individuals and thereby alter selective pressures that will allow the selection of new coincident variations:

But here too [Delage and Goldsmith reflected upon the classical example of deer antlers] this [Lamarckian] heredity is only apparent: in reality, it is as a result of an innate variation that, in some individuals, the antlers were once more developed than in others; then, ontogenetic adaptation brought about in them the corresponding increase in the muscles of the neck and shoulders that made the innate variation of the antlers useful. Conversely, in one of the following generations, it may happen that an innate variation that strengthens the muscles arises; *it would have been useless and would have faded away if the previous development of the antlers, an innate variation preserved thanks to an acquired variation, had not assigned it its place in the functioning of the organism in advance.* Natural selection therefore protects it, and the two variations thus continue to grow in parallel.[13]

As already emphasized, such an understanding of the causality involved in organic selection (OS2) progressively grew during the first half of the 20th century and Delage and Goldsmith's chapter pointed in that direction. It is also worth noting that they never connected this issue to Gulick's (1905) theoretical positioning, where the idea that plasticity polarizes natural selection is fully endorsed, as we have seen. In 1909, they already knew Gulick's work on Hawaiian snails, to such an extent that the following chapter is entitled "Segregation" and is devoted to a discussion of Wagner's, Romanes', and Gulick's conceptions about geographical isolation and speciation. Yet, they referred only to Gulick's 19th-century articles and not to his 1905 book. Despite Delage and Goldsmith's significant level of erudition, they were then unaware of Gulick's final shift, when he eventually promoted OS2 as a pivotal mechanism in speciation. Such a treatment seems to be the common case: before Hall's 2006 paper, the fact that Gulick appropriated and developed organic selection in his personal understanding of evolution went remarkably unnoticed.

After this brief and inexhaustive account of the way organic selection became available in French biology – there is little doubt other texts also contributed to introducing this new mechanism in the French-speaking community during the first years of the century – I will focus on a zoologist who was one of the most central figures in French science during period 1900–1950, Lucien Cuénot (1866–1951). Cuénot remained famous both because of his pioneering work in animal genetics[14] and because he developed, at the end of his career, a nuanced finalist view of life[15] that was in contradiction with the nascent Modern Synthesis and molecular biology.[16] Cuénot is important here for two reasons. First, his work and books exemplify the disappearance of organic selection in the debate about the mechanisms of adaptive evolution. Even if Cuénot (like Piaget; see Section 2.2) had every reason to be interested in this mechanism, he only very rarely mentioned it in the thousands of pages he published about adaptation and species formation. Second, he developed a very influential conception of the relationship between adaptation, ecology, and evolution, namely "preadaptation", according to which adaptation was, at least for non-sessile organisms, not so much a consequence of natural selection (or the inheritance of acquired characters) but rather the chance meeting between an already formed structure and an ecological niche where it can be of some use. As we shall see, Cuénot's concept of preadaptation was the starting point of Hovasse's theory of "postadaptation", that is, a reworked and developed version of organic selection.

In contrast to most of his French colleagues, Cuénot is remarkable in that he never endorsed a Lamarckian reading of evolution. During his career, he changed his mind several times, from Darwinism and natural selection to mutationism,[17] and from mutationism to preadaptation.[18] Despite working in Nancy, quite far away from Paris, he was a highly influential and respected figure in French science, and his books achieved a wide readership. Like most of his colleagues, he was engaged in a long-lasting reflection on the issue of adaptation. Since Weismann formulated his neo-Darwinian stance, adaptation was the main theoretical issue for French biologists. During the inter-war period, because of the repeated failure to demonstrate the inheritance of acquired characters, the issue became even more problematic. For instance, it was in this context of growing skepticism that zoologist Etienne Rabaud contested the very idea of morphological adaptation: in his view, adaptation was only about minimal and strictly metabolic adequacy between an organism's protoplasm and his milieu.[19]

Because the issue was then so confused and debated, Cuénot deemed it necessary to devote a whole book to biological adaptation. This book is mainly a documented reflection upon the possible relationships between adaptive plasticity ("accommodation") and adaptation; thus, it should have represented the perfect platform to devote some space to organic selection and the like. But, despite Cuénot's dissatisfaction with both neo-Lamarckian and neo-Darwinian explanations, and despite the fact that he was obviously looking for a third way that he was then unable to find,[20] he did not.

This quasi-absence is made all the more remarkable by the fact that he paid special attention to the specific case of the development of calluses in various species (dromedaries, ostriches, warthogs, etc.) – a case which would have a central role in Waddington's own argument in the early 1940s. Such spectacular cases seemed to necessarily imply a Lamarckian account: given the perfect parallelism between the direction of accommodation and the one of adaptation, it was "attractive"[21] to think that the latter was the direct consequence of the former because of the cumulative effect of the inheritance of acquired characters. Nevertheless, Cuénot remained unconvinced and formulated several criticisms against such an understanding.

It is at the end of this chapter (chapter 10, "Can the "plastic response" [*Accommodat*] become hereditary over time?") that the only reference to an adaptive process close to organic selection is made. This is how Cuénot briefly termed the issue, in the specific case of plant adaptations to various environments (maritime and alpine):

> It can happen that plants . . ., simply accommodated, present over the course of the ages mutations which *coincide* with their needs; experiments show indeed that among plants of the same environment, but of very diverse localities, there are some which are at the accommodated stage, and others at the mutation stage, although their external aspect is almost identical.[22]

First, it must be emphasized that Cuénot is very elliptic about the conceptual content of the mechanism at stake, and only a very attentive reader might recognize here something that resembles organic selection. His formualtion of it appears merely incidental, and he never develops the idea more consistently, in this book or in other works, for instance, his 800-page *La genèse des espèces animales* (1911).[23]

Second, he never mentions Baldwin, Morgan, or even Delage, to such an extent that it remains very difficult to ascertain whether he was even aware of the seminal debate or if he in fact came close to re-discovering organic selection following his own theoretical path. Third and most importantly, even though there is little doubt that these rare passages came close to organic selection, it is obvious that Cuénot never understood that such a mechanism might be relevant to start building a third way, that is, exactly what he was looking at. For instance, in chapter 12, he admits that he does not know how somatic accommodations might be passed on to the germ plasm and that he is unable to explain how plasticity could lead to adaptation.[24] Thus, it seems that he never conceived organic selection as a significant process in adaptive evolution. This is also consistent with the fact that the previous quote is more a description of stages than an actual process exemplifying a causal relationship between adaptability and adaptation.

Another alternative for Cuénot, which he regarded much more positively, was his concept of preadaptation. In his work, the idea dated back to 1901, at least in print, but he only coined the term in 1914. Because neither the Darwinian nor the Lamarckian mechanisms were satisfactory solutions to him, he came to renounce the idea that morphological structures were the adaptive consequences of the way organisms interact with their environments. For Cuénot, adaptation was no longer a process but only the recognition of a match between structures and ecological niches. For mutation reasons – a rather common view among early Mendelians – he thought that morphological features were formed independently of external demands and that, afterward, there existed a kind of "ecological filtration" by the milieu. Populations and their preformed characteristics were distributed in various environments until those features eventually (but not necessarily) matched a specific niche. Gayon documented how Cuénot's concept of preadaptation was highly discussed at that time, including by some of the founders of the Modern Synthesis such as Simpson and Wright.[25] He also pointed out that, decades later, Gould and Vrba also referred to Cuénot's work when they developed their own concept of exaptation. What was true on the international scale was even more true on the French scale: Cuénot's conceptions were indeed inescapable during the interwar period and a necessary starting point in the adaptation debate. In the third section of the present chapter, I will show how Raymond Hovasse explicitly rooted his concept of "postadaptation" in a critique of Cuénot's preadaptation.

## 2.2 The complex case of Jean Piaget's early zoological work and its subsequent reinterpretation

Jean Piaget (1896–1980) is one of the founders of developmental psychology and is also remembered today as an epistemologist who had a long-lasting influence on didactics.[26] In this section, I will not be interested in Piaget's better-known legacy and will focus on his theorization of the relationship between plasticity and adaptation within evolutionary theory. There are at least two reasons to consider Piaget-the-zoologist in the present narrative. The first is that he is still sometimes acknowledged as a key figure in the history of the Baldwin effect and genetic assimilation. For instance, ethologist Patrick Bateson, in a 2004 critical review of Weber and Depew's edited volume on the Baldwin effect, praised Piaget for having provided an "early example of how the process [of the Baldwin effect] might work". More importantly, Bateson argued that Piaget "was much influenced by Baldwin".[27] There is a common assumption that Piaget was aware of the concepts formed by Baldwin, including organic selection, when he undertook his important work of measuring the torsion of the shells of certain species of freshwater mollusks in the 1920s.[28] As we shall see, both assertions are incorrect. In the interwar period, like for Cuénot, while there were many reasons why Piaget should have been aware of Baldwin's ideas,[29] he was not, and at the time he was unable to conceptualize any alternative to Lamarckism and Darwinism. The second reason is that Waddington himself, in his 1975 *The Evolution of an Evolutionist*, devoted a short chapter to "Genetic Assimilation in *Limnaea*" in which he referred to Piaget's early zoological work as the "most thorough and interesting study of genetic assimilation under natural conditions".[30]

Piaget's place in the history of the Baldwin effect and genetic assimilation is indeed complex and debatable. First because of the strong temporal discontinuity in his concern for organic selection. In the 1920s, he conducted an enormous amount of empirical work on various species of gastropods. But it was not until the late 1960s, that he reinterpreted his results within the framework of organic selection and genetic assimilation. Second because his subsequent reinterpretation was highly confused, to such an extent that he distanced himself from both Baldwin's and Waddington's conceptions – which did not completely escape Waddington.

Let us begin by examining Piaget-the-zoologist.[31] Piaget appears to have been very early highly interested in classical zoological work with respect to mollusks, especially systematics and bio-geographical issues (he published his first scientific paper in 1911, at the age of 15). From 1911 to 1921, before he turned to developmental psychology, he published 25 scientific articles and, in 1918, defended a PhD dissertation devoted to the malacology of a Swiss region. From this early descriptive work, it is clear, as Gayon has emphasized,[32] that Piaget was especially concerned with the question of the evolutionary origin of morphological adaptations, being dissatisfied with standard Lamarckian and Darwinian accounts. Even though he moved to psychology in the 1920s, he managed to perform substantial empirical work on the phenotypic variation and geographical distribution of *Limnaea stagnalis* in several places in Switzerland. This work was published as a 270-page monograph in the 1929 volume of the *Swiss Journal of Zoology*[33] [*Revue Suisse de Zoologie*]. It is this impressive sum of data that was later praised by Waddington as the first and most significant study of genetic assimilation in natural conditions.

In the early 1920s, Piaget decided to gain a better understanding of the debated and controversial issue of adaptation on a rigorous empirical ground. For years, he patiently worked in two complementary directions. In his fieldwork, he performed a statistical and biometrical analysis of the forms of shells in various environmental conditions. Piaget was a good naturalist and a rigorous observer, and he knew from experience that the variety *lacustris*, which inhabits the more turbulent environment of Swiss lakes, has a more contracted shell than other forms (Figure 2.1).

What he was looking at was a statistical confirmation of this intuition. For this, he collected approximately no less than 80,000 specimens from several populations. Each population was then characterized by a frequency curve of a contraction index, H/A, where H measures the total length of the shell and A the length of the opening. Thereby a shell elongation is directly proportional to index H/A. The basic idea was quite simple: the more a population inhabits a milieu where the water flow is strong, like the turbulent shores of the biggest lakes, the more its shell is contracted, the more the index (H/A) is small. The statistical analysis perfectly fits with Piaget's hypothesis (Figure 2.2).

In parallel with this massive field investigation, Piaget also performed numerous experiments in the laboratory that were designed to show the

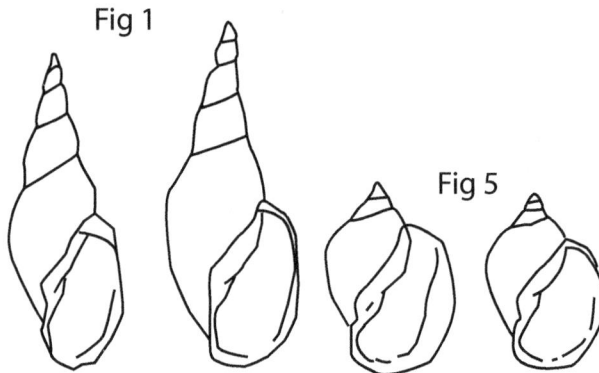

Figure 2.1 *Phenotypes of two different populations ("races" in Piaget's wording) of* Limnaea stagnalis. *"Fig 1" represents what Piaget considers as the "primitive" phenotype, which is typically elongated (H/A = 1.89). "Fig. 5" shows the morphology of individuals of the Neuchatel lac which exhibit the more contracted forms (H/A = 1.43). Reproduced from: J. Piaget, "Les races lacustres de la* Limnaea stagnalis *L. Recherches sur les rapports de l'adaptation héréditaire avec le milieu",* Bulletin biologique de la France et de la Belgique, *1929, 63, pp. 424–454. With permission from the Jean Piaget Foundation.*

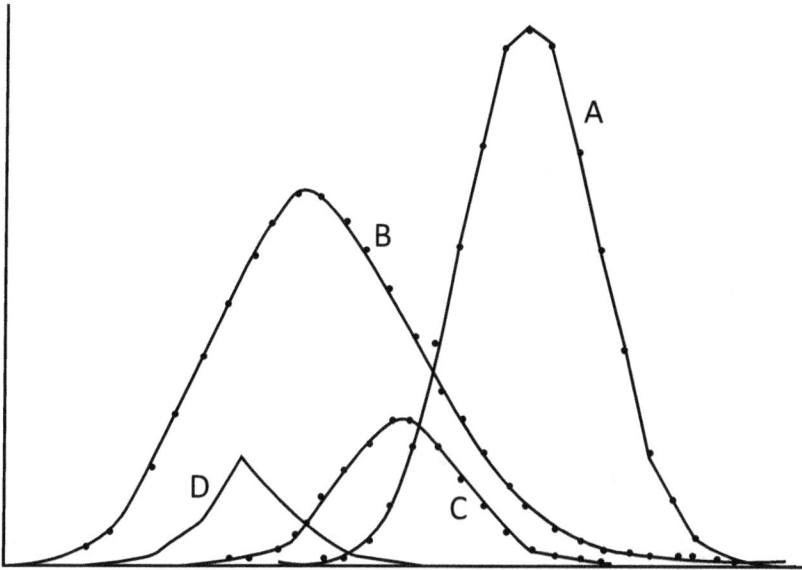

*Figure 2.2 Four distinct populations of* Limnaea stagnalis. What is measured on the x-axis is the shell contraction: the more the index H/A is high, the more the shell is elongated. The A curve represents the distribution of 8.000 "non-lake" (i.e., typical) individuals, whereas the B curve represents the distribution of 8.000 "lacustris" (i.e., contracted) individuals. Reproduced from: J. Piaget, "Les races lacustres de la *Limnaea stagnalis* L. Recherches sur les rapports de l'adaptation héréditaire avec le milieu", *Bulletin biologique de la France et de la Belgique*, 1929, 63, pp. 424–454. With permission from the Jean Piaget Foundation.

morphological impact of individual plasticity. In controlled conditions, in aquariums, he submitted hundreds of young individuals of the various populations he had identified to different flows. Here again, the results were telling: the index of contraction was significantly smaller when individuals were raised in an agitated environment (Figure 2.3).

These experiments ascertained that the shell contraction was also a phenotypic adaptation resulting from the organism's movements in response to its environment. Piaget theorized this aspect as early as 1929, openly using Edward D. Cope's already-dated notion of "kinetogenesis".[34] It was the organism's behavior and muscular reaction against the water that progressively developed its shell into a specific contracted morphology. Here, Piaget directly and frequently mentioned the famous "use and disuse" law,[35] but this did not imply that he went to favor a Lamarckian interpretation.[36] His results supported two other meaningful characteristics of this accommodative process:[37] (1) it was unable to immediately (i.e., in a single generation) change one population into another (in other words, the differential morphology had an obvious hereditary basis), and (2) it was not cumulative (i.e., there was no inheritance of acquired characters). This is why Piaget eventually concluded

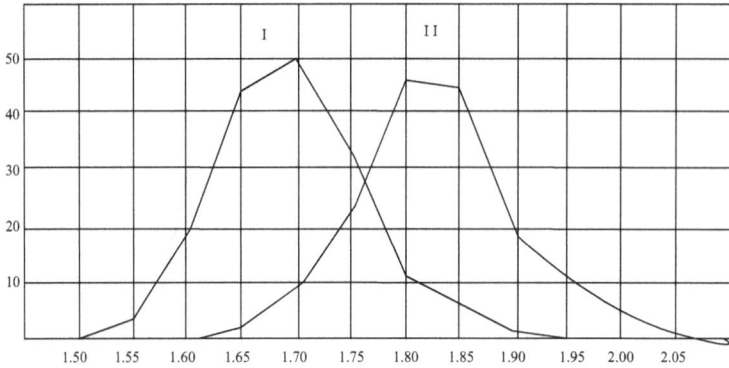

*Figure 2.3  Graphs showing the distribution of individuals according to their shell contraction index (H/A). The first curve (I) represents the distribution of individuals raised in an agitated milieu. The second curve (II) represents the same distribution in a quiet environment. Reproduced from: J. Piaget, "Les races lacustres de la Limnaea stagnalis L. Recherches sur les rapports de l'adaptation héréditaire avec le milieu", Bulletin biologique de la France et de la Belgique, 1929, 63, pp. 424-454. With permission from the Jean Piaget Foundation.*

that adaptive phenotypes were the additive results of both "accommodation" (plastic phenotypical reaction) and "adaptation" (hereditary characteristics of "races").[38] The next point at issue was now to understand the possible causal relationship between adaptive plasticity and adaptation. Lamarckism was no longer an option because the repeated work in aquariums showed no cumulative effects. Piaget framed the issue as follows:

> The problem is thus this: How can we explain the appearance of genotypes adapted to a certain environment when these genotypes seem to constitute the extension, by their structure and behavior, of phenotypic accommodations due to kinetogenesis?[39]

Darwinism was also discussed at length by Piaget, who opposed this possibility on several grounds, including the argument that plasticity prevents the efficiency of natural selection.[40] It must be emphasized here that Piaget's understanding of the workings of natural selection was rather schematic. Like others, he thought of natural selection from a typological perspective: the various gastropods forms were seen as direct products of one single major mutation, in de Vries' or Bateson's sense, and only secondarily maintained or eliminated by selection.[41] We shall see that, on other occasions, Piaget also promoted simplistic accounts of molecular concepts.

Like so many others at the time, Piaget remained unsatisfied by the alternative between Lamarckism and Darwinism for explaining adaptive evolution in natural populations where organisms are endowed with the ability to plastically react to their environment. None of these standard accounts seemed able to explain the causal relationship at stake within the documented example of *L. stagnalis*. What he was looking for was a third way, beyond Lamarckism and Darwinism,[42] that is, another kind of causal link between plasticity and adaptation.[43] In other words, had he known about it, Piaget would have been tremendously interested in a mechanism like organic selection. During those years, nowhere in his printed work did he make any reference, even distant and allusive, to organic selection or to related concepts. This is a telling example of the eclipse of organic selection taking place during the interwar period. Despite its significant introduction in French biological literature at the beginning of the century, and despite the fact that Piaget was obviously in demand of such a conception, he remained unable to formulate it. This was made even more surprising by the fact that Baldwin, in the late 1910s, became a close friend of Swiss psychologist Edouard Claparède (1873–1940), who was himself one of Piaget's mentors during the early 1920s.[44] As we have seen before, many of Baldwin's major writings were translated into French and therefore very easily accessible. Moreover, according to several accounts, Baldwin's psychology seems to have been a major resource for Piaget.[45] And, still, Piaget seems to have been unaware of the concept of organic selection back in the 1920s.

It was not until the mid-1960s that Piaget finally understood that Baldwin's and Waddington's theorizations offered the third way that he was looking for. At that time, he successively published three speculative books devoted to the interrelations between evolution, behavior, and intelligence: *Biology and Knowledge*,[46] *Vital Adaptation and Psychology of Intelligence: Organic Selection, and Phenocopy*,[47] *Behavior, Motor of Evolution*.[48] From the titles it was clear that organic selection was now at the forefront for him.

As far as I know, Piaget first mentioned organic selection, broadly speaking, as a promising third way, in a 1965 zoological paper published in the *Revue suisse de zoologie*. In this short article he admitted that the third way he had long been hoping for had been elaborated by Waddington.[49] From 1965 onward, Piaget blended Waddington, Baldwin, and his understanding of the molecular knowledge of the time into an idiosyncratic and cryptic synthesis that in no way clarified what was at stake. His three aforementioned books do not differ substantially in their general argument. Therefore, for the sake of convenience, I will mostly draw upon the English translation of the last one, which was published in 1978 – two years before Piaget's death – under the title *Behavior and Evolution*.[50] My aim is not to give an exhaustive picture of Piaget's complex views but to emphasize two main aspects that are especially relevant for the topic of the present book.

## (a) Conceptual vagueness and heterodoxy

What is striking in Piaget's writings is the way he gave standard concepts very unusual and unorthodox meanings, sometimes deliberately (e.g., for the concept of phenocopy) and sometimes seemingly without realizing it (e.g., genetic assimilation). Because his wording is especially confusing and vague, it is also often hard to understand to what extent he was aware of the conceptual gap between his account and more standard ones. For instance, he sometimes seems to use Baldwin's and Waddington's wording interchangeably,[51] whereas, on other occasions, when he considers his own conception, he uses a meaning of "organic selection" that has nothing to do with Baldwin's (see (b)). This is even more striking in his understanding of the molecular biology of the time, which seems to have been very weak, to say the least.[52] For instance, he refers to Whyte and Britten's regulatory model writing that this model considers "the possibility of a 'regulation of mutations'".[53] This is not true: the model was about the regulation of genetic expression and had nothing to do with the occurrence of new mutations. Many examples testify that his understanding of molecular biology[54] and, to a lesser degree evolutionary biology, was at best superficial, if not incorrect.

## (b) A "structuralist" understanding of organic selection

Because of (a), it is not an easy task to fully understand the ideas developed in Piaget's books. Already in 1975, Waddington noticed that Piaget's understanding of genetic assimilation was unusual and might even mask a Lamarckian framework because of some sort of retroaction of the phenotype on genetic mutations.[55] It is true that Piaget was less interested in the ecological dimension of organic selection (i.e., the way selection is impacted by plastic responses) than in a structuralist/internal/endogenous mechanism that he chose to term organic selection. Was his mechanism Lamarckian (as Waddington suggested)? Not in a way one can think at that time. Repeatedly, Piaget opposed a Lamarckian reading of his conception. In particular, he objected to the possibility of feedback of genetic information flow from proteins to RNA and from RNA to DNA. In other words, he largely stuck to the so-called central dogma of molecular biology. Nevertheless, he did not give up the possibility that the expression of genes might impact the mutational process (he spoke of "canalized" mutations, once again using a term (canalization) in a very unusual sense). All in all, and bearing in mind that Piaget remains cryptic and hard to understand, his mechanism might be summed up as follows:[56]

- New environmental conditions produce a phenotypic perturbation, which might be adaptively compensated during an organism's lifetime.
- This "imbalance" [*déséquilibre*], at first strictly phenotypical, if persistent, might spread within the different levels of the organization and reach the internal environment.
- If this imbalance is not compensated, in the end, it might produce a reorganization of the whole genome, leading to a new hereditary structure. It is this very genomic reorganization under the constraint of a disturbed metabolism that Piaget called "organic selection".[57]

Piaget's conception is interesting not because of its scientific soundness or because of its conceptual rigor but because it helps characterize the fitness-based dimension of the original account of organic selection. In contrast, Piaget was looking for a more "structuralist" mechanism, given that he never gave up on the idea that, at a strictly individual level, there might exist a causal link (yet not Lamarckian) between the phenotype and the genotype. As such, this late conception was perfectly congruent with his early Mendelian and typological understanding of both mutation and selection. Piaget never reflected upon these issues in a populational framework. One might also notice that, at least in its general working, Piaget's underdeveloped intuition is rather close to the recent epigenetic model supported for a few years by Laura Fanti, Sergio Pimpinelli, and their colleagues.[58] In this so-called pseudoassimilation model, epigenetic modifications might produce DNA rearrangements in corresponding loci in the germ line, and thus it exemplifies a kind of "structuralist" non-Lamarckian causal link that Piaget might have endorsed.

All in all, what can we conclude about Piaget's place in the history of organic selection and genetic assimilation? First, that his impressive empirical and experimental work of the 1920s should have been the starting point of the empirical turn documented in the next chapter. Yet, as we have seen, this was not the case: Piaget simply ignored organic selection in those years and was unable to conceive a third way. This demonstrates how strongly this concept was eclipsed during the interwar period, especially given that Piaget was most probably directly sensitized to Baldwin's ideas in psychology. Second, it is only retrospectively, and indeed very late in the 20th century, that Piaget reinterpreted his work on *L. stagnalis* in the framework of genetic assimilation and organic selection. As such, Piaget did not have any causal role in the history of these concepts. Third, even in the late 1960s and 1970s, Piaget did not – to say the least – have a clear understanding of these concepts, and his account of organic selection was especially heterodox and misleading.

It also remains doubtful that Piaget's results could be seen as decisive evidence in favor of genetic assimilation, as Waddington claimed in 1975. What Piaget patiently demonstrated is that (a) *Limnaea* populations ("races") are hereditarily adapted to their environment, (b) individuals are endowed with the plastic ability to respond to a change in their environment, and (c) they are not able to hereditarily transmit their accommodations. This work was without a doubt an unpreceded clarification of the debate, especially regarding the possibility of a Lamarckian account. Nonetheless, given the methodology used, it cannot show that (b) was causally responsible for (a). This was – and largely remains – a difficult and central issue: how to experimentally tackle the problem of the possible causal relationship between plasticity and adaptation? In the next chapter, we shall see that the Russian school also struggled in this aspect during the late 1930s and 1940s, even if Soviet biologists went much further than Piaget in the direction of an experimentally grounded demonstration.

### 2.3   From *preadaptation* to *postadaptation*, Raymond Hovasse's adaptationist synthesis

The eclipse of organic selection ended during the second half of the 1930s, as the Modern Synthesis was moving to the fore. In the French context, and, at the international level, Raymond Hovasse's work and writings participated in this renewal. Hovasse contributed to the reappraisal of what would come to be known as the Baldwin effect in two books, published in French and never translated. The first is a 75-page essay, *From Adaption to Evolution through Selection*. Apparently, the book was ready in 1939, but its publication was delayed because of the war and became available only in 1943.[59] The second is a 136-page more detailed synthesis published in 1950 with a less enigmatic title, *Adaptation and Evolution*. In this second opus, Hovasse situates himself explicitly within the growing literature devoted to organic selection. These two books were considered important enough for George Gaylord Simpson to discuss them in his 1953 seminal article on the Baldwin effect. Their theoretical content is the main issue of the present section.

Raymond Hovasse (1895–1989), a contemporary of Jean Piaget,[60] was, for most of his career, a descriptive zoologist who specialized in protistology.[61] For almost three decades, he held the chair of Professor of Zoology at the Clermont-Ferrand University, until he retired in 1965. Most of his work is typical of mid-20th-century descriptive protozoology, focused primarily on identification and classification. Hovasse managed to identify numerous new species and new subcellular structures, especially within flagellate groups.[62] Therefore, most of his published work did not directly concern organic selection or even evolution. Out of his approximately 160 papers and books, under 10 publications primarily concerned evolution, and the majority of these are mostly general articles published in popularizing journals. In contrast to his Soviet contemporaries and to Waddington, and even in contrast to what Piaget was doing already in the 1920s, Hovasse's interest in the process of organic selection did not take the form of a concrete research program. His books offered a general reinterpretation of adaptive evolution in the light of organic selection and not a quantitative or experimental approach to the issue. Nonetheless, as I have already stressed, these books remain remarkable because, some 60 years before Mary-Jane West-Eberhard's 2003 volume, they constitute a non-Lamarckian attempt to bring together the entire process of adaptive evolution based on the plastic abilities of living beings. Hovasse's synthesis lacked empirical support, and he never expanded it into a research program. Yet, he might have been the first to try an alternative synthesis in which Darwinian evolutionary change is fueled, not by chance variations but by plastic accommodations.

Two years after he had already written his first book, in 1941, Hovasse published an article titled "Adaptation and Change of Milieu: Preadaptation or Postadaptation". In this text and in the books, Hovasse challenged Cuénot's famous preadaptation hypothesis. According to Hovasse, in the late 1930s, Cuénot's evolutionary mechanism was gaining momentum and had "conquered the favor of many biologists".[63] From the start, Hovasse's quest was about contesting the causality promoted by Cuénot's concept. This is to say that Hovasse did not start his reappraisal by reflecting upon the scientific literature devoted to organic selection. In fact, it appears that he simply ignored – as did so many others – this body of work at least until the mid-1940s. In his 1941 paper, the only authors cited are three French zoologists: Cuénot, Rabaud, and Guyénot. Even in 1943, in his first book, Hovasse entrenched his ideas about adaptation exclusively within the French context and never referred to Baldwin, Lloyd Morgan, or others. Only in 1950 did he position his work within the international debate resurrected by Soviet biologists and in the context of the nascent Modern Synthesis. Like with the cases of Cuénot and Piaget, this shows that these ideas were forgotten in the interwar years, and how often these mechanisms had to be reinvented and rediscovered since the end of the 19th century.

It appears that Hovasse himself did not completely re-elaborate the concept of organic selection from scratch. In his 1941 article he explicitly referenced[64] Cuénot's short and elusive account, first published in *L'Adaptation* (1925) and then reproduced in the third edition of *La genèse des espèces animales* (1932). In 1950, he explicitly acknowledged that it was Cuénot's account which had offered him a starting point.[65] While Cuénot was never interested in organic selection, Hovasse thought it very promising and decided to build on it. Because in this view adaptation was the end result of plastic reactions, he chose to label this process "postadaptation", both to explicitly challenge Cuénot's notion and to highlight the temporal dimension at stake in natural conditions. On a strictly phenomenal level, postadaptation mimicked the Lamarckian process of the inheritance of acquired characters, but the causality involved was significantly different. At the end of the article, Hovasse detailed how he conceived the succession of the different steps:

If the advantageous variations acquired by the soma of the parent do not directly reach its own germline, and are not transmissible to the immediate descendants, they may leave the germline of these descendants the time to acquire, by mutation of its genes, the possibility of corresponding hereditary variations which will be substituted for the somations. A somation will not be transformed into a corresponding mutation, but over time a previous somatic variation may be replaced by a germline variation that is not necessarily identical, but simply equivalent.[66]

Two things must be highlighted about this 1941 formulation of organic selection. First, like Baldwin and Lloyd Morgan (though most probably independently of them), Hovasse also primarily put the emphasis on the "breathing space" dimension of organic selection: what he called "somations" [*somations*], that is, plastic phenotypical responses, are important because they allow the persistence and survival of organisms. Second, as we shall see, he did not conceive an adaptation as a developmentally stabilized form of accommodation (like Waddington would) but only as a trait which confers an enhanced fitness to its bearer; hence, the final hereditary outcome was only "equivalent". Taken together, these characteristics of Hovasse's early account tend to show that he promoted an ecological view of organic selection, where plastic responses are evolutionary significant because they interact with natural selection. This would be confirmed in his two following books.

In *From Adaption to Evolution through Selection*, Hovasse did not add much to his 1941 article, which makes sense since the book was completed before the article was written. Nonetheless, at least three things are worth mentioning. First, Hovasse did not hesitate to present himself as a Darwinian, which remained rare in mid-20th-century France. Second, Hovasse claimed that his Darwinism was the consequence of his daily interactions with population geneticist Philippe L'Héritier when both of them worked at Strasbourg University in the 1930s.[67] At that time, L'Héritier, in close collaboration with Teissier, had already designed his famous "population cage" to test natural selection on *Drosophila* populations in controlled conditions.[68] L'Héritier and Teissier were early actors of the Modern Synthesis[69] and Hovasse benefited from these contacts. The third and most important point is that, though Hovasse was not perfectly explicit about this yet, he understood "postadaptation" as something more significant than just a secondary mechanism restricted to some specific cases of adaptive evolution. As early as the late 1930s, Hovasse believed that the selective process that starts with plasticity and ends with adaptation was no less than the *primum movens* of the entire evolutionary process.

What had only been a suggestion in 1943 became a fully endorsed credo by 1950. Hovasse was looking for a general evolutionary model in which plastic changes were seen as the initiating event.[70] This is why he concluded his book by arguing that "the application of this principle can lead to a *general explanation* of Adaptation".[71] Such a general understanding of the evolutionary consequences of organic selection was, according to him, what made his contribution original in comparison to many others.[72] Now aware of the existing literature, he made numerous references to Russian-speaking biologists (like Gause, Lukin, Schmalhausen, and Kirpichnikov), Lloyd Morgan, Baldwin, Delage, and even Waddington. He attested that what he interchangeably called "Baldwin principle" or "parallel selection" had experienced a durable eclipse for almost 30 years and had survived only in occasional sentences lost in voluminous books.

If so-called postadaptation was the standard evolutionary process, it would mean that rather than emerging from the creativity of natural selection, like in the usual neo-Darwinian view, evolutionary novelties arise from the non-predetermination of the embryological development. For Hovasse, and in this he differed significantly from Waddington (see Part II), development was (mainly) an open process which was not hereditarily pre-programmed and which included the environment as a key causal actor.[73] Hovasse was especially concerned with this central issue and distinguished between two forms of plastic responses[74] equivalent to what Schmalhausen, at the same period, termed "morphoses" and "modifications" (see Part II). Hovasse called the first kind of accommodations "indirect accommodats": in some cases, because of previous evolutionary history, plastic responses are evolved automatic devices encoded in the genome that cannot be involved in the production of evolutionary novelties. He called the second kind of accommodation, which was much more relevant for progressive evolution, "direct accommodats". In this case, the unusual interaction between the environment and the developing organism brings into being embryological disruptions, that is, true phenotypical novelties, which might later be genetically stabilized because of "parallel selection". On some occasions, Hovasse even came close to West-Eberhard's recently formulated slogan according to which genes are only followers, not leaders, in adaptive evolution.[75] Such a view was fully consistent with the idea that organic selection was above all an ecological theorization and not a developmental one as it was for Waddington. In Hovasse's understanding, like for nearly all the supporters of organic selection in the 20th century, final genetic adaptations do not need to be stabilized outcomes of the same developmental pathways that were at first recruited in the plastic phase. As long as fitness is maximized along the way, postadaptation was supposed to be effective:

> The mutation substituting the somation [plastic response], producing the provisional adaptation, *may only copy it on the functional level*, since natural selection can act only physiologically [i.e., in relation to fitness]. A provisional adaptation, achieved by a particular morphological device, may be replaced by a definitive adaptation, only parallel to it, realized by another morphological device, but that functions in the same way.[76]

Hovasse's radical understanding of organic selection as an almost universal adaptive mechanism also allowed him to draw upon an old evolutionary idea entrenched in 19th-century thinking,[77] that each phyletic group goes through two distinct and successive phases: first a progressive one during which structures remain highly plastic, followed by a regressive one during which most of the plastic "potential" has already been exhausted.[78] For instance, according to him, most insects are only able to produce "indirect accommodats", which means that their progressive evolution is nearly over. For *Drosophila* especially, as classical genetics has strongly documented, genetic determinism is very strong, and no true freedom appears to be active within the system[79] (this is precisely the reason why the concept of epigenetic landscape was designed by Waddington on the basis of the causal role of genes during the development of *Drosophila*'s wings (see Part II)). On the contrary, according to Hovasse, at least some groups of vertebrates are still highly plastic and capable of true innovations, at both the developmental and evolutionary time scale.[80]

To conclude, one last point deserves special attention. As we have seen in the previous chapter, the common notion (at least for Baldwin) that plasticity first prevents the action of natural selection sometimes coexisted with the idea that plastic accommodations also polarize new selective pressures and that this is why genetic adaptations progressively replace phenotypic variations. As I have already argued, this second causal role devoted to plasticity, at first very cryptically sketched in initial formulations of organic selection – with Gulick's 1905 book as a remarkable exception – tended to be more and more explicitly acknowledged and supported over the course of the 20th century. Hovasse's writings directly contributed to such a conceptual change. If Hovasse fully endorsed the "breathing space" dimension of organic selection, he also increasingly emphasized that plastic responses were evolutionary relevant because they causally transform ecological interrelations and thus polarize natural selection. It is in the conclusion of his 1950 book that Hovasse was the most explicit about this difficult issue. On several occasions, he clearly stated that the "accommodat . . . polarizes natural selection",[81] or that natural selection is thereby "polarized in definite direction".[82] Already at the beginning of the book, he asserted that the "selective filter is oriented 'as needed' [comme il convient]"[83] because of phenotypic variations. Even though he never expanded on this crucial and neglected aspect, he very clearly understood what was at stake.

Hovasse was interested in an adaptationist evolutionary synthesis which would be compatible with the Darwinian credo of the Modern Synthesis but which put the emphasis on natural selection acting on phenotypic variations in altered environmental conditions. In a sense, he pushed organic selection to the limit by characterizing it as the central evolutionary mechanism able to produce most, if not all, adaptive evolution. In so doing, he participated in consolidating the ecological dimension of organic selection, where plasticity is causally relevant inasmuch as it actively modifies selective pressures, either by reducing (OS1) or by reorienting them (OS2). Such an unreasonable ambition did not escape Simpson's notice, and this was why he opposed Hovasse's conclusions. "By making the Baldwin effect virtually all-powerful in adaptation", Simpson wrote, Hovasse "weaken[s] the whole case for the over-all importance of the Baldwin effect".[84] Indeed, Hovasse might have been way too confident in his daring theorization, and there is no doubt that his adaptationist view strongly lacked empirical support. Nonetheless, such bold speculation helps understand the ecological dimension of organic selection, a core aspect that Simpson almost entirely missed, as we shall see in the third part of the present book.

## 2.4   Chapter's conclusion

The eclipse of organic selection remains somewhat of a mystery. It has often been argued that the explosive setting up of Mendelian genetics may explain the disappearance of what was at that time a highly speculative hypothesis.[85] Hovasse himself supported such a reading of the history.[86] This might indeed be one of the reasons, but this cannot be the whole of it. If genetics had been the main causal factor, then organic selection should have been discussed at length in the French context given that (a) classical genetics was barely an issue in the first half of the century,[87] (b) organic selection was substantially introduced in French journals and books around 1900, and (c) a significant number of zoologists (like Cuénot and Piaget) were looking for a third way beyond Lamarckism and Darwinism in the interwar period. And yet, as the present chapter has documented, organic selection was only very rarely mentioned, if not totally ignored, until Hovasse developed his own account of "postadaptation" in the late 1930s. Thus, to some extent, there remain historiographical issues to tackle in the discontinuous fate of organic selection during the first half of the 20th century.

Another related aspect which deserves emphasis is that the highly discontinuous nature of this history makes it difficult to ascertain to what degree later actors re-conceptualized organic selection from scratch and to what extent they were informed of the 1900 seminal debate. As we have seen, it seems that both Cuénot and especially Hovasse, at least at the beginning, simply ignored the historical dimension of the concepts they were using. This is a constant difficulty within the history of organic selection and genetic assimilation. As we will see with other cases (e.g., ethologist W.H. Thorpe's early work; see Chapter 3), these concepts were repeatedly rediscovered more than they were elaborated building upon the existing literature. Waddington represents another remarkable instance of such a recursive process, which necessarily contributes to the conceptual mess that is a significant part of the history at stake. Because so many scientists were not aware of the deep history of these concepts, they started their own conceptual paths with new terminologies and new empirical frameworks, thus creating theoretical tensions between what they were looking for and what their predecessors had already done. This remains an issue today, which highlights why the history of science matters for science in the making.

## Notes

1 Lull (1917); Herbert (1919).
2 Bowler (1983); Loison (2010, 2011).
3 Baldwin (1897c, see chapters 7 and 16).
4 Baldwin (1897c, p. 181).
5 Here Marillier directly quoted Baldwin: Baldwin (1897c, p. 181).
6 Marillier (1897, pp. XII–XIII). My translation.
7 Baldwin is using exactly the same phrase on page 181.
8 Baldwin (1897c, p. 187). See also p. 440.
9 L. Defrance (1902, "L'Année Biologique", *année*, pp. 414–417).
10 Defrance (1902, p. 417).
11 Y. Delage, M. Goldsmith, 1912, *The Theories of Evolution*, London, Frank Palmer. The book was translated by André Tridon.
12 Delage & Goldsmith (1909, p. 290).
13 Delage & Goldsmith (1909, pp. 286–287, my emphasis).
14 Burian et al. (1988).
15 Cuénot (1941).
16 Loison (2020b).
17 Limoges (1976).
18 Gayon (1995).
19 Rabaud (1922).
20 Cuénot (1925, p. 360).
21 Cuénot (1925, pp. 242, 255).

22 Cuénot (1925, pp. 258–259), my translation and emphasis.
23 In the 1932 edition of this book he repeated his 1925 equivocal formulation: Cuénot (1932, p. 267).
24 Cuénot (1925, p. 360).
25 Gayon (1995, pp. 340–341).
26 For a comprehensive survey, see Müller et al. (2009).
27 Bateson (2004, p. 287).
28 Messerly (2009).
29 Piaget was a regular visitor to Paris in the 1920s where he regularly met Paul Janet but apparently not Baldwin directly (Bennour & Vonèche, 2009). It should be noted that, in the 1910s and 1920s, Baldwin seemed to have completely set aside his theory of organic selection. He says nothing about it, for example, in his imposing treatise of *Genetic Logic*, published in three volumes (1906–1911), and which was translated into French. If Piaget became aware of Baldwin's theses in developmental psychology through this reading, he could not encounter the concept of organic selection.
30 Waddington (1975a, p. 92).
31 Gayon (1989); Messerly (2009).
32 Gayon (1989, p. 149).
33 A few weeks before, Piaget published a detailed summary in the French (initially openly Lamarckian) journal "*Bulletin biologique de la France et de la Belgique*". The title was designed to attract the attention of a Lamarckian readership: "Lake races of *Limnaea stagnalis* L. Research on the relationship of hereditary adaptation with the environment" [Les races lacustres de la *Limnaea stagnalis* L. Recherches sur les rapports de l'adaptation héréditaire avec le milieu] (Piaget, 1929a).
34 Piaget (1929a, p. 447; 1929b, pp. 455–456, 486–487).
35 Piaget (1929b, pp. 455–456).
36 Piaget (1929b, p. 456).
37 Piaget (1929a, p. 435).
38 Piaget (1929a, pp. 444–445, 1929b, p. 447).
39 Piaget (1929b, p. 495).
40 Piaget (1929b, pp. 511–512).
41 Piaget (1929b, p. 503).
42 Piaget (1929b, p. 518).
43 Piaget (1929a, pp. 454–455).
44 Bennour & Vonèche (2009).
45 Müller et al. (2009). This is also the view supported by the Piaget foundation (page consulted on November 17, 2022): www.fondationjeanpiaget.ch/fjp/site/presentation/index_auteur.php?PRESMODE=1&auteurID=9
46 Piaget (1967).
47 Piaget (1974).
48 Piaget (1976).
49 Piaget (1965, p. 781). My translation.
50 Piaget (1978).
51 Piaget (1974, p. 45).
52 Gayon & Mengal (1992, p. 54).
53 Piaget (1978, p. 42).
54 See, for example, Piaget (1974, pp. 23–24 and 60).
55 Waddington (1975a, p. 95).
56 See also Messerly (2009, p. 97).
57 For a rather developed passage where Piaget tried to sum-up his view, see, for instance, Piaget (1978, pp. 21–22).
58 Fanti et al. (2017); Pimpinelli & Piacentini (2020).

59 Hovasse (1943, p. 74).
60 Piaget directly cited Hovasse on several occasions in his late books. Note that after his academic retirement, Hovasse moved to Switzerland, and it is not impossible that he was then in direct contact with Piaget.
61 Hovasse's career remains poorly known. For the main lines, one can refer to De Puytorac (1965, 1990).
62 De Puytorac (1990).
63 Hovasse (1941, p. 412).
64 Hovasse (1941, p. 416).
65 Hovasse (1950, p. 10).
66 Hovasse (1941, p. 419). My translation.
67 Hovasse worked in Strasbourg from 1934 to 1938, whereas L'Héritier arrived there in 1938. Therefore, they were colleagues during only a few months. Yet L'Héritier seems to have exercised a profound influence on Hovasse.
68 Gayon & Veuille (2001).
69 Boesiger (1980).
70 Hovasse (1950, p. 10, 72–73).
71 Hovasse (1950, p. 127), my translation and emphasis.
72 Hovasse (1950, p. 128).
73 Hovasse (1950, p. 32).
74 Hovasse (1950, p. 39).
75 Hovasse (1950, p. 107).
76 Hovasse (1950, p. 17). Emphasis in the original, my translation.
77 See, for instance, the work of paleontologists Alpheus Hyatt and Edward Drinker Cope.
78 Hovasse (1950, see especially p. 128).
79 Hovasse (1950, p. 23).
80 Hovasse (1950, pp. 29–33).
81 Hovasse (1950, p. 158).
82 Hovasse (1950, p. 125).
83 Hovasse (1950, p. 12).
84 Simpson (1953, p. 115).
85 Simpson (1953).
86 Hovasse (1950, p. 14).
87 Burian et al. (1988).

# 3 The experimental turn, from plastic change to speciation

As we have already seen with the case of Raymond Hovasse, the eclipse of organic selection ended in the second half of the 1930s, for reasons that still have to be ascertained. This movement of positive reconsideration was global: it happened in various local contexts and research traditions, like in France or England, and was remarkably strong in Soviet science, where several important biologists were instrumental in resurrecting organic selection in the form of "substituting selection" or "coincident selection". Before examining the contribution of the Russian school and William H. Thorpe's research program, I will start by providing a basic understanding of what happened during the period 1935–1945. In the chapter's conclusion, we will come back to this picture in a more refined way, isolating and emphasizing four major features specific to what can be seen as the "experimental turn" in the history of organic selection.

The first and most prominent characteristic of this renewed interest in organic selection was thereby its experimental dimension. Whether it was Thorpe in England or Gause in the USSR, these works shaped the experimental turn toward organic selection. Until the late 1930s, and despite Piaget's huge empirical effort, organic selection had remained a speculative issue expressed only in abstract propositions. Around 1935, this was about to change, and most of the second generation of biologists interested in organic selection favored an experimental take on the issue (with of course exceptions, e.g., Hovasse). In the present chapter, I focus on the two research programs that seem to have played the most central role in this process of reconsideration, the one led by work of Lukin, Gause, Alpatov, Kirpichnikov, Naumenko, and others in the USSR (Schmalhausen's case will be dealt with in the second part, alongside Waddington's), and the more restricted one performed in the UK by ethologist and entomologist William H. Thorpe. Other scientists in other countries may also have been involved in similar conceptual and empirical endeavors at the same time, but it was the Russian school's experiments on the one hand, and Thorpe's results on the other, that were the most discussed (including by the architects of the Modern Synthesis, as we shall see), and these were the works that left the most significant mark on the scientific literature.

DOI: 10.4324/9781003422990-5

The second remarkable characteristic of this experimental turn is that it primarily (but not exclusively) focused on a specific evolutionary step, namely, speciation. It was thought (especially by Thorpe) that adaptive plasticity was evolutionarily relevant because it causally impacts gene flow between individuals. Thereby, the issue at stake was different from what was previously central for psychologists and ethologists like Baldwin or Lloyd Morgan, who were concerned with the path that goes from adaptive plasticity to evolutionary adaptation. This emphasis on diversification rather than adaptation is congruent with what Gulick had already expressed in detail as early as 1905. It is also perfectly in accordance with the hypothesis that organic selection was fundamentally an ecological concept that makes sense in a geographical understanding of the milieu where competitive relationships between individuals are the main concern. This ecological view was already substantial around 1900, gained momentum in the cases studied in the present chapter, and, eventually, culminated in the early 1960s in a few writings that opposed the gene-centered view of the Modern Synthesis. This aspect will be given special attention in the chapter's conclusion.

The third and last point that needs to be emphasized in this introductive sketch is, once again, the obvious conceptual and terminological uncertainty in the articles and books published in the mid-20th century. This vocabulary was far from being stabilized, especially before Simpson tried to bring clarity to it in his 1953 landmark paper published in *Evolution*. This requires special attention to capture what the concepts mean despite the huge terminological confusion. As already stressed in the conclusion of the previous chapter, this is in direct connection with the fact that the history of organic selection was – and remains – particularly discontinuous. Most of the actors reinvented the conceptual framework of organic selection from scratch rather than consciously engaging with earlier contributions. Baldwin or Lloyd Morgan were not absent from the literature, but their names usually came rather late in the process, at a stage where concepts were already formed and engaged in experimental designs. For the most part – and this might have been true also for Waddington – mid-20th-century biologists did not start with Baldwin, Morgan, or Osborn but met them at some points in their own idiosyncrasies.

## 3.1 The Russian school: adaptive plasticity as an ecological factor in speciation

The remarkable contribution of Soviet biology to evolutionary theory in the 20th century is still undermined in standard (Western) historical narratives. Despite Mayr and Dobzhansky's interest,[1] and despite the impressive body of work put together by the historian Mark B. Adams,[2] the names of Sergei Chetverikof, Nikolaï Dubinin, Nikolaï Timofeeff-Ressovsky, among others, are barely remembered, especially in comparison to those of Fisher, Haldane, and Wright. Yet the Soviet contribution was significant[3] in paving the way for experimental population genetics and on a more general level in its attempt to connect naturalist disciplines with experimental biology within a synthetic framework. In a slightly different historical configuration, that is, by erasing Lysenko, a Modern Synthesis – or something close to it – could have emerged first in USSR rather than in the US and the UK. While Lysenkoism directly opposed scientific work that pertained to evolutionary biology and genetics, there is another explanation for the obscurity of the research that was done: it was mainly published in Russian and in scientific journals that were very often unavailable to Western scientists.

The same was true for Soviet research programs related to organic selection from the mid-1930s onward. As we will see, the movement toward organic selection was a general trend that involved many biologists in several institutions. Unfortunately, only a small fraction of the dozens of articles and books were published in English, and these form the basis of my historical reconstruction in this section. Leaving Schmalhausen aside from the moment, I know of seven publications in English-speaking journals, with a total of approximately 113 pages (most of these were by Gause and his collaborators). This limited amount of literature might still be enough to gain a minimal understanding of what happened in USSR more than ten years before Waddington started his own research program on genetic assimilation. In addition to these publications in English, I was able to obtain[4] an article in Russian summarizing Naumenko's experimental work on *Drosophila* strains and published in the *Reports from the USSR Academy of Science* in 1941. This text has been translated,[5] and its content, which is extremely informative, is examined in the third and last subsection devoted to the Russian school.

What clearly emerges from this limited number of primary sources is that the two prominent features of the "Russian style" in evolutionary biology, namely, its ecological-naturalist and experimental focus,[6] were also representative of the way Soviet biologists resurrected the issue of the evolutionary role of plasticity. In several illuminating papers, Adams addressed the puzzling fate of Russian evolutionary biology during the interwar period, which was remarkably fecund while being under threat of the growing Lysenkoist agenda.[7] Adams emphasized the key role played by the Koltsov Institute of Experimental Biology, based in Moscow, in framing the experimental and "synthetic" turn in the so-called "Russian school".[8] In the first years after the revolution, the civil war, and World War I, the Soviet government invested massively in science, which allowed for the rapid rise of several institutions, including the Koltsov Institute. It was there that Chetverikof turned to population genetics and was able to find the stimulating intellectual environment that led him to publish, in 1926, "one of the first theoretical papers synthesizing biometric, naturalist, and genetic approaches to evolution in a Darwinian framework".[9]

How precisely organic selection came to attract the interest of several young scientists within Ukrainian and Russian biology remains to be elucidated.[10] As far as can be told from the published sources available in English, it seems that Ukrainian zoologist Efim Lukin played a pivotal role in bringing the hypothesis back into view.[11] This, of course, could be properly investigated only by Russian-speaking historians of biology, and I hope this book will encourage such a work to be produced in the near future.

### 3.1.1 Gause's experimental research program

Out of the seven articles published in English, five were published by Gause, either as the sole author or in collaboration with N.P. Smaragdova and W.W. Alpatov. These papers appeared in quick succession, between 1941 and 1947. The sixth article was a brief development published in 1944 in *Nature* by Alpatov in which he explained the mathematical treatment that he and Gause used to contribute to the demonstration of what they termed the inverse relation between "acquired" and "inherent" characters (see the following text). Finally, in 1947, geneticist Sergeevich Kirpichnikov (1908–1991) – a former student of Chetverikov who was trained at the Institute of Experimental Biology – published, in the *Journal of Genetics*, a very dense review article in which he summarized some aspects of what had been done on "coincident selection" during the past ten years in Russia. Here I start by giving an overview of the experimental research performed by Gause before turning to more theoretical issues regarding organic selection based on Gause's and Kirpichnikov's 1947 detailed reviews.

Georgyi Frantsevich Gause (1910–1986) was trained as an evolutionary biologist at Moscow State University, which he entered in 1927.[12] In the early 1930s, he performed a famous series of experiments on the struggle for existence[13] using micro-organisms and insects. This work was at the root of his well-known competitive exclusion principle,[14] for which he received worldwide recognition. His experiments on the struggle for existence among micro-organisms led him to start working in the emerging field of antibiotics, and from the early 1940s onward he devoted much of his time to the development of theoretical bases for the search of new antibiotics and the industrial organization of their large-scale production.

Apparently, Gause's interest in organic selection was never central, and it seems that this research emerged as the natural extension of his experimental work on the struggle for existence using micro-organisms. As a member of the Institute of Zoology in Moscow, he focused on a few experimental systems among ciliates. Three species of *Euplotes* (*patella*, *vannus*, and *elegans*) were used, at first, to test the phenotypic and evolutionary consequences of increased salinity of the medium. Then he turned to three other species, *Paramecium aurelia*, *caudatum*, and *bursaria* to address similar issues and to investigate the evolutionary origin of geographic differences.[15]

Experimental results were rapidly obtained and published in 1941 and 1942. The 1945 paper was more a review of what Gause thought he had demonstrated and his 1947 52-page "Problems of Evolution" was a general synthesis designed to attract the attention of evolutionary biologists on the neglected causal role of adaptive plasticity in genetic evolution. It must be emphasized that it was only in this last article that Baldwin and organic selection were mentioned. Before that, only his Soviet colleagues – like Schmalhausen or Lukin – and alternative vocabulary – especially "stabilizing selection" and "substituting selection", later "coincident selection" – were cited and used. Once again, this is a case of spontaneous reconceptualization rather than a conscious elaboration on previous theoretical paths. As already stressed, this reconceptualization within the Soviet context brought about a strengthening of the ecological focus. Adaptive plasticity was evolutionarily relevant because of its ecological implications: it explained how a single species could be split into several ecotypes in relation to different environmental conditions. Plastic accommodation was then thought of as a possible first causal step in the process of speciation. Gause was himself clear about this aspect, when for example he wrote that the "study of the mechanism of the adaptive process is essentially an ecological problem"[16] and that stabilizing selection might have a role "in the origin of geographic differences".[17]

His work on *Euplotes* started in 1939 and was based on a well-documented observation: body size in different species was closely correlated to salinity. There was a regular decrease in body size from freshwater species (*E. patella*) to species living in media of increasing salinity (*E. vannus* and then *E. elegans*). What was striking was the remarkable parallelism between individual acclimatization to salinity and the hereditary adaptations observed in the natural environment: for reasons related to osmotic regulation, in both cases, body size was decreasing. In controlled conditions of increased salinity (up to 5%), Gause started to cultivate various clones of *E. patella* and *E. vannus*, and these experiments easily showed a reduction of the body surface.[18] At the same time, taking advantage of the conjugation abilities of *Euplotes*, Gause demonstrated that individual accommodations left no trace in genetic heredity: in the absence of selection no variation was detected in comparing the average size of a great number of offspring clones to the mean of the parental ones.[19]

In a final series of experiments, Gause tackled the key issue of what would happen in experimental conditions allowing both acclimatization *and* natural selection to proceed (a step not present in Piaget's experimental designs). To answer this question, he used different clones that he cultivated both independently (i.e., with no significant genetic diversity) and after two episodes of crossing and conjugation (i.e., with substantial genetic variation). Each clone cultivated independently (thus with almost no genetic variation available for natural selection) was successively acclimatized to increasing salinity (2.5% and then 5%). The same conditions were applied in media where mass conjugation operated. It was shown that the average body surface was 114.2[20] in the first case and only 107.0 in the second. On this basis, Gause was convinced to have produced the first experimental demonstration of what he did not yet call organic selection:

> In other terms, natural selection in a mixed population works in the direction of strengthening the adaptive modification. Under the effect of increased salinity all infusoria in pure clones diminish in size (adaptive modification); at the same time in mixed population relatively smaller clones are selected. It appears that we are dealing here with the case of so-called stabilizing selection, which improves the adaptive modification by selection of casual inherent variations in the same direction.[21]

Around the same time, closely related protocols were designed and tested on several *Paramecium* species. At first, they seemed to confirm the work done on *Euplotes*, the main difference being that Gause hoped to shed some light on the evolutionary origin of recorded geographic differences in this genus.[22] But on closer inspection, this work also served as the basis for a strongly entrenched theoretical commitment that Gause believed to be his second main contribution to the issue (in addition to the experimental demonstration of organic selection), the so-called "principle of compensation" between "inherent" and "acquired" adaptation. From the start, Gause was especially interested in the fact that the total adaptation of a given organism was the sum of the acclimatization gained during its development and its hereditary abilities. He summarized what was first an empirical finding in stating that "the total magnitude of adaptation . . . can consist of two components: initial adaptation and acquired adaptation".[23]

This theoretical aspect was important for Gause for at least two contradictory reasons. In the end, it eventually helped him clarify the fact that genetic mutations are selected only inasmuch they allow for a global improvement in "the magnitude of adaptive character" (see the next section). Yet, from the early 1940s onward, and still in 1947, Gause also supported a different account: in his view, there was a kind of balance between plastic abilities and hereditary abilities. On both empirical and theoretical grounds, Gause argued that, usually, populations with high plastic capacities tended to show low genetic adaptation and vice versa. This was the reason why, in the first place, he engaged with a second round of experimentations using several species of *Paramecium*. His goal was then to show that the most common case which should be the rule was the "constancy of total adaptation in some adaptive characters".[24] To support his claim, he focused on the resistance to the salinity of various clones, a feature that he chose to define by the concentration of salt killing 50% of the individuals per 24 hours. He then patiently measured the initial resistance and the final resistance (after acclimatization) of numerous clones of *P. caudatum* and demonstrated that the "acquired resistance" was in a linear and negative relationship with the "initial resistance". More importantly, the coefficient of inverse proportionality was, in most cases, equal to unity, which showed the complete "compensation" between accommodation and adaptation.

On these empirical bases, Gause was confident that adaptability and adaptation were in inverse proportion for each genetic clone.[25] This was central because, in complete accordance with what was at the core of Waddington's and Schamlhausen's conception (Chapter 4), he thought, at least in the early 1940s, that the very same developmental and physiological "system" responsible for the plastic response was also the one recruited in genetic evolution and thus impacted by coincident mutations. In such a theoretical framework, where the final genetic adaptation is seen as the progressive reinforcement of the same initially plastic accommodative path, the results Gause obtained with *Paramecium* showed that this sort of general balance between accommodation and adaptation was seen as experimental evidence proving the case.[26]

Nonetheless, already in 1942, Gause was ready to consider other possibilities. For example, when acclimatization was sped up, it appeared that a higher level of adaptation was reached by the most plastic clones, a selective configuration which favored the evolutionary reinforcement of plasticity but not the genetic fixation of accommodations. But even when he considered these alternative cases, Gause eventually remained confident in his idea of a general and overall stable balance between what accommodation and adaptation can do: such cases were exceptions to the rule, in the sense that they showed "non-proportionate compensation",[27] that is, a compensation that felt to be completed.

In the end, he firmly stated that "for some physiological reasons" (i.e., the homology of the developmental mechanism responsible for plastic and hereditary adaptations) "such combinations as that of strong adaptability with the genotypically strong initial adaptation turn out to be impossible".[28] In the mid-1940s, Gause and Alpatov thought that the inverse relation between inherent and acquired properties of organisms that they had experimentally revealed was something close to a general principle in evolution, like the competitive exclusion principle. This is why they summed up their findings and results in an article published in *The American Naturalist* in 1945, in which they did not hesitate to link their principle to the idea of "compensation" that was already promoted in the early 19th century by French zoologist Etienne Geoffroy Saint-Hilaire. According to them, the logic was the same ("a redistribution of the limited amount of matter in a biological system"[29]) but was effective at the level of species or even genus and not, in this case, at the level of an individual organism.

### 3.1.2   *Plastic enough but not too plastic: how to conceptualize the causal role of plasticity in adaptive evolution?*

The efficiency of organic selection as an evolutionary mechanism rests on a minute adjustment of plasticity, which must be neither too strong nor too weak. If plasticity is too strong, then the new adaptive peak is reached immediately by developmental adjustment alone, which prevents natural selection from taking place. In such cases, there are no selective pressures to start genetic evolution. Conversely, if plasticity is too weak for a new environmental challenge, then either the population will go extinct, or evolution will take place only by standard selection of chance mutations. The complex theoretical issue of the range of plasticity that makes organic selection possible and efficient was explicitly addressed by Gause.

In the early 1940s, he performed several experiments designed to test what would happen when the new environment is substantially different and does not correspond to what the species might have experienced in the past. This is why he submitted various strains of *P. caudatum* to quinine, a chemical factor "never met . . . in their natural habitats".[30] As expected, results showed that adaptability was rather low and did not play any significant role in the process of adaptation. In such cases, evolutionary adaptation is mostly about the "selection of the casual initial strength of resistance",[31] that is, standard Darwinism.

The reverse situation, when initial plasticity is already strong enough to phenotypically reach the adaptive optimum, was seriously considered by Gause later in the 1940s and he only openly addressed it in his 1947 essay. The problem was that this contradicted the "compensation principle", which he still firmly supported in the essay. Gause never mentioned this theoretical tension – perhaps he never saw it. In "Problems of Evolution", he still argued that, besides the first experimental demonstration of organic selection, what he had brought to the debate was his "principle of inverse relation between phenotypic plasticity and genotypic specialization".[32] He repeatedly emphasized the significance of the so-called principle, to such an extent that the last section (out of four) was exclusively devoted to it. Yet, despite the fact that the principle still held a central position in his theoretical framework, this time, in opposition to what he initially thought, he no longer endorsed the idea that the initial accommodation and the final adaptive character relied on the same developmental path or "physiological system". The reason for that was firstly theoretical: if organic selection was supposed to produce evolutionary (i.e., genetic) adaptations, then it follows that coincident mutations, to be selected, must lead to a phenotypic advantage compared to what plasticity can do. And for this to be the case, it is more likely that coincident mutations do not impact the phenotype in exactly the same way. According to Gause:

> It is unlikely on *a priori* grounds that the resemblance of genoadaptations to phenoadaptations which they substitute will be a far-reaching one. As far as substituting or organic selection is based upon a *greater adaptive value* of possible genovariations as compared to modifications, and not upon the inheritance of the latter, it is probable that the likeness between these two types of acquirement will be limited to a superficial resemblance only. It is hardly reasonable to expect that genoadaptation will imitate all particulars of a physiological response.[33]

He thought the issue important enough to deserve an experimental investigation. This is why he designed new protocols, using the small housefly *Fannia canicularis* L. as a new system. Apparently, the first results were published as soon as 1941, in Russian, which indicates that Gause might have had an early interest in this theoretical problem. In any case, the basic idea was to precisely measure and compare "inherent" (geographical) and "acquired" (induced) differences in two strains of flies. He thus collected insects in the vicinity of the Karadag Biological Station (Crimea) and in Moscow. As expected, the southern strains were smaller, and when reared in the same laboratory conditions (in Moscow), a significant difference persisted, showing that "southern flies appear to be inherently smaller than the northern ones".[34] Reared at high temperature (28.5°C), the Moscow strains became smaller, once again as expected. What was highly significant, for Gause, was that the decrease of body size by means of modification and genetic geographic variation was not proportionally equivalent; that is, "proportions between different organs change in two different manners". This meant that there was "a difference in the mechanism of the adaptive adjustment of an organism by means of modification and of genoadaptation".[35]

Thereby, most probably for the first time in the history of organic selection,[36] not only the issue of the fixation of mutations was acknowledged as a problem that could not be taken for granted, but a precise hypothesis was elaborated and then submitted to a careful empirical test. In organic selection, for the plastic response to only be a transient phase that causally leads to genetic adaptation, accommodations have to be surpassed, in terms of fitness, by what mutations can do. In other words, if the plastic response has to be adaptive, that is, if it moves the phenotypic space toward the optimum, it also has to be not adaptive enough to reach this optimum, which then creates selective pressures that will promote genetic evolution. As we shall see, a comparable argument was independently produced, also in the mid-1940s, by ethologist W.H. Thorpe (Section 2.2).

However, while Gause endorsed such an evolutionary causality, he had to renounce the structural homology between accommodations and adaptations, an aspect central to Schmalhausen and Waddington and that he himself had supported in the early 1940s, as we have seen. In so doing, what was emphasized here was the ecological perspective of organic selection: during most of its history, organic selection was a fitness-based argument not committed to any developmental hypothesis. This historical episode perfectly exemplifies the nascent difference between Gause's conceptualization within the framework of organic selection and what Schmalhausen and Waddington, independently, were about to theorize.

Such a crucial distinction was also at work in the way Kirpichnikov formulated his argument in his 1947 review "The Problem of Non-Hereditary Adaptive Modifications (Coincident or Organic Selection)". Taking as a fact the renewed interest in organic selection and the significant amount of experimental work done in Ukraine and Russia in the past ten years, he thought that a "review of these papers may be of general interest".[37] Kirpichnikov's text was by far the best clarification then available regarding the conceptual content of organic selection and its empirical basis. It clearly shows the superiority of Soviet biology in this domain, just before Lysenko definitively took preeminence and almost destroyed genetics and evolutionary theory in the USSR. This review thereby constitutes a valuable resource for today's historians. Here, I will only briefly focus on the issue of the developmental relationship between plastic accommodation and genetic adaptation. This aspect was also central to Kirpichnikov and seems to have motivated the terminology he endorsed: following Gulick, he favored "coincident selection". He was not satisfied with Schmalhausen's "stabilizing selection", because it seemed to imply a general decrease in the plastic abilities of organisms during adaptive evolution. He was not convinced by Gause's "substituting selection" either. "Indirect selection" could have been an option, but it already meant something else.[38] In the end, he conceived "coincident selection" as the more appropriate terminology because mutations and adaptive modifications, in this theoretical framework, must be only "more or less similar phenotypically",[39] that is, with respect to fitness but not in their developmental bases. What was important for mutations was to "coincide" with the adaptive niche opened by initial plastic accommodations, no matter how developmentally they might impact the phenotype.

Such a reading is supported by the fact that, as early as 1947, Kirpichnikov, was perfectly aware of the difference of perspective between organic selection on the one hand and Schmalhausen's (and later Waddington's) conception on the other. He understood that what was not yet termed genetic assimilation was a developmental-based argument that primarily aimed to explain the progressive "automonization" – a term favored by Schmalhausen – of the epigenetic building of the phenotype with respect to environmental cues. In such a theoretical framework, one predictable consequence is that one must observe an "increased dominance of the 'wild type'",[40] an argument that, as we shall see, was central for Schmalhausen and Waddington too. In complete contrast, organic selection's causality was blind to development, with only the ecological relationships between adult phenotypes and their environment at stake.

Last but not least, it must be emphasized that for Kirpichnikov, as for Gause, what ultimately had to be explained was not simply adaptation but how this process of coincident selection could lead to ecological diversification and ultimately to speciation. Remarkably, he seemed fully aware of Gulick's earlier contributions, to whom he explicitly refers (as well as to Thorpe's), when considering how adaptability to divergent environments might "hamper free cross-breeding".[41]

### 3.1.3 Naumenko's early experimental work: contrasting organic selection and genetic assimilation

It is indeed the "non-developmental" logic typical of the concept of organic selection that we find at the heart of the experimental work led by V.A. Naumenko in the early 1940s. For the moment, I have not been able to access biographical information concerning him or other existing works, and I am here simply putting into perspective the contents of a series of experiments published in Russian in a single article entitled "Stabilization of Specific Mutations in the Artificial Selection of Respective Modifications". The article, "presented" by Schmalhausen, was published in the *Reports from the USSR Academy of Science* on May 3, 1941.

Naumenko, who may have been a student of Schmalhausen,[42] was well aware of the scientific literature concerning the problem of the genetic stabilization of plastic variation and cites Kirpichnikov, Lukin, and Schmalhausen (among others) in his article. What is particularly remarkable is that in many respects his experimental design seems to anticipate what would be the basis of Waddington's extremely famous work in Edinburgh in the 1950s. Naumenko used a specific strain of *Drosophila melanogaster* (named "Nalchik", probably in connection with the locality of origin of the flies) which he subjected to the following treatment:

• During the development, at a "critical period" (which is not specified in the article), the *Drosophila* are submitted to significant X-ray irradiation (4000 r).
• This leads to the production of "morphoses" (phenotypic variations) and in particular to the expression of phenocopies (the term is used by Naumenko) of the "rough eyes" and "divergent wings" types.
• In each generation, there is a selection of the "phenodeviants" obtained, which are then used as breeders.
• The steps are repeated during 12 successive generations.

Each of the 12 generations involved several hundred flies (from 132 to 538), so this constituted a quite substantial empirical work. In both the "rough eyes" and "divergent wings" phenotypes, Naumenko obtained at each generation a very significant proportion of morphoses, of the order of 30% (but also extremely variable, which according to Naumenko can be explained by the fluctuation of the X-ray doses received). In both series of experiments (the series where the "rough eyes" phenotype is selected and the series where the "divergent wings" phenotype is selected), at the 12th generation (but not before), Naumenko observed the sudden appearance of mutants showing the "divergent wings" phenotype (the phenotype is then inherited in a stable way, without treatment during embryogenesis).

It is important to note on this point that the experimental work of Naumenko radically differs from that which Waddington would lead a decade later. In Naumenko's work, there is no structural/developmental relationship between the plastic variation and the mutations which, in the end, produce an analogous but not equivalent phenotype. Nothing at all is progressively developmentally reinforced by artificial selection. This is not an aspect that would have escaped Naumenko, on the contrary:

There can be no doubt that the mutations have appeared without any connection to fly selection based on morphosis, as in the first case the selection was conducted based on a different trait ("rough eyes"), and although in the second case there was a mutation which in certain aspects was similar to the selected kinds of morphosis, nevertheless the character of its heritability still differs drastically from the expression of morphosis (it is expressed solely in females, and the males carrying this mutation die).[43]

If selection has not led to a progressive reinforcement of plastic variation before the accumulation of standing mutations, up to a threshold, makes it heritable (which will be the case with Waddington, see Chapter 4), what is the mechanism at work here? This implies on the one hand the ability of X-rays, not only to produce morphoses, but also *de novo* mutations, and on the other hand the repetition of the procedure until the "good" mutations, that is, those "coinciding" with the desired phenotype, eventually occurs. This is again the core of the logic of organic selection, in the specific form of OS1, in which plasticity plays a role in that it allows for the transient survival of the population until genetic mutations can take over. Naumenko had no doubt about the plausibility of such a mechanism in natural conditions, where plastic variations would have been adaptive to environmental perturbations:

Although in my work there was a selection of non-adaptive modifications, the obtained results point towards a possibility for non-random selection and stabilization in the presence of certain modificational changes of their analogical mutations; in natural circumstances this might lead to adaptive modifications being replaced by respective mutations in the process of natural selection.[44]

Naumenko's results were sufficiently well known to be cited by Kirpichnikov in his 1947 review and much later (1961) by Waddington himself. Kirpichnikov saw them as an authentic experimental demonstration of organic selection, on par with the one obtained by Gause.[45] Predictably, Waddington was more critical. The environmental perturbation here was so extreme (massive doses of X-rays) that any response to selection was in fact impossible, and only the accidental occurrence of advantageous mutations could allow the fixation of the right phenotype. Therefore, to Waddington, this represented a "limiting case" that might be non-existing in natural conditions.[46]

Whatever the actual plausibility of this mechanism, Naumenko's work helps us distinguish certain aspects of organic selection from that of genetic assimilation. As we will see in detail in the second part of this book, Waddington did focus on the response to selection following environmental treatment, and, in his experimental designs, artificial selection had a cumulative role in enabling genetic assimilation. In other words, the epigenetic landscape was progressively altered during the selective process until the new adaptive path was canalized. This required nothing more than the use of the initial standing variation. The logic of organic selection, perfectly grasped by Naumenko, was different. There was here a complete causal independence (in the embryological sense) between the plastic variation and the genetically fixed variation. The former did not act as the outline of the latter, which was necessary to consolidate, but it suspended the action of natural selection and thus offered the time necessary for the occurrence of the "good" mutations (OS1).

The Russian school, which was already prominent in experimental population genetics in the 1920s, was also instrumental in the resurrection of organic selection in the late 1930s and early 1940s. In comparison to what Hovasse did approximately at the same time, differences are obvious. First and foremost, Soviet biologists, primarily Gause and Naumenko, took issue with developed experimental bases: several series of experiments were launched and run parallel during those years. Second, as Gause and Kirpichnikov's 1947 detailed reviews exemplified, the conceptual richness was unprecedented. Soviet biologists without question went further in the conceptualization of the mechanism of organic selection than any previous or contemporary Western counterparts ever did. The reader must keep this in mind when we turn to Waddington in the next chapter, because, to emphasize his originality and the specificity of genetic assimilation, he would minimize what Soviet biologists brought to the picture. Third, the ecological dimension of organic selection gained momentum, in at least two respects: adaptive plasticity was given a key role in speciation and the developmental architecture of the phenotype was, eventually, acknowledged as irrelevant to the process.

However, the basic logic of the concept remained close to what had already been proposed by Baldwin and Lloyd Morgan at the beginning of the century. It was thought that plasticity was an evolutionary factor because it interacts with natural selection, that is, because it modifies competition in a given environment. It enables the survival of the population in new conditions long enough for genetic evolution to eventually proceed. As Gause put it in 1947,

> modifications [plastic variations] repeated for a number of generations may serve as the first step in evolutionary change, not by becoming impressed upon the germ-plasm [Lamarckism], *but by holding the strain in an environment where mutations tending in the same direction will be selected and incorporated into the constitution.*[47]

Yet, beyond this apparent constancy in wording, my aim was to show that the Russian school brought the concept to another level of scientific development.

## 3.2   W.H. Thorpe's research program: behavioral plasticity and speciation

The only experimental agenda that, in the late 1930s and early 1940s, was comparable to what was done in USSR was William H. Thorpe's work on insect conditioning. Strangely enough, despite the fact that Thorpe's results and interpretations were already widely known and discussed at that time, including by some of the founders of the Modern Synthesis (see Chapter 7), until very recently, Thorpe has remained largely ignored by historians of science interested in the history of the Baldwin effect.[48] Indeed, not until Gregory Radick's 2017 monographic article was Thorpe reinstated as a significant actor in this history.[49] Radick's paper is tailor made for my own argument, because, as its title indicates ("Animal Agency in the Age of the Modern Synthesis: W.H. Thorpe's Example"), it focuses on the issue of animal agency exactly at the time when the nascent Modern Synthesis considered it evolutionary meaningless. My aim is to build upon Radick's concerns in two directions. First, in an effort to compare Gause and Thorpe's research programs, I pay special attention to the experimental dimension of Thorpe's work on animal conditioning, which also started in the mid-1930s. Second, Thorpe's results led him to promote the concept of sympatric speciation while Mayr became the uncontested champion of geographic speciation. A careful examination of this opposition will help us understand how Mayr himself evolved on the subject, an aspect that is usually undermined and will be dealt with in Chapter 7. As a necessary prelude, I will first focus on the way Thorpe articulated his experimental results and his ecological account of sympatric speciation.

For the most part, Thorpe's ideas independently followed the same ecological trend that was already at work within the Russian school (yet with some significant differences, particularly with regard to what the experimental designs were supposed to show). For both Thorpe and his Soviet counterparts, plasticity – in Thorpe's case, behavioral plasticity – was evolutionarily relevant mainly because it causally reduced gene flow between ecotypes, which is the first step that might lead to speciation and evolutionary diversification. This is the basic idea that Thorpe supported in the early 1940s, and this is how he was understood by Huxley, Mayr, Simpson, and, later, Waddington himself, whom he knew for years.[50] Thorpe's case presents a complication and refinement in the conceptualization of the causal process of organic selection, close to the one already underlined in the previous section for Gause and his colleagues.

William Homan Thorpe (1902–1986) was a British ethologist who specialized in ornithology and entomology.[51] He remains one of the important figures – behind Lorenz and Tinbergen – of the mid-20th-century ethology, and his main book, *Learning and instincts in animals* (1956), was considered a significant contribution to the scholarship.[52] Thorpe, who was a naturalist since his childhood, was first trained in the agricultural sciences and, from there, moved to entomology. In 1929 he was awarded his PhD at Cambridge and spent most of his career there as Professor of Animal Ethology. As Radick and Gillespie[53] have both emphasized, Thorpe was an unusual figure in the landscape of 20th-century science: he endeavored to tie his religious beliefs and his scientific work as an ethologist together in his own natural theology. This will not however be the focus of the present section: beyond his motives, I will examine the way he articulated evolutionary concepts related to organic selection in connection to his substantial experimental work on insect conditioning.

The chronology and the general pattern of this episode run parallel to what we have noted in the Soviet case. Thorpe started his experiments in 1936–1937 formulating at first very cautious hypotheses which were unrelated to the previous history. In the series of papers that resulted from these experiments, Thorpe mentioned neither Baldwin nor organic selection. He was simply trying to extrapolate what might be the most significant implications of his results. In the mid-1940s he turned more explicitly to theoretical issues and it was only then that he became aware of the significant history that started in the late 1890s. Like Gause, he then tried to reconnect the historical thread, positioning his own contribution within this history. For the sake of clarity, this section is divided following the same dichotomy: first, it deals with the experiments Thorpe performed in the second half of the 1930s and, second, it addresses the conceptual content of his elaboration on the process of sympatric speciation.

### 3.2.1   Olfactory conditioning in insects: a model system for sympatric speciation?

As early as 1930, only a few months after he defended his PhD, Thorpe published a comprehensive review article on "Biological Races in Insects and Allied Groups" where he reflected upon the issues that would later serve as the basis of his research program. As the title indicates, Thorpe was interested in biological races, that is, populations of the same locality that are most of the time indistinguishable by morphology alone but that are nonetheless isolated, for instance, by food preferences.[54] The paper's aim was rather straightforward: to promote the idea that behavioral differences might lead to speciation.[55] Thorpe's ecological perspective was transparent from the outset and would remain a constant guideline for the rest of his career. Retrospectively, two ideas in this early publication stand out that proved central to Thorpe's subsequent work. First, he was already referring to "Walsh's host selection principle", a theoretical principle that he designed his research program to test. This principle refers to the observation that many adult insects demonstrate a preference for the host species on which they themselves developed as larvae.[56] Second, as soon as 1930, Thorpe considered that behavioral plasticity could accommodate significant environmental variation, thus directly producing divergent ecotypes, a step that would be the first in the speciation process. In doing so, Thorpe distanced himself from the standard Lamarckian explanation of the evolutionary relevance of behavior and favored a framework closer to organic selection. With no reference to the existing literature, Thorpe formulated the mechanism as follows:

> These experiments [. . .] are of very great interest, and are perhaps the clearest and most satisfactory case known among insects, of the

artificial production of a biological race. However, the interpretation put upon these results by Harrison in stating that the new habit was germinally fixed, seems open to question. He does not seem to have considered the possibility that his results may be due to some sort of "larval memory," the adult having a tendency to seek out for oviposition a plant of the kind upon which she had fed as a larva. While perhaps unlikely, such an explanation is by no means impossible, and has been invoked before now to explain such cases as the provisioning instincts of the hunting wasps. Indeed, such a theory is suggested by the very words of Walsh's "host selection principle," in its original form. If such host plant changes *are* of any significance as initiating evolutionary divergence, the postulation of a "mnemonic" preference keeping the nascent race to its new host, perhaps for a long period, till germinal differences appear, seems difficult to dispense with. It would also be of great interest to know if mating preferences were developed along with the change in oviposition response. As we have seen, they are known to occur in natural biological races, but it is hardly conceivable that these could be *immediately* developed as a direct result of food plant change. If, however, a preference for a particular plant was handed on to the adults from the larvae the individuals attached to one plant would, one might suppose, be more likely to mate together than with members of another host race merely because of their proximity; and it is not difficult to imagine how a mating preference might then in time be built up as an indirect result of the food plant change.[57]

This passage was not written by a biologist aware of the existence of organic selection and its already long history but by a young field entomologist trying to find alternatives to Lamarckian heredity to understand how behavior might be considered a causal factor in the evolutionary process. Even if the wording is difficult, if not cryptic at some points, some of the basic ingredients of organic selection are already there: plasticity accommodating environmental change (a new host) that is maintained long enough during successive generations to allow for "germinal differences" to appear. Thorpe mostly remained faithful to this first theoretical sketch which he would never stop elaborating and enriching and which would form the framework of his experimental work from late 1936.

This series of experiments led to three successive papers on olfactory conditioning, published in 1937, 1938, and 1939 in the *Proceedings of the Royal Society*. They first made clear that Thorpe was trying to experimentally address the host-selection principle[58] (and, secondarily, the possibility of a form of Lamarckian heredity). For this to be the case, it was necessary, for experimental workability, to use a parasitic insect that chooses its host on the basis of a single parameter only: odor. This is why he focused on a small ichneumonid, *Nemeritis canescens*, in which the sense of smell resides in the antennae and which develops at the expense of two species of flour

moths, *Ephestia kühniella* and *Ephestia elutella*. This organism completes its
life history in a relatively short period of 3–4 weeks and could be very easily
reared in the laboratory. As an endoparasitic species, *Nemeritis'* eggs are laid
inside the larva, where they develop at the expense of the host tissues. Thorpe
started to cultivate individuals of this species from 1934 onward and worked
on dozens of successive generations. Moreover, in the course of experimental
work, he discovered that *N. canescens* could also be reared on a third species,
the small wax moth *Meliphora grisella*. Thereby, Thorpe had at his disposal
a very suitable experimental system to test the host-selection principle. His
question was basically the following: to what extent is host choice for oviposition determined by olfactory conditioning imposed during the larval stage
of the parasitic insect?

To tackle the issue on an experimental basis, Thorpe used an "olfactometer" (see Figure 3.1). The general principle was rather simple. Young

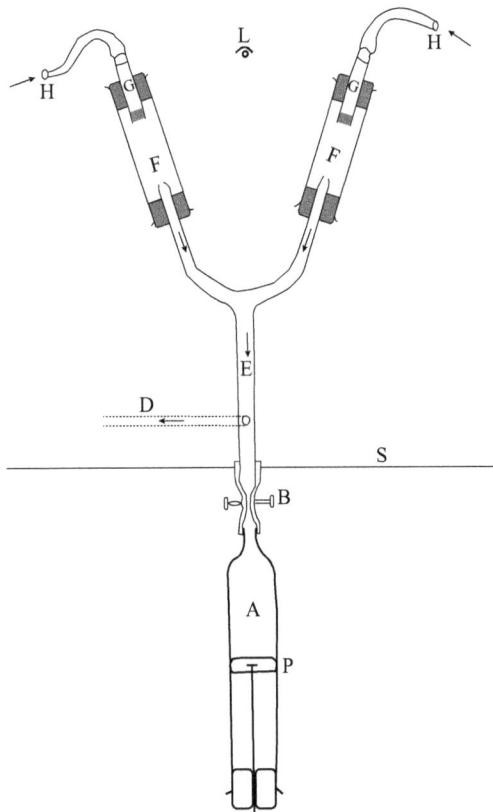

*Figure 3.1 Diagram of Thorpe's olfactometer.* Arrows indicate the direction of
airflow. For explanation, see the text. Reproduced from: W.H. Thorpe,
F.G.W. Jones, "Olfactory conditioning in a parasitic insect and its relation
to the problem of host selection", *Proceedings of the Royal Society*, B,
1937, 124, pp. 56–81.

*Nemeritis* adults were placed in compartment A and when clip B was open, they had to make a choice between two options to reach the light (L), either go left and eventually be trapped in compartment F or right and be trapped in the other compartment F. The only difference between those two compartments was the odor released by an air flow from H to E: in one of them, larvae were placed in G and their smell reached E, while the second one served as a control tube, with no specific odor.

This experimental design was used repeatedly by Thorpe in several configurations. For the sake of concision, given the huge number of experiments performed and the quantity of data obtained, I will discuss only the most significant results. First, insects reared on their normal host *Ephestia* show no reaction to the smell of *Meliphora*. Thorpe interpreted these results in terms of a "strong germinally fixed tendency"[59] to follow the normal odor. Nonetheless, it was also shown that, when reared this time on *Meliphora* (the non-usual host), *Nemeritis* young adults positively responded to the smell of *Meliphora*, even if they still preferred their normal hosts:[60] the proportion choosing *Ephestia* was then reduced from 85% to 65.8 %. It was also noted that this type of conditioning did not increase in successive generations, therefore, there was "no indication of any inheritance of acquired characters".[61]

This first series of results reinforced Thorpe's conviction that behavior might be an evolutionary factor even when Lamarckism is no longer an option. Already in 1937, he opposed the geographic account of speciation and favored the ecological one. Like in his 1930 review, he concluded by formulating a mechanism that was close to organic selection, even if, once again, he focused especially on speciation and gene flow instead of adaptation:

> Now it must be realized that under natural conditions two hosts of a phytophagous or parasitic insect would hardly ever be equally available and were the conditions such that the alternative host became more readily available, then even a change of smaller magnitude than that described in the present work might well be sufficient to isolate a population effectively over a long period. During this period physiological or other modifications of genetical origin might arise and be conserved by the ecological barrier provided by the conditioned food preference, and so the establishment of a new race genetically distinct from the parent form could be produced.[62]

The next series of experiments was designed to confirm and clarify the results published in 1937. It was definitively demonstrated that only the odor was at stake in the conditioning process. For this purpose, a "conditioner" was constructed which prevented direct contact with *Meliphora* larvae during conditioning: only the smell remained. More importantly, Lamarckism was, once again, experimentally defeated, and "this particular line of investigation" was then "dropped".[63] In the conclusion of his 1938 paper, Thorpe engaged again with the evolutionary relevance of such phenomena and linked his own results and hypotheses with the new notion of "isolating mechanisms", recently developed and promoted by Theodosius Dobzhansky in his 1937 *Genetics and the Origin of Species*.[64] He tried to give empirical substance to Dobzhansky's abstract grasp on the issue by concretely showing how an isolating mechanism of behavioral type might function.

Finally, Thorpe also questioned the developmental stage at which the conditioning might be efficient. In 1939, he came back to this problem making use of a new experimental system, namely *D. melanogaster*.[65] The results obtained showed this time that a pre-imaginal conditioning was at stake,[66] whereas previous results on *Nemeritis* rather supported a conditioning on the young adult still enclosed in the cocoon which includes part of the skin of the host.[67]

As we shall see in the next section, all this experimental work grounded Thorpe's theorization on sympatric speciation and his ecological account of organic selection. As already stated, the parallel between his research program and Gause's is striking in several aspects. Both Thorpe and Gause were interested in understanding the issue through controlled and rigorous experimental bases. Both programs were launched in the second half of the 1930s, in total ignorance of the seminal 1900 debate. Both submitted organisms to unusual environmental conditions and quantified the observed consequences. Nonetheless, there were also significant differences when comparing their work.

First, even though Gause was not at first aware of what had been done initially by Baldwin, Lloyd Morgan, and Gulick, he was conscious from the outset that he was engaged in the theoretical problem of the causal relationship between plastic accommodation and genetic adaptation. His references were only Russian and Ukrainian, but they allowed him to position what was at stake in his work within the correct conceptual framework. Things were much less clear in Thorpe's case, because in the 1930s he had no colleagues who could have played the role of Schmalhausen or Lukin. In other words, Thorpe was more isolated than Gause, and this might be the main reason why his initial formulations were sometimes not as clear. Second, and more importantly, the demonstrative scope between the two research lines was not equivalent. In Gause's case, the new environment was not only a new developmental environment able to change the phenotypes, but it also worked as a new selective environment which caused the evolutionary transformation of the populations of ciliates. In the end, ciliates evolved new adaptations that paralleled initial plastic responses. This is the reason why Gause thought he had experimentally demonstrated a case of organic selection. Thorpe, using a less favorable material, never engaged in a similar way. In more modern terms, what he was able to show was that the initial plastic response *could* have trans-generational effects not because of some sort of Lamarckian inheritance but because of a self-reinforcing ecological shift: it *might* induce a lasting change in host for the generations that follow the initial conditioning. Yet, given his experimental setting, he could only hypothesize the long-term evolutionary consequences of olfactory conditioning. In short, Gause had substantial reasons – later opposed by Simpson (see Chapter 7) – to be convinced that he had produced an experimental demonstration of organic selection, whereas Thorpe did not, both because he was initially unaware of the concept and because his experiments were not primarily designed to address the possible interference between plasticity and natural selection.[68]

### 3.2.2 Behavioral plasticity, sympatric speciation, and the evolutionary efficiency of organic selection

In the mid-1940s, Thorpe significantly expanded the implications of his results. One of the reasons is easy to understand: between his last experimental paper published in 1939 and his 1945 theoretical and important articles devoted to animal learning and habitat selection, Julian Huxley published his famous *Evolution, The Modern Synthesis* (1942). As we will see in Chapter 7, Huxley, as an ethologist, was especially interested in Thorpe's work and immediately understood the connection between Thorpe's results and the old idea of organic selection. In an important passage of the book, to which I shall refer again later, Huxley drew the line between Baldwin, Lloyd Morgan, and Thorpe as follows:

> To use Thorpe's own words, "the theoretical importance of such a conditioning effect is that it will tend to split a population into groups attached to a particular host or food-plant, and thus will of itself tend to prevent cross-breeding. It will, in other words, provide a non-hereditary barrier which may serve as the first stage in evolutionary divergence". We have here a beautiful case of the principle of organic selection (p. 523), as enunciated by Baldwin (1896, 1902) and Lloyd Morgan (1900), according to which modifications repeated for a number of generations may serve as the first step in evolutionary change, not by becoming impressed upon the germ-plasm, but by holding the strain in an environment where mutations tending in the same direction will be selected and incorporated into the constitution. The process simulates Lamarckism but actually consists in the replacement of modifications by mutations (see also Osborn, 1897).[69]

Huxley fully embraced Thorpe's ecological perspective and made him conscious of the significance of his experimental work. Still in 1940, in a chapter published in Huxley's edited volume *The New Systematics*, Thorpe was apparently unaware of the history of organic selection.[70] Not until Thorpe read *Evolution, The Modern Synthesis*, did he eventually make the connection, in a couple of papers published in 1945, now better equipped to situate his own achievement and to emphasize what was at stake on an evolutionary level. Thorpe did not just follow in Huxley's footsteps, he actually went further than Huxley in conceptual refinement and even further than Gause to some extent. In "Animal Learning and Evolution", a short article published in *Nature* in July 1945, he referred to the way Huxley linked his previous work to the subject of organic selection and argued that he wanted to push the idea further:

> Huxley suggests (ref. 6, p. 524) that the organic selection principle might be expected to account for the replacement of non-heritable variations by "mutations" (presumably merely on account of the increased adaptation to the niche thus conferred), and that "where the modifications are extensive the process of their replacement by mutations may closely simulate Lamarckism". The object of the present communication is to suggest that recent developments in genetics and in the field of animal learning, particularly the concept of imprinting due to Lorenz and others (refs. in Thorpe [1944]), make the principle at once more probable and easier to understand.[71]

Besides attesting to Huxley's importance, this passage shows two other remarkable aspects of Thorpe's evolving conceptualization. First, it was Thorpe and not Huxley who proposed that genetic fixation was the result of the higher fitness provided by mutations ("increased adaptation to the niche thus conferred"). This pivotal idea is indeed absent from Huxley's writings, and as already argued, was a missing part of the founding debate even if Baldwin seems to have come close (see Chapter 1). Thereby, it seems that it simultaneously emerged in USSR (and in particular in Gause's writings) and in Western science (at least in Thorpe's publications). Second, Thorpe wanted to expand on the subject on the basis of recent findings both in genetics and in ethology: his ambition was to go beyond a mere application of organic selection. In the mid-1940s, he began to argue that the initial plastic response, in splitting populations, created new ecologies that resulted in new selective pressures (OS2). These new selective pressures, in the end, would lead to complete speciation. In other words, there is no doubt that Thorpe started to elaborate a process that was close to the one depicted in Niche Construction Theory. This is how he summarized his evolutionary scheme in "Evolution and Learning":

> Suppose for the sake of argument that the initial basis of separation is a host-plant preference based on olfactory conditioning, or a " 'locality imprinting" holding the animal to a restricted locality or environment, renewed afresh in each generation. Suppose also that this is strong enough and has continued long enough significantly to reduce the intensity of selection for relational balance. There will thus be a definite selective advantage for such new variants as favour more complete isolation. Among new germinal variants of equal magnitude, those which

are of the same nature and direction as the phenotypic learned response already operating will, besides resulting in closer adaptation to the niche, be the most effective in furthering isolation, and will therefore be most strongly favoured by natural selection. Thus the learned or conditioned response of the animal to the environmental situation will, besides tending to reinforce and make effective slight topographical and geographical barriers, give momentum to, and set the direction for, the selective processes tending to bring about genotypic isolation. These selective processes will thus bring about the reinforcement and perhaps the eventual replacement of non-heritable modifications by genetic modifications, and will thus closely simulate a Lamarckian effect.[72]

As we have seen, the fact that plasticity interacts with natural selection in altering the orientation of selective pressures was to some extent already present in a few passages of Baldwin and Lloyd Morgan's texts. This aspect however remained difficult to read. This idea, on the other hand, was already remarkably explicit and central in Gulick's 1905 book. With Thorpe, a new stage was reached: the idea was not only now clearly formulated, as the previous quotation shows, but it was at the basis of a much more refined model of sympatric speciation, which had at the time no equivalent in the scientific literature. Even Soviet biologists and Gause (at least, as represented in the sources available in English) did not go as far in that direction. If they also saw organic selection as an important causal process in speciation, they did not theorize the fact that plasticity was evolutionary relevant in altering the selective environment.

In his second 1945 paper, "The Evolutionary Significance of Habitat Selection", Thorpe continued along the same theoretical line: learning and conditioning, even without any Lamarckian effect, are of evolutionary significance in making a "new niche"[73] inhabitable, an ecological process that will "give momentum to and set the direction for the selective processes tending to bring about genotypic speciation".[74] Over ten years later, he would use the exact same idea and wording in his ambitious 1956 book, *Learning and Instinct.*[75] Therefore, once clearly conceptualized in the mid-1940s, Thorpe probably remained faithful to this ecological understanding of organic selection which, like Gulick had 40 years before, remarkably anticipated some core conceptions of the more recent Niche Construction Theory. As we will see in the conclusion of Part I, and despite the fact that these ideas would be quickly and strongly opposed within the Modern Synthesis framework (Chapter 7), this growing momentum in emphasizing the causal link between plasticity and selective pressures did not end with Thorpe. Others, like his colleague in Oxford zoologist Alister Hardy, would go even further in that direction in the early 1960s, at a time when the Modern Synthesis had already crystallized, even "hardened".

## 3.3   Chapter's conclusion

As we have seen, after the eclipse of organic selection there was a substantial reappraisal of the concept in different countries and research traditions like in France (Hovasse), Russia (Gause), and England (Thorpe). To conclude the present chapter, I would like to summarize the main findings of this survey to emphasize the features of this general (though limited) movement of positive reconsideration that was initiated in the mid-1930s.

(1) *Historical discontinuity.* What we have here is a series of independent re-creation of the concept rather than conscious developments and enrichments of the seminal debate. None of these biologists, in the beginning, knew about Baldwin, Lloyd Morgan, or "organic selection". At a moment in the history of biology when standard Lamarckism had been dismissed, they were all looking for alternative explanations to understand how plasticity might be given an evolutionary role. They all reinvented organic selection from scratch, at first independently from one another. This is the reason why so many different names flourished at the time, like "postadaptation" (Hovasse) or "substituting selection" (Gause). The historical thread was broken and restored only retrospectively. In the meantime, concepts originated in a scientific context different from the one of the 1900 debate.

(2) *Conceptual difference.* Concepts were re-formed in a more openly ecological perspective, where the main concern was not adaptation per se, but rather geographic specialization, ethological isolation, and eventually speciation. At least for Gause and Thorpe (but not for Hovasse), the 1940 concept of organic selection was thus not only a more elaborated version of the 1900 version (though it was) but was also a slightly different one in the place it occupied within evolutionary theory. It came at a time when speciation was once again becoming a central issue,[76] and when gene flow between ecotypes was given special attention, especially in the wake of Dobzhansky's 1937 *Genetics and the Origin of Species.*

(3) *Conceptual refinement.* While a strong ecological commitment characterized the 1940 concept of organic selection, it also brought about a clarification of the causality of the process following two distinct lines. First, Gause (and most probably other Soviet biologists) and Thorpe present a clear and conscious understanding that genetic fixation requires an additional hypothesis, namely, that the new adaptive peak cannot be reached only by plastic accommodation and needs genetic mutations to maximize fitness. Second, and this was especially explicit in Thorpe's publications, adaptability is not only responsible for survival under new environmental circumstances (OS1, the most central aspect around 1900 and still in Hovasse's treatment) but is also – and perhaps even mainly – evolutionarily relevant because in creating a new ecology, it modifies the orientation of selective pressures and then the direction of the evolutionary process (OS2).

(4) *Experimental turn*. Finally, the most obvious feature in this 1940 revival is that, in most cases (but not in all, see Hovasse), biologists favored an experimental take on the issue. At a time when evolutionary theory definitively entered the realm of experimental science (partly in the wake of experimental population genetics), organic selection made no exception. This hypothetical evolutionary mechanism was not seen as only a speculative concept, a thought experiment, as in Baldwin's or Morgan's early writings, but also as a refined and articulated set of specific hypotheses that Gause, Naumenko, Thorpe, and probably others tried to submit to experimental investigation in different material contexts.

## Notes

1 Dobzhansky (1980).
2 Adams (1968, 1970, 1980a, 1980b, 1988).
3 Wake (1986).
4 With the assistance of Russian historian Sergey Shalimov.
5 Thanks to the work of Aleksandra Traykova.
6 Dobzhansky (1980).
7 See especially Adams (1968, 1980b).
8 The term "Russian school" was regularly used by Adams. For matter of convenience, I am doing the same here. This does not imply that every scientist involved in this history was Russian (several were Ukrainian) but that there was something like a "school of thought" that revolved around Moscow (and to a lesser extent Kharkiv and Kiev) where ideas and concepts were discussed on a regular basis, whether directly or in Russian-speaking journals.
9 Adams (1980b, pp. 242–243).
10 According to Gause, the fact that Lamarckism (i.e., a physiological process of the inheritance of acquired characters) was no longer an option in the 1930s made necessary to rethink the causal link between adaptability and adaptation (Gause, 1947, p. 64).
11 Lukin (1936, 1940).
12 Most of the biographic information given here comes from the obituary published in July 1987 by Brazhnikova.
13 Gause (1932).
14 Gause (1934). This principle still considered valid today, and sometimes called "Gause's law", stipulates that two species competing for the same limited resource cannot remain at stable levels over time: when one species has even the slightest advantage over another, it will lead, in the long run, to the extinction or ecological exclusion of the weaker species.
15 Gause et al. (1942).
16 Gause (1942, p. 99).
17 Gause et al. (1942).
18 Gause (1941, pp. 89–92).
19 Gause (1941, p. 92).
20 "Body surface" was expressed in $\mu m^2$. Gause multiplied half-length by half-width to obtain these results (the product was not further multiplied by $\pi$ since this constant is the same throughout).
21 Gause (1941, p. 93).
22 Gause et al. (1942).
23 Gause (1942, p. 101).

24  Gause (1942, p. 101).
25  Gause (1942, p. 103).
26  Gause (1942, pp. 103–104). My emphasis.
27  Gause & Alpatov (1945, p. 479).
28  Gause (1942, p. 107).
29  Gause & Alpatov (1945, p. 478).
30  Gause (1942, p. 107).
31  Gause (1942, p. 109).
32  Gause (1947, p. 19).
33  Gause (1947, p. 37). My emphasis.
34  Gause (1947, p. 38).
35  Gause (1947, p. 39).
36  According to Kirpichnikov, Lukin already emphasized this issue in the mid-1930s (Kirpichnikov, 1947, p. 168). As we have seen in Chapter 1, Baldwin, at least in one occasion (in his 1902 reply to Delage's critic), already briefly considered such a theoretical possibility.
37  Kirpichnikov (1947, p. 164).
38  Kirpichnikov (1947, pp. 167–168).
39  Kirpichnikov (1947, p. 168).
40  Kirpichnikov (1947, p. 173).
41  Kirpichnikov (1947, p. 165).
42  Waddington (1961a, p. 288).
43  Naumenko (1941, p. 78).
44  Naumenko (1941, p. 78).
45  Kirpichnikov (1947, p. 172).
46  Waddington (1961a, p. 289).
47  Gause (1947, p. 21). My emphasis.
48  For example, he is absent of the 2003 rich edited volume *Evolution and Learning, the Baldwin Effect Reconsidered* (Weber & Depew, 2003). In his detailed 39-page article on Thorpe, Neal C. Gillespie devoted only a few lines (p. 30) and a foot-note (n. 87) to the issue of the Baldwin effect (Gillespie, 1990).
49  Radick (2017).
50  Both were trained in Cambridge in the late 1920s, and both were impressed, at least at the beginning, by Whitehead's processual philosophy.
51  For a biographical survey and an exhaustive list of references, see Hinde (1987).
52  Hinde (1987, p. 629); Radick (2017, p. 46).
53  Gillespie (1990).
54  Thorpe (1930, p. 177).
55  Interestingly, Thorpe started by citing an example of ethological isolation in two closely related populations of cuttlefish (*Sepia officinalis*) documented by Lucien Cuénot. On this occasion, he included a small quotation directly in French in his paper (Thorpe, 1930, pp. 178–179).
56  This principle, later called Hopkins' Host-Selection Principle in the literature (HHSP, Thorpe himself favored this terminology in his following publications), is still debated today in insect ethology. For a recent review, see Baron (2001).
57  Thorpe (1930, p. 202). Emphasis in the original.
58  Thorpe & Jones (1937, p. 56).
59  Thorpe & Jones (1937, p. 80).
60  Thorpe & Jones (1937, p. 70).
61  Thorpe & Jones (1937, p. 78).
62  Thorpe & Jones (1937, pp. 78–79).
63  Thorpe (1938, p. 390).
64  See Dobzhansky (1937, chapter VIII, Isolating Mechanisms, pp. 228–258).

65 The strains used were given to him by Waddington (Thorpe, 1939, p. 425).
66 Thorpe (1939, p. 429).
67 Thorpe & Jones (1937, pp. 70–71).
68 In fact, it would be over half a century before this kind of work on the conditioning of egg-laying behavior in insects included artificial selection procedures. The results obtained then pointed in the direction of the Baldwin effect, without, once again, establishing a link with the conceptual issue at stake: Mery & Kawecki (2002).
69 Huxley (1942, pp. 303–304).
70 Thorpe (1940).
71 Thorpe (1945a, p. 46).
72 Thorpe (1945a, p. 46).
73 Thorpe (1945b, p. 68).
74 Thorpe (1945b, p. 69).
75 Thorpe (1956, pp. 256–257).
76 Cain (2009).

# Conclusion of Part I

From Weismann to Lloyd Morgan, from Baldwin and Gulick to Gause and Thorpe, the aim of this first part was to explore the complex history of half a century of controversies about the debated hypothesis of organic selection. The aim here was not exhaustivity. Indeed, several actors, especially from the Soviet context, are most probably missing from the narrative proposed. The aim was more modestly to present a general panorama that allows a minimal understanding of the original concept of organic selection.

Despite differences between what was theorized by the biologists involved, these conceptions, often developed independently from one another, nonetheless shared a common basis in the idea that plasticity is evolutionary relevant because it modifies the way natural selection takes place. In more modern terminology, this means that organic selection was in the first place a fitness-based argument. In such a framework, there is no need for the final (genetically encoded) adaptation to be structurally homologous to the initial plastic accommodation. What matters is maximizing fitness according to a defined function from a competitive-ecological perspective. In other words, organic selection is not concerned with anatomical structures, physiological mechanisms, or embryological development but focuses only on the way plasticity transforms the "ecology", that is, according to Thorpe, "the relation of the organism to its whole environment".[1] Here again, we are faced with the ancient alternative in biology between structure and function, between "internalism" (constraints of structure) and "externalism" (needs of function). Organic selection was without a doubt on the side of function and externalism: what drives evolutionary change is function, that is, the competitive performance in a given niche (as we shall see in the next part, Schmalhausen and Waddington made a different choice in focusing on the other side of the alternative). In such an externalist-ecological perspective, two related and contiguous lines of theorization were asymmetrically explored or at least started to be worked out during the period 1890–1950.

The first, which was dominant at least until the early 1950s, held that plasticity allows populations not to go extinct in new environmental conditions and thus genetic evolution to catch up (OS1). This "buying time"[2] or

DOI: 10.4324/9781003422990-6

"breathing space" role was explicit from the outset and was already fully acknowledged and emphasized in Baldwin and Lloyd Morgan's articles and books and later in most of the subsequent work done at the time of the reappraisal of this evolutionary mechanism in the late 1930s and early 1940s. This causal role, to be evolutionarily effective, implies auxiliary assumptions that define a range in which plasticity could pave the way for subsequent genetic evolution. If plasticity is too weak, then the time available for genetic evolution is too short and the population will go extinct (or evolution will proceed by the direct selection of genetic variants). Conversely, and more importantly, if initial plasticity is too high, then the new fitness peak is immediately reached, which prevents natural selection from taking place. At that time, whether around 1900 or in the 1940s, biologists did not formulate these auxiliary assumptions as clearly. Nonetheless, they were already struggling, in one way or another, with precisely these issues. Repeatedly, and as early as in the late 1890s, an objection against organic selection was raised: if plasticity enabled sufficient adaptation, then there would be no more genetic evolution because there would be no more differences in fitness between genotypes. A genuine progress was made in such a debated problem when it was acknowledged, first only briefly and incidentally by Baldwin himself, and then in a more explicit and argued manner in the late 1930s and early 1940s (by Gause and Thorpe), that genetic mutations must not only mimic plastic accommodation but, to be selected, they must enhance the adaptation to the new conditions: somehow fitness must be maximized for organic selection to be an evolutionary mechanism.

Another way to think of the evolutionary role of plasticity from an ecological perspective is to assume the idea that plasticity is evolutionarily creative in opening new avenues for natural selection to take place. In this view (OS2), plasticity is not so much about the accommodation of pre-existing and fixed selection pressures externally imposed on the system but is rather about the production of these pressures themselves. In such an account, the phenotypic space is enlarged, and a new selective regime emerges from this initial plastic exploration/creation. In short, organic selection is more about creating a new ecology than buying time. These two possibly distinct (but not independent) causal roles were already conceptualized in the first half of the 20th century, even if OS2 was often much more cryptically expressed and only became visible in the articles and books of the time after careful analysis. This second evolutionary role of plasticity was to some extent partly envisioned by Lloyd Morgan and Baldwin, as we have seen, but neither made clear what exactly they had in mind on this important aspect. It was Gulick, as early as 1905, who, with remarkable hindsight, explicitly acknowledged this understanding in his book *Evolution, Racial and Habitudinal*. The idea was later partly endorsed here and there, like in Hovasse's books, but it was not until 30 years later that Thorpe made the idea that animal agency should be a topic of interest for evolutionary theory central again. Historically speaking, it therefore seems that there was something like an erratic and discontinuous trend toward an increasing awareness that plasticity – and especially behavioral plasticity – not only accommodated fixed selection pressures (OS1) but also shaped the selective environment (OS2).

Despite the rise of the gene-centered Modern Synthesis (that will be dealt with in Part III), such a trend did not end with Thorpe. In the 1950s and 1960s, a few zoologists still explicitly supported the view that living animals are evolutionary agents that actively produce their own selective regime.[3] Perhaps the most prominent and well-known example of a zoologist who, in the mid-20th century, openly endorsed that behavioral plasticity actively determines subsequent morphological evolution was English marine zoologist Alister Hardy (1896–1985). Professor of zoology at Oxford, remembered today for what is now called the "aquatic ape hypothesis", Hardy was especially interested in the way behavioral changes might drive Darwinian evolution without any need for Lamarckian heredity. This was the main topic of his 1963 Gifford Lectures that were subsequently published as a book entitled *The Living Stream*. It is in the sixth and seventh chapters ("Behaviour as a Selective Force" and "Habit in Relation to Bodily Structure") that he expressed the clearest argument in favor of such a theoretical positioning. Dissatisfied with the dominant interpretative framework of standard population genetics,[4] relying on several occasions on Thorpe's experimental results,[5] Hardy argued that "habit [is] fostering a form of selection"[6] and that evolutionary creativity, that is, the genesis of evolutionary novelties, has to be found in "the new habit",[7] which is "the real initiating agent in the process".[8] Aware of what had been done in the past by Lloyd Morgan and Baldwin,[9] Hardy stressed, even more than Thorpe did, organisms' agency in drawing a distinction between two forms of natural selection, which seems remarkably equivalent to Gulick's 1905 distinction between "passive" and "active" (or "endonomic"; see Section 1.5) selection:[10]

> I would, however, make another kind of division between the forms of selection. I would distinguish all the foregoing kinds under a super heading of *external* selective agencies, meaning those acting from outside the organisms concerned, *i.e.* the selective forces acting from both the animate and inanimate environments; and in contrast to these I would place an *internal* selective force due to the behavior and habits of the animal itself.[11]

Thereby, at least verbally, Hardy was still pointing in the very same theoretical direction as Gulick and Thorpe before him. Such conceptions (OS2) are indeed very close, if not basically identical in the causality involved, to our modern Niche Construction Theory (NCT). NCT emerged in the late 1980s as, it was thought, an alternative to standard Modern Synthesis[12] and promotes the same theoretical framework based on an acknowledged evolutionary role of organisms' agency. In NCT, plastic organisms transform their environment to such an extent that "acquired characters . . . become evolutionary significant by affecting selective environments in systematic ways",[13] a modern definition that could have been perfectly endorsed by Gulick, Thorpe, or Hardy.

In my reading, the concept of organic selection framed in two successive episodes during the first half of the 20th century already encompasses the evolutionary mechanism (OS2) that will only later be clearly individualized in the NCT framework. In other words, the core idea promoted in NCT, that is, that organisms "transform some of the selection pressures in the environments that subsequently select them"[14] largely predates the 1980s. What was not envisioned in these early works, neither in 1896–1905 nor in the 1940s, is the case when the selective regime is altered because organisms are *engineering* their material environment. In short, for Gulick, Thorpe, and others (like Waddington; see Part II), what was at stake was more habitat selection than habitat construction (in the material sense of the term). Nonetheless, on a formal level, this distinction is meaningless because it does not affect the way fitness is impacted by plasticity; that is, it does not change how such a phenomenon can be mathematically modeled.

My point here is not to undermine the significant amount of work, both theoretical and empirical, which has emerged in these past 30 years or so within NCT. In many aspects, NCT is now a much more refined framework than OS2 ever was. What was especially missing in the 1940s was a population genetic theory of organic selection that would have allowed to model how habitat selection could drive evolutionary change. This was obviously never Thorpe's intention nor competence. Such a bridge between OS2 and nascent population genetics could have been achieved in the USSR, but, unfortunately, it was not.

Moreover, because today NCT is a hotly debated topic, it also proves fruitful in raising many theoretical problems that were never envisioned before.[15] In this concluding section, I simply aim to show that the basic idea that plasticity alters selection pressures was not invented by Richard Lewontin in the 1970s, nor even by Waddington back in the 1950s,[16] but indeed had already been framed as part of the early history of organic selection. The fact that NCT seemed *new* to most biologists back in the 1990s is in itself highly telling: a historical thread was somehow broken again, so to speak, between the early 1940s (Thorpe) and the late 1980s (Odling-Smee). In the last part of this book, I investigate what happened then, but the basic idea can already be briefly sketched. During the 1950s and 1960s, during the so-called hardening of the Modern Synthesis, the complex set of concepts grouped under the term "organic selection" was reframed (and drastically simplified) into Simpson's Baldwin effect, which (barely) survived as a footnote in the Modern Synthesis framework. This is the second major discontinuity within the complex history under scrutiny. It started when plasticity was confined to a "proximate" mechanism, which long precluded any possibility of attributing it a substantial evolutionary role.

## Notes

1 Thorpe (1940, p. 341).
2 Pennisi (2018).
3 For instance, paleontologist and ethologist Rosalie F. "Griff" Ewer explicitly argued that "this somewhat grudging use of the concept of survival value is, however, not sufficient. It leaves out the live animal and concentrates too much on what it is, too little on what it does. It would appear to be glaringly obvious that what an animal does, or tries to do, can determine what characters are of survival value, *i.e.* can decide the direction of natural selection – and yet even today this indirect action is frequently ignored and the only possible effect of an animal's activity which receives consideration is the direct Lamarckian one" (Ewer, 1960, p. 163).
4 Hardy (1965, p. 153).
5 Hardy (1965, pp. 161–162, 172–173).
6 Hardy (1965, p. 185).
7 Hardy (1965, p. 172).
8 Hardy (1965, p. 172).
9 Hardy (1965, p. 161).
10 Gulick was never mentioned by Hardy, who seemed to be unaware of his work.
11 Hardy (1965, pp. 171–172). Emphasis in the original.
12 The term "niche construction" was first coined by British biologist John Odling-Smee in 1988 (Odling-Smee, 1988). Usually, the 1996 article "Niche construction" published in *The American Naturalist* is seen as the seminal paper (Odling-Smee et al., 1996).
13 Laland et al. (2016, p. 192).
14 Odling-Smee (2010, p. 176).
15 For instance, the distinction between two different impacts organisms might have on their environment: "mere effects" or "adaptive engineering". See especially Sterenly, 2005 and Dawkins, 2004. For a general philosophical treatment of NCT basic assumptions, see Pocheville (2010).
16 Odling-Smee (2010, pp. 175–176); Laland et al. (2016, p. 192).

## Part II

# Waddington and Schmalhausen

Thinking plasticity developmentally

# Introduction of Part II

I was looking at evolution very definitely from the point of view of a developmental biologist.

— Waddington, 1975

As we have seen, while the history of the debate surrounding organic selection has mostly been forgotten,[1] this debate represented an active field of research within evolutionary theory, first around 1900 and then from the mid-1930s onward. Several prominent biologists, in various local and national contexts, contributed to the issue, and (sometimes in complete isolation from one another), as time passed, concepts and questions were refined, and experimental programs were eventually launched. In short, even though the history at stake is remarkably discontinuous, from the 1890s to the 1940s, the issue has followed, by and large, a progressive path.

In the 1940s and the 1950s the history of the conceptual relation between adaptability and adaptation should have been substantially transformed by Schmalhausen's and Waddington's involvements. In the 20th century, their contributions constituted by far the most impressive body of work on the causal role of plasticity in the production of evolutionary adaptations. Neither before nor after did any other biologists come close to what Schmalhausen and Waddington have done. Both reframed the issue in new and stimulating ways, coined new terms for new concepts, colligated known facts that supported their views, and, at least for Waddington, started experimental programs designed to show how significant plasticity could be as an evolutionary factor.

In contrast to others, Waddington has never been forgotten. Nonetheless, what biology currently owes – or could owe – him remains an open question. Were his many concepts – "genetic assimilation", "canalization", "epigenetics", and so on – only new fashionable ways of repackaging rather standard ideas, or were they endowed with genuine heterodox content? This aspect is especially problematic and has remained debated.[2] One reason for this ambiguity is that Waddington regularly moved from one subject to another, without necessarily deepening the issues at hand.[3] In other words, it might be

DOI: 10.4324/9781003422990-8

the case he did not deliver on all his promises. Another reason is that some of his ideas became embedded within the nascent philosophy of biology from the 1980s onward. For many philosophers of biology, Waddington remains an attractive thinker because he supposedly challenged the orthodoxy of the gene-centered view of Modern Synthesis and molecular biology.[4] The existence of such literature is sometimes a helpful resource because it provides a conceptual analysis of some of Waddington's core notions.[5] Yet, it might also constitute an obstacle because the philosophy-of-biology Waddington might obscure other aspects of his work.

For obvious reasons, Schmalhausen's case is different. If he remains known by some biologists (mainly developmental biologists interested in evolution[6]), he never became a central figure within theoretical biology or philosophy of biology, even though he was seen as one of the scientists who participated in the foundation of the Modern Synthesis by at least Dobzhansky and Simpson. Most of his published work – not to mention unexplored archives in Moscow – still wait for a substantial historical examination. The main reason is that Schmalhausen wrote almost exclusively in Ukrainian and Russian. It seems that only three of his numerous writings were published in English. The first is his important 1949 book *Factors of Evolution. The Theory of Stabilizing Selection*,[7] which presents a detailed account of his view. The second is a 16-page paper published in *Evolution* in 1960 titled "Evolution and Cybernetics". As the title makes clear, Schmalhausen tried, at the very end of his career, to show that his conceptions worked well in the new framework of cybernetics.[8] The third is his book, *The Origin of Terrestrial Vertebrates*, published posthumously in 1968 (five years after Schmalhausen's death).[9] Unfortunately for our purposes, this work, typical of his last period when he was marginalized under Lysenko's reign,[10] is mostly a detailed empirical study[11] and does not tell us much about Schmalhausen's views on evolutionary causality. For someone unable to read Russian, the problem is thereby the same as the one already encountered in Chapter 3: past conceptions have to be rebuilt on fragile grounds, even if the 300 pages of *Factors of Evolution* are very informative. This is why I had one of his main articles translated, his 1941 "Stabilizing Selection and Its Place among the Factors of Evolution".[12] These limited sources do not allow us to understand how his ideas originated in the context of the Soviet evolutionary morphology of the interwar period;[13] nonetheless, they provide enough to ascertain that, despite the fact that the translation of *Factors of Evolution* is far from excellent, it is reliable enough and can be used as a general account of Schmalhausen's views, at least as they crystalized in the 1940s.

Given this imbalance, the present reconstruction necessarily relies on Waddington more than on Schmalhausen. In accordance with what previous historians[14] and the actors[15] themselves thought, there is a remarkable convergence between their views, which makes the issue of access to Schmalhausen's writings slightly less problematic. Terminology aside, they asked the same questions, within the same general framework and developed almost the same concepts. The few differences that can be found remain secondary. Thereafter, this part's argument rests first and foremost on Waddington and only secondarily on Schmalhausen. When necessary, differences between their views will be explicitly stressed and explained. A lot of work remains to be done on Schmalhausen, at least in English.[16] I hope the present book will serve as an encouragement for other colleagues to devote some research to this still-neglected 20th-century biologist.

### A conscious choice? Development rather than ecology

In contrast to the scientists studied in the first part, Waddington and Schmalhausen focused on the developmental aspect of plasticity rather than on the ecological aspect. In other words, they were interested in the embryological connection between plastic adaptability and genetic adaptation, and not primarily in the ecological interaction between plasticity and natural selection. This was partly a conscious choice: both were to some extent aware of most of the theoretical dimensions that grounded the work of Baldwin, Lloyd Morgan, Thorpe, Gause, and others. As trained developmental biologists, they simply did not want to go in that direction. That does not mean that they did not consider ecology. They sometimes did, and in detail.[17] But when it came to plasticity, they were much more interested in the way embryological development could be (selectively) altered or even rebuilt to transform a plastic and phenotypic accommodation into a genetic adaptation. In short, the central argument of the present part is that whereas organic selection is a fitness-based argument, "genetic assimilation" (in Waddington's famous terminology[18]) is a mechanism-based argument, and the aim is to show how Waddington and Schmalhausen went to conceive embryological restructuration under the creative force of canalizing (Waddington) or stabilizing (Schmalhausen) selection.

Waddington's emphasis on development rather than ecology can be seen in the (uneven) way in which he worked on the different sub-systems that made up, according to him, the evolutionary process. He distinguished four, namely the "genetic system", the "natural selective system", the "epigenetic system", and the "exploitive system".[19] The former two were at the core of the Modern Synthesis, whereas, for Waddington at least, the latter two had been mostly neglected. If, as is well known and acknowledged,[20] Waddington tried, for decades, to produce new knowledge about the epigenetic system, he was far less interested in the exploitive system, that is, basically, niche construction.[21] He felt that this field was promising, but he did not work much in that direction.

He only devoted one publication,[22] seemingly inspired by Thorpe, to the elucidation of a component of the exploitive system, habitat selection. In 1954 in collaboration with B. Woolf and M.M. Perry, he designed an experiment in which different stocks of *Drosophila melanogaster* were able to choose between various chambers in which temperature, light, and humidity were assigned different values.[23] The experiment showed that flies were active in selecting their environment and that they usually chose the environmental conditions that maximized their fitness. As promising as they were, these results never launched further research. Waddington remained focused on the epigenetic system and its evolutionary formation.[24]

Whether this was a fully conscious decision on Waddington's part is open to discussion. It must be pointed out that his partial lack of interest in the exploitive system might have played a role in his depreciation of organic selection. Explicitly in their writings, both Waddington and Schmalhausen distanced themselves from organic selection.[25] According to them, this evolutionary mechanism was both different from their own and most probably not very efficient under natural circumstances. This issue will be dealt with later (see Chapter 4's conclusion) and a single aspect needs to be stressed here. When Waddington wanted to distinguish his genetic assimilation model from the one Simpson had just termed the "Baldwin effect", he was especially concerned with the evolutionary role of the plastic response. In traditional organic selection, as we have seen, the plastic response is evolutionarily relevant in that it blocks and/or polarizes natural selection. It both "buys time" (OS1) and creates a new ecology with an altered selective regime (OS2). For Waddington, this was not central, to say the least. He never considered the "breathing space" hypothesis to play a causal role,[26] and he never really elaborated on the fact that the plastic response could shape new selective pressures.

In short, neither Waddington nor Schmalhausen ever embraced the ecological dimension that was at the root of organic selection and that would be instrumental in the rise of the modern Niche Construction Theory. They saw plasticity and the connection between adaptability and adaptation through the eyes of developmental biologists, and most of their work, both empirical and experimental, was driven by such a view. So, while natural selection was part of their global framework, they treated it as something closed to a fixed, external parameter (Waddington's experiments are here especially telling; see Section 4.2). They were interested in the way canalizing/stabilizing selection enables the transformation of a reaction norm,[27] that is the transition from a conditional accommodation to a canalized adaptation.

This part aims to give a reliable and critical account of their developmental perspective. A complete description of their theoretical commitments goes beyond the scope of the present book.[28] I will focus almost exclusively on genetic assimilation, that is, what they viewed as an alternative evolutionary mechanism that causally connects adaptability to adaptation.

Today, it is still unclear what precisely this mechanism consists of, and whether or not it is supported by certain experimental results. Many reasons can account for this state of uncertainty: the vagueness that systematically surrounds Waddington's formulations, the intrinsic complexity of the concepts involved, and so on. The main hypothesis that will be discussed throughout the three chapters of this part is, however, the following: two different concepts have coexisted under the name of "genetic assimilation". One is quite easily reducible to the framework of quantitative genetics; the other is probably more radical and heterodox. According to this hypothesis, the history of genetic assimilation is above all the history of a conceptual ambiguity whose meaning we shall try to reactivate here.

## Notes

1 To such an extent that contemporary scientists active in the field, like Samuel M. Scheiner, could write that Baldwin's 1896 seminal paper was "mostly ignored by biologists at the time" (Scheiner, 2014, p. ii).
2 As we shall see, Ernst Mayr thought Waddington's wording was unnecessary. More recently, developmental biologist Adam S. Wilkins was still wondering if genetic assimilation was something new within the evolutionary synthesis (Wilkins, 2003, p. 24).
3 According to his colleague Alan Robertson: "With [Waddingtons's] width of vision he could see clearly what should be done and he was good at persuading funding organizations to provide support. But having got support for specific projects, he seemed often to lose interest in their subsequent progress, perhaps a result of intellectual impatience" (Robertson, 1977, pp. 612–613).
4 This is especially true for philosophers involved in so-called Developmental System Theory (DST), like Susan Oyama. Waddington is also praised as a prominent figure in the history of organicism (see, for instance, Gilbert & Sarkar, 2000 and Peterson, 2016).
5 Griffiths' 2006 interpretations are especially valuable (Griffiths, 2006).
6 For instance, Adam S. Wilkins dedicated his classic textbook *The Evolution of Developmental Pathways* "to the memory of Ivan Schmalhausen (1884–1963), a great, but neglected, pioneer" (Wilkins, 2002).
7 The book was written during the Second World War (in Borovoje, now in northern Kazakhstan), mostly in isolation with respect to Western scientific literature. A first draft was ready as soon as 1943 (Levit et al., 2006, p. 92). The original version was published in Russian in 1946 (the date 1947 is sometimes mentioned) and translated into English in 1949 by Isadore Dordick, under the supervision of Theodosius Dobzhansky, who wrote a short foreword. In comparison to the original text, the English translation is shorter. According to Dobzhansky, mostly pages on standard genetic notions were eliminated (Dobzhansky, 1949, p. xvii). According to historian Mark B. Adams, Dobzhansky would in fact have reworked the original text more substantially (Adams, 1988, p. 283). The book was reprinted in 1986 with a new foreword by evolutionary morphologist David B. Wake. It is this 1986 version that I use.
8 According to Levit and colleagues, "in the last period of his scientific career, Schmalhausen was occupied by the idea of explaining evolution in terms of cybernetics" (Levit et al., 2006, p. 98).
9 The Russian version was also published posthumously, in 1964.
10 Gans (1968, p. viii); Wake (1986, p. v); Levit et al. (2006, p. 92). Note that Schmalhausen's situation had improved after 1953 and Stalin's death (Levit et al., 2006, p. 93).
11 Wake (1986, p. vii).
12 Here, I owe again special thanks first to Sergey Shalimov, who provided the original text, and moreover to Aleksandra Traykova, who translated it into English.
13 The links between morphologist Alexei Nikolaevich Severtsov (1866–1936) and his student Schmalhausen have already been studied and emphasized. In English,

see especially Adams (1980a) and Levit et al. (2006). I briefly come to the issue in Section 4.1.

14  Gilbert (1994); Levit et al. (2006, pp. 100–102); Wilkins (2008, p. 230). An insightful and almost systematic comparison between Schmalhausen and Waddington could be found in a highly stimulating PhD work that, unfortunately, has been neglected so far: Paul D. Lewin, *Embryology and the Evolutionary Synthesis: Waddington, Development and Genetics* (Leeds, 1998). The whole text can be downloaded here: https://core.ac.uk/download/pdf/43017.pdf.

15  For instance, this is how Schmalhausen ended his preface of *Factors of Evolution*: "[Waddington's] solution is very close to ideas I had previously evolved in a series of books and articles beginning in 1938. The difference between Waddington and myself amounts to a somewhat different terminology. I employ the terms 'autoregulation' and 'autoregulating mechanism' in approximately the same sense as Waddington uses the term 'canalization' of development" (Schmalhausen, 1949, p. xxii). Waddington cited Schmalhausen minimally but admitted their theoretical proximity, although he especially emphasized what he thought were significant differences (Waddington, 1961a, pp. 273–274). At the end of his life, he was more likely to acknowledge their proximity: "Schmalhausen . . . was thinking along lines very similar to those I was following, at about the same time and quite independently" (Waddington, 1975a, p. viii).

16  It seems that Russian historian of science Yakov Gall has published extensively on Schmalhausen and other actors of the Russian school, unfortunately only in Russian.

17  As for example Schmalhausen did in his late cybernetic reading of the main steps of a microevolutionary process (Schmalhausen, 1960).

18  "Genetic assimilation" might be the only Waddingtonian term that does not have any equivalent in Schmalhausen's framework (see Section 4.2, for a more detailed discussion). Such an absence, among other things (including Waddington's priority quest), made Waddington skeptical about the fact that Schmalhausen went as far as he did in conceptualizing genetic assimilation (Waddington, 1961a, p. 273).

19  Waddington (1959a, p. 400).

20  Peterson (2016); Fabris (2018); Nicoglou (2018).

21  Despite the fact that he is regularly presented as the grand-father of Niche Construction Theory, with Richard Lewontin as the father (Odling-Smee, 2010; Laland et al., 2016).

22  According to what seems to be a complete list of references stored in Edinburgh's archives (EUA IN1/ACU/A/5/4), Waddington published no less than 465 texts, all included, from 1927 until 1978 (he died in 1975). An examination of the titles suggests that his 1954 paper was his only work related to the exploitive system.

23  Waddington et al. (1954).

24  Note that at the end of his career Waddington showed a renewed interest for the ecological impact of plasticity. In his 1969 text "Paradigm for an evolutionary process", he encouraged his young colleagues to work into the direction of Niche Construction Theory: "I have myself suggested that the parameter which is continually increased in evolution would have to express ability of members of the system to find some way or other of keeping alive and leaving offspring. This is perhaps not very far from MacArthur's suggestion but involves the possibility that the organisms may not simply become more efficient at using available resources but may begin exploiting new resources. The way to give more precision or more penetration to such ideas is, I suggest, a new investigation of the logical structure of the evolutionary process" (Waddington, 1969b, pp. 114–115).

25  Schmalhausen (1941, pp. 308, 327; 1949, p. 204; see also Levit et al. (2006, p. 102). Waddington (1953b, 1961a, pp. 287–289).

26 See, for instance, Waddington (1953b, p. 386).

27 The idea that the reaction norm of a genotype can be altered, which plays a central role in Schmalhausen's *Factors of Evolution* in particular, can already be found in the seminal paper by Richard Woltereck (1877–1944), in which Woltereck lays the foundation for the modern concept of the reaction norm (see the fourth and final section of his text, Woltereck, 1909). Nevertheless, in this article at least, Woltereck considers this alteration without ever mentioning a selective process and refers instead to the very vague and ambiguous term of "acclimatization" of Daphnia to new conditions of nutrition. It is therefore difficult to know whether the mechanism he had in mind could in certain aspects anticipate that of genetic assimilation.

28 I refer the reader to Lewin's PhD for an attempt into that direction (Lewin, 1998).

# 4 "Autonomization", "canalization", and "genetic assimilation"

## Toward a developmental perspective on the relationship between adaptability and adaptation

For decades, facts of so-called parallelism provided the strongest evidence in support of the Lamarckian account of the evolutionary process. When organisms were placed in unusual environmental conditions, they developed plastic accommodations over the course of their lives that closely, but often less completely, mimicked the hereditary adaptations of related species. In the 19th century and still at the beginning of the 20th century, what seemed to be the most rational explanation was the inheritance of acquired characters: somehow, plastic accommodations ("acquired characters") were able to retroact on the hereditary substance (Weismann's germplasm) and produced in the next generation an innate variation that was already directed toward adaptation. From generation to generation, if the new conditions were maintained, it would supposedly lead to hereditary reinforcement, until complete fixation was achieved. The standard Lamarckian scheme was individual and physiological: for reasons that remained to be understood, an organism was endowed with the ability to physiologically transform its heredity (i.e., to bypass "Weismann's barrier") according to what it experienced during its life.

Waddington and Schmalhausen were primarily concerned with the same class of facts. Like their predecessors, they were highly interested in the numerous documented examples of parallelism and were convinced that "acquired characters" were important causal factors in organic evolution.[1] But unlike their predecessors, they strongly denied the standard Lamarckian hypothesis any explanatory power.[2] What they sought to do was to replace the old Lamarckian concept of inheritance of acquired characters with a new one that would be fully compatible with a Darwinian framework and Mendelian genetics.[3] The concept of inheritance of acquired characters therefore had to be entirely reframed. There was no need for a single organism to be able to directly transform its heredity. But what would happen if, in a population, organisms plastically and phenotypically accommodated new environmental constraints?

For Waddington and Schmalhausen, plasticity starts a microevolutionary process that will indeed end up transforming and remodeling embryological pathways but because of natural selection acting at the population

DOI: 10.4324/9781003422990-9

level. Waddington was especially explicit in this regard.[4] He continued to use the term "inheritance of acquired characters" well into the 1950s and the 1960s,[5] even though he knew that it would raise ambiguity and strong opposition at the time of Lysenkoism. What he had in mind was a renewed and populational understanding of the concept.[6]

For both biologists, explaining how acquired characters might become part of the hereditary repertoire at the population level required a reconceptualization of the causal relationships between the developmental process, natural selection, and heredity. If, over the course of generations, a population subjected to natural selection becomes less and less dependent upon environmental cues in the production of the newly adapted phenotype (that will tend to become the "wild type"), then it necessarily implies that embryological development has become insensitive and automatic. Such developmental robustness had great evolutionary consequences for our biologists, as we will see. Waddington termed it "canalization", whereas, Schmalhausen independently called it "autonomization", a term which would not have the same legacy.

The present chapter is meant as an introductive characterization of their developmental interpretation of the relationship between adaptability and adaptation. In contrast to most of the other biologists that were studied in the previous part, Waddington and Schmalhausen remained primarily focused on the embryological consequences of a special form of natural selection, that they termed canalizing (Waddington) or stabilizing (Schmalhausen) selection. Above all, they wanted to understand how a developmental pathway could be transformed or even rebuilt to produce a new and stabilized reaction norm. This was in contrast with previous thinkers – Baldwin, Lloyd Morgan, and Gulick in the years around 1900 on the one hand, Hovasse, Thorpe, and Gause in the early 1940s on the other hand – who never thought of this issue as central. Within the framework of organic selection, the embryological production of accommodations and adaptations was not at issue. But Waddington and Schmalhausen decided to focus on the embryological aspect of the genetic reinforcement of phenotypic accommodations.

## 4.1   Schmalhausen's and Waddington's paths, an overview

It is difficult to reconstruct Schmalhausen's work and career without having access to Russian sources. Beyond a few biographical notes,[7] only two documented articles are available in English.[8] This is hardly enough to produce a satisfactory narrative of how Schmalhausen, as a morphologist, came to conceive stabilizing selection, in other words, how in the 1930s he progressively moved from morphology and embryology to evolutionary theory and genetics. Nevertheless, before turning our attention to Waddington, it is necessary to provide some minimal biographical information about Schmalhausen.

### 4.1.1    Schmalhausen: how did he come to stabilizing selection?

Ivan Ivanovich Schmalhausen (1884–1963) was a generation older than Waddington. This is important because he sometimes appears to have been more familiar with 19th-century evolutionary morphology (like Haeckel's) than with the interwar population genetics. This might also explain why, in contrast to Waddington, Schmalhausen never tried to launch an experimental program on stabilizing selection. Another reason is that he was strongly impacted by the rise of Lysenko in the USSR during the late 1940s and 1950s, to such an extent that he was highly limited in what he could do at the time when Waddington started his own experimental work in Edinburgh.

Schmalhausen was born in Kiev, at the time still part of the Russian Empire. His father, Johannes Theodor Schmalhausen (1849–1894), of German origin, was a renowned paleobotanist and one of the founders of the discipline in Russia. Importantly, he was committed to Darwinism already in the 19th century.[9] Very early in his training as a biologist, most probably around 1902, Ivan Schmalhausen became acquainted with evolutionary morphologist Alexey Severtsov (1866–1936), who would become his mentor and later his close collaborator. According to Adams, Severtsov was the leader of the Russian school of morphology[10] and was himself highly interested in the causal relations between evolution and development, as some of his books testify.[11] Severtsov's work was still closely related to the biogenetic law, the relation between ontogeny and phylogeny, and most of the issues that had been pivotal in the rise of evolutionary morphology since the 1860s. As regards evolutionary causality, he was already reluctant to Lamarckism and endorsed the primacy of natural selection, even if he remained much more interested in phylogeny and macroevolution than in the mechanism of evolutionary change.[12]

Under Severtsov's supervision, Schmalhausen worked as a PhD student in Moscow and at Naples' zoological station from 1912 until his defense in 1916. His early work was typical of Severtsov's own interests: Schmalhausen studied the morphology of amphibian limbs from an evolutionary perspective.[13] He then started an academic career at Kiev University. It seems that until the mid-1930s, Schmalhausen's work "closely mirrored his teacher's",[14] in that he confined himself to questions of evolutionary morphology. His change of perspective, his new interest in genetics and in the mechanisms of evolution, dated back to his final move to Moscow. In 1935, Schmalhausen was elected to the Soviet Academy of Sciences, and in 1936, following Severtsov's death, he took over the direction of the "Severtsov Institute of Evolutionary Morphology".[15]

Schmalhausen only started to elaborate a synthesis of his own between morphology, natural selection, and genetics in the second half of the 1930s, when he was already in his fifties. In 1938–1939, he published in rapid succession two important books (never translated into English): *The Organism as a Whole in Individual and Historical Development* and *Trends and*

*Laws of the Evolutionary Process.* As far as we know,[16] it is in this second book that, for the first time, he introduced the concept of stabilizing selection that would later become the centerpiece of *Factors of Evolution* (1946, 1949a, for the English translation). From then on, Schmalhausen stayed in Moscow (except for a few years during the war), even though, because of Lysenko's rise, he was institutionally marginalized during the late 1940s and early 1950s. Nonetheless, it seems that Schmalhausen's last few years were quite productive when his position improved after Stalin's death. According to Levit and colleagues, he was planning new books when he died on October 7, 1963.[17]

As the title of his 1938 book makes clear, Schmalhausen came to favor a holistic approach to the organism. Like Waddington and other supporters of the organicist movement,[18] he opposed the reductionist and atomist view that was central to genetics and population genetics during the 1930s. As we shall see, this theoretical commitment was still at the root of *Factors of Evolution*. In 1941, he published a 43-page detailed article entitled "Stabilizing Selection and Its Place among the Factors of Evolution". This paper already offered a refined account of stabilizing selection. Thus, it seems that the concept of stabilizing selection, which had absolutely no equivalent In Severtsov's own morphological take,[19] emerged rather quickly during the period 1939–1941. Where did it come from? This question remains unresolved and would require the ability to read Russian. According to Levit and colleagues,[20] Schmalhausen himself credited Gulick for his previous distinction between "balanced" and "unbalanced" selection. Such a distinction predated Schmalhausen's between "dynamic" and "stabilizing" selection (see Chapter 6). It might be the case that Gulick inspired Schmalhausen. Yet, given the very weak elaboration of the distinction in Gulick's 1905 book between balanced and unbalanced selection,[21] this distinction at best participated in stimulating Schmalhausen's reflections, but it could hardly have played a major role. In *Factors of Evolution*, Schmalhausen did refer to Gulick[22] but only to mention the possible influence of accidental processes in microevolution.

### 4.1.2 Waddington: from canalization to genetic assimilation

Waddington's path to genetic assimilation has been much better documented, and several in-depth studies are already available.[23] The present section is only an overview of the personal trajectory of Conrad Hal Waddington (1905–1975), and I refer the reader to existing literature for more detailed accounts. In the late 1930s, as an experimental embryologist, Waddington progressively elaborated the concept of canalization that led him to hypothesize, in the early 1940s, the mechanism of genetic assimilation. In short, in Waddington's case, his involvement in developmental biology ended with a new concept that launched his research program as an evolutionary biologist.

As a graduate student in Cambridge, Waddington started a PhD in paleontology on certain species of ammonites. Ammonites interested Waddington because their shells are reified processes, each part corresponding to a developmental stage. For the young Waddington, this aspect was especially stimulating as it "forces on one's attention the Whiteheadian point that the organisms undergoing the process of evolution are themselves processes".[24] Yet, because he wanted to be involved in active experimental science,[25] Waddington progressively lost interest in paleontology[26] and never completed his doctoral dissertation, becoming a doctor only in 1938 on the basis of his published work in causal embryology,[27] a field he entered in the early 1930s. Causal embryology was then very attractive for Western young scholars, in the wake of Spemann and Mangold's breakthrough results on embryological induction. They were able to demonstrate how a group of cells, called the "organizer", was responsible for the induction of the neural tissues during development in various amphibian embryos. This discovery allowed for many experimental possibilities and set the stage for experimental embryology for decades. In his recent book on 20th-century organicism, Peterson recalls how understanding the mechanism of induction "became something of a scientific gold rush in early 1930s European embryology and biochemistry".[28]

Waddington was no exception. With a few friends and colleagues from Cambridge – especially Joseph and Dorothy Needham – they started the famous "Theoretical Biology Club" that was first devoted to understanding the organizer. For almost ten years, in collaboration with others (mostly the Needhams and Jean Brachet), Waddington worked as an experimental biologist, performing many researches related to "epigenetics" in birds and mammals. One of his main achievements was to succeed in developing new techniques for the study of the avian egg (especially of the chick) and the process of induction, a material much more complex than the amphibian embryo.[29] As a developmental biologist, he published around 53 papers on the subject, from 1930 to 1943,[30] which earned him recognition in the field.[31] Apart from the work done in Cambridge, where he tried to develop a laboratory of experimental embryology, Waddington also stayed in Berlin, in Otto Mangold's laboratory, where he was invited in 1933.

After years of intense work in several laboratories in Europe and Japan, no one was able to identify the chemical nature of the organizer, which was, at the time, far beyond the scope of available techniques. Waddington himself was soon embarrassed by the results obtained, yet he stood out in his persistence to build on the repeated failure to identify the inducing factors. It can be argued that the formation of the concept of canalization was largely the direct consequence of these repeated failed attempts. Waddington progressively convinced himself that we should transform our understanding of embryogenesis.[32] Given that an exceptionally large set of substances could bring about induction in embryos, he thought that rather than understanding development and differentiation as being "caused" by the organizer, a better framework might be to look at them as the end result of a "complex system of actions and interactions".[33] To put it briefly, the specificity of developmental paths was transferred from the chemical substance produced by the organizer to the system able to react to it, the developmental system. He put forth the concept of competence to stress that what was important was the faculty of the developmental system to be responsive, whatever the nature of the inducing signal. For Waddington, competence was an evolved property of embryonic tissues and the main cause of the specificity of development. The organizer's chemical product, the so-called evocator, was progressively downgraded to only a mere stimulus able to push the tissue into one of the developmental paths already available.

This theoretical move culminated in his 1940 seminal book, *Organisers and Genes*, in which Waddington drew the consequences of his 10-year effort in experimental embryology. In this book, he tried to offer a synthesis of his own, a synthesis between developmental biology and genetics,[34] in the wake of Thomas Morgan's challenge.[35] In early 1939 he went to Caltech and worked for a few months in Morgan's laboratory. He devoted a lot of energy at that time to trying to elucidate the genetic determinism of wing patterns in *Drosophila*[36] and became convinced that the next theoretical step would be to produce a synthesis between causal embryology and classical genetics. For Waddington, there was no doubt that competence was under genetic control, especially given the results he managed to obtain at Caltech.[37] Yet, to him, even if genes were ultimately the cause of development, it did not imply that a precise phenotypic character was the consequence of any individual allele. In contrast, Waddington most usually favored a systemic understanding of the genome, where all the genes were collectively responsible for the developmental process.[38] We will see that this would have pivotal consequences for his second account of genetic assimilation (see especially Chapter 6).

Waddington's processual commitment inspired by Whitehead's philosophy,[39] and his special ability to think diagrammatically, was crystallized in *Organisers and Genes* in the famous representation of the "epigenetic landscape" in which developmental paths are well-defined discrete entities which pre-exist the running down of the ball that symbolizes the course of cell differentiation (see Figure 4.1). Even though the term "canalization" is absent from *Organisers and Genes*, the concept already appeared on a few occasions.[40] It was in a 1942 article published in *Nature* that Waddington coined the term "canalization" which he defined as the "[adjustment of developmental reaction], as they occur in organisms submitted to natural selection . . . so as to bring about one definite end-result regardless of minor variations in conditions during the course of the reaction".[41]

During the war, when Schmalhausen was writing *Factors of Evolution* in almost complete isolation, Waddington had to distance himself from his scientific work as he became involved in operational research in the Royal Air Force and was eventually appointed scientific advisor to the Commander in Chief of Coastal Command from 1944 to 1945.[42] After World War II, he looked for a permanent position and was offered a professorship in animal genetics at the University of Edinburgh, where he settled in 1947. It was in Edinburgh that Waddington developed his concept of genetic assimilation

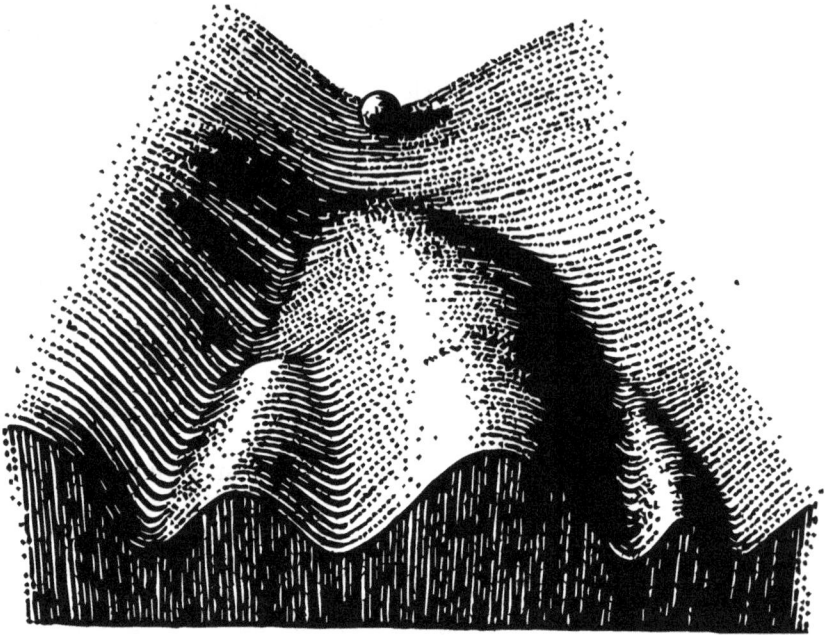

*Figure 4.1 Waddington's most usual representation of an epigenetic landscape.* Reproduced from: C.H. Waddington, *The Strategy of the Genes*, London, Allen and Unwin, 1957. With permission from Taylor & Francis.

and several series of experiments designed to support his view. These experiments will be given special attention in the following section.

In the Institute of Animal Genetics, Waddington succeeded in creating a very active school of genetics and attracted many prominent scholars[43] in the 1950s and 1960s. It must already be noted that most of the members of his department, for instance, Douglas S. Falconer and Alan Robertson, did not follow him in promoting canalization as the cornerstone of the process of genetic assimilation (see Chapter 8). In the late 1950s and early 1960s, at the peak of his career at the Institute,[44] Waddington's genetic assimilation model stabilized in its definitive form, as testified in his 1961 in-depth review.[45] Waddington repeatedly argued that the Modern Synthesis was incomplete and unsatisfactory,[46] but he was unable to renew and expand his criticism and what he wrote in the 1960s and early 1970s therefore did not add much to what he had already published.

His last detailed thoughts on these subjects were published in the four volumes he edited under the title *Towards a Theoretical Biology*. These volumes were the result of a series of meetings hosted at the Villa Serbelloni, in Bellagio near Lake Como, over several consecutive summers from 1966 onward. Under Waddington's patronage, each meeting gathered between 15 and 20 scientists from different backgrounds, from evolutionary biology (Ernst Mayr), to mathematics (René Thom) and biochemistry (Stuart Kauffman), who were all interested in abstract issues. Unsatisfied with the reductionist approach of contemporary molecular biology,[47] Waddington attempted to do for biology something akin to what theoretical physics had done for physics.[48] At that time, and until his death in 1975, he saw genetic assimilation as an original evolutionary model that was in contradiction with at least some of the core conceptions of the Modern Synthesis and standard population genetics.[49]

## 4.2 The problematic issue of empirical basis

The remarkable equivalence between Schmalhausen's and Waddington's conceptions, as already pointed out, has been acknowledged for decades and first by Schmalhausen and Waddington themselves. Because these conceptions were formed in isolation from one another, the terms they coined differed but it is mostly easy to draw a correspondence between them. For example, Waddington's "canalizing selection" is the same as Schmalhausen's "stabilizing selection". One ambiguity remains: as previously noticed, it seems that there is no equivalent in Schmalhausen's vocabulary for Waddington's "genetic assimilation". Some scholars might disagree because Schmalhausen's "autonomization" is usually seen as corresponding to Waddington's "genetic assimilation".[50] In my reading, "autonomization" and "autoregulation" (autoregulation being a weaker/incomplete form of autonomization) are closer to "canalization" than to "genetic assimilation".

That this ambiguity exists is in the literature is telling: for both Waddington and Schmalhausen canalizing/stabilizing selection produces canalized/ autonomous reaction norms that, eventually, allow genetic assimilation to occur. Thus, there is indisputably a close connection between canalization and genetic assimilation, a causal connection that does not exist within the framework of organic selection and that is usually missing in the scientific literature.[51] Canalization and genetic assimilation are both evolutionary products of canalizing selection. But canalization is a property of a single developmental system, whereas genetic assimilation is a populational process that consists in the evolutionary stabilization, in a population, of a new reaction norm. In short, in Mayr's standard terminology,[52] whereas canalization is mostly a proximate mechanism, genetic assimilation is an evolutionary one. Development, under certain circumstances, can become canalized. This was exactly how Schmalhausen used "autonomization", he always thought of a "progressive autonimization of development".[53] Autonomization, like canalization, is achieved when "the organism ceases to respond through formative reactions to certain changes in factors from the external environment".[54]

As is now acknowledged, a developmental path can become insensitive to intrinsic (genetic) and extrinsic (environmental) perturbations. In contemporary biology, this is why genetic canalization is distinguished from environmental canalization. This distinction was already present in Waddington's writings but without this exact wording. The same distinction can be found also in Schmalhausen's texts,[55] leaving little doubt that his use of "autonomization" was equivalent to Waddington's canalization. Finally, Schmalhausen himself, in the preface of *Factors*, explicitly recognized this correspondence.[56]

What is missing in Schmalhausen's repertoire is a synonym for genetic assimilation. The most probable interpretation for this absence is that Schmalhausen, who tended to coin a lot of new terms (most of them totally forgotten), thought it unnecessary to add a specific one somewhere in between stabilizing selection and autonomization. If stabilizing selection was correctly understood as a different form of selection able to produce specific developmental effects, there was no need for another term. Waddington, in contrast, chose to emphasize the specificity of the process and even supported the view that Schmalhausen "did not . . . formulate in any precise form the process which has been referred to here as genetic assimilation".[57]

A remarkable convergence also existed on the empirical basis that grounded this view of the evolutionary process. As we shall see, Waddington performed several experiments on genetic assimilation in the 1950s, when Schmalhausen was struggling with Lysenkoism. But before that, they both independently emphasized the same class of phenomena that were supposed to support the idea of progressive developmental autonomization: parallelism between adaptive ecotypes and plastic accommodations[58] (including ostriches' calluses[59]), the fact of allelic dominance understood as an evolved property, high frequency in natural conditions of the "wild-type" phenotype despite obvious genetic polymorphism.[60] To these facts, Schmalhausen added Gause's recent results (see Chapter 3) which, to him, constituted substantial evidence in favor of stabilizing selection.[61]

The ubiquity and stability of the wild type in natural populations were especially significant to them because they often favored the idea that biological heredity was not an intrinsic property of genes but instead a systemic phenomenon progressively produced by the gradual action of canalizing/stabilizing selection. In contrast to the view that was instrumental in the setting up of molecular biology, that is, in opposition to Schrödinger's notion of the "aperiodic crystal", intergenerational resemblance was not based on supposed intrinsic stability and permanence of a molecular structure (later identified to DNA) but on the evolved robustness of a highly integrated system. As soon as 1941, Schmalhausen left no room for ambiguity on the subject:

> At the same time, the stability of organization is tied precisely to the existence of complex systems of correlations of a regulative character and is not, of course, "explained" with the strength of the genotype (as done by Mashtaler, 1939, 1940, in full compliance with formal genetics). "Hereditary endurance of organization is based on the complexity of the historically established system of correlations, and not on the strength of hereditary . . . and their genes" (Schmalhausen, 1938, p. 138), and *the relative strength of organic forms in its deepest*

*foundations is based not on the genotype's strength, but on the complexity of the systems of correlations* which tie the organism together into a single whole and do not allow for any significant impairments without lethal consequences for the organism itself (Schmalhausen, 1939, p. 39).[62]

In standard population genetics, natural selection acts on hereditary variants. In Schmalhausen's and Waddington's favored view, at least in some of their writings (see Chapter 6), heredity is instead an evolutionary product of stabilizing selection and does not precede the working of selection.[63] As a matter of fact, their conception was much closer to the theoretical framework of quantitative genetics than to that of population genetics. This is why the usual instability of mutants, especially in *Drosophila*, was also an argument in support of this "constructivist" understanding of biological heredity[64] where the response to selection was the main issue.

For various reasons, Schmalhausen remained on the side of the consistency of his theory of stabilizing selection with known facts as they already existed in various fields, like morphology, embryology, and genetics. In contrast, Waddington launched several investigations to experimentally test at least some of the predictions derived from his model of genetic assimilation. During the 1950s, Waddington tackled two types of questions involving selection procedures in *Drosophila* populations. One was of course the experimental demonstration of the process of genetic assimilation itself. These experiments, carried out from 1952 to 1959, remain well known today. They are essentially the ones that are presented in this section. Nevertheless, in addition to this series of experiments during which Waddington coined the term "genetic assimilation", he carried out a smaller number of experiments on one specific aspect of his model: the fact that canalizing selection can produce developmental canalization. It was mostly in reaction to certain criticisms that Waddington wanted to show that the degree of canalization was accessible to experimentation. As we will see in Chapter 8, several other colleagues such as John M. Thoday, Curt Stern, Kenneth Mather, James M. Rendel, R.B. Dun, and A.S. Fraser (among others) worked on roughly the same subject, though mostly within a different interpretative framework. Most of them also performed experiments on these issues in the 1950s and early 1960s. Waddington was thus only one member of a network of researchers who sought to understand how developmental stability could emerge from specific forms of selection. This is worth noting because most of the time Waddington's experiments are presented in isolation, as if Waddington were a lonely figure at the margin of the Modern Synthesis, which is obviously a simplification. Therefore, I will expand on his experimental work devoted to canalizing selection in Chapter 8, where the context of what happened in quantitative genetics will be dealt with. For the sake of simplicity, I limit myself here to his work on genetic assimilation.

Waddington's 1950s series of experimental works on genetic assimilation in various *Drosophila* strains are still rather well known. But, from the outset until nowadays, the interpretation of his results remains a subject of controversy. This represents a textbook example of the underdetermination of the theory by data. The question of what Waddington's experiments actually show remains a largely open question, and in the current scientific publications, opinions strongly diverge depending on the author.[65] In the history of science, some experiments become, and remain, famous for settling a question. These are usually called *experimentum crucis*. But there is a second category of experiment, probably just as important but not as well identified from an epistemological point of view: experiments that have not increased the amount of knowledge available as such but that have continuously produced research through the questions they raise. Undoubtedly, Waddington's work belongs to this second category and has a remarkable ripple effect, which can still be felt today.[66]

Even Waddington, it seems, hesitated between at least two main interpretations. In his writings from the 1950s onward, he was equivocal enough to legitimize two divergent accounts. An especially contested aspect was the need to include canalization as a necessary component of the explanatory framework. For many of his contemporaries, and still today, simpler, and much orthodox interpretations should be favored (see Chapter 8). In this section, I focus on the experimental designs and the results obtained. I will turn to the interpretations produced in the following section and thereafter in Chapters 6 and 8.

Waddington's experiments on genetic assimilation can be seen as an extension of the numerous works of Richard Goldschmidt carried out between the end of the 1920s and the mid-1930s, which were at the basis of the concept of phenocopy. Within the framework of a vast research program on homeosis,[67] Goldschmidt had indeed shown that most of the mutations classically identified in *Drosophila* could be mimicked, phenotypically, if the embryonic development of the individuals was adequately disturbed (see Figure 4.2). In 1935, he gathered his results in a synthesis article in which he coined the term "phenocopy".[68] In his experimental work, Goldschlmidt systematically used a heat shock (35°C to 37°C) applied at a precise stage of larval development. He emphasized that for such a procedure to be operative the stage at which the development was effectively sensitive to heat shock needed to be precisely identified, an aspect that was also central in the protocols later established by Waddington. In later writings, Goldschmidt was confident that all the known mutant phenotypes could be obtained through such experimental procedures.[69]

With such a background in mind, Waddington performed three main sets of experiments during the 1950s, where he added to Goldschmidt's teratological methodology specific selective procedures. The first two, by far the most developed, were based on the same general design. Waddington and his collaborators submitted numerous *Drosophila* larvae (or eggs) to a strong environmental perturbation at a precise developmental stage to produce altered phenotypes and then selected both the more responsive and less responsive genotypes. The more responsive genotypes, after environmental treatment, exhibited a specific phenotype that could be easily characterized and that in some cases was identical to a "Goldschmidtian" phenocopy. The treatment/selection procedure was repeated several dozen times until the new phenotype became "assimilated", that is, automatically produced without any need for environmental stimulation. What was at first an "acquired character"[70] was in the end part of the genetic repertoire of the experimental population.

Waddington's work involved very large populations of flies of at least several hundred, sometimes several thousand. The important workload involved in carrying out the manipulative procedures and especially in counting the adult individuals should therefore not be underestimated.[71] On these aspects, Waddington was largely assisted by the technicians of his laboratory, particularly Evelyn Paton, whom he thanked very sincerely in his publications.[72]

The first series of experiments, performed in 1952, is the most famous and involved a heat-shock treatment that produced, among other phenotypic outcomes (which included "dumpy" or "miniature"[73]), a break in one of the wing cross-veins of a proportion of *Drosophila* individuals (the so-called crossveinless phenotype). After a series of exploratory experiments, Waddington also focused on this phenotype for practical reasons: it was relatively easy to obtain at least for some *Drosophila* strains. After the publication of preliminary results,[74] he published in 1953 a more developed article in *Evolution* titled "Genetic Assimilation of an Acquired Character" in which the term "genetic assimilation" appeared in print for the first time.

Waddington and his collaborators submitted pupae of a wild Edinburg strain (S/W5) to a temperature shock (4 hours at 40°C) 17 to 23 hours after puparium formation. "Crossveinless" phenotypes (see Figure 4.2) were positively ("upward" selection) and negatively ("downward" selection) selected at each generation (Figure 4.3). It was shown that quite rapidly (in some lines after only 14 generations of selection), the new phenotype was indeed constitutively produced in a small fraction of the experimental population, without any need for unusual environmental parameters.[75] Importantly, Waddington stressed that the phenotype of interest, crossveinless, was purposively chosen not only for practical reasons but also because it could not be an adaptation to high temperature. Such a design excluded from the outset any Lamarckian interpretation of the results, an aspect that was crucial to him. The induced phenotype was then submitted to artificial selection.

As early as 1952, several crossing experiments allowed Waddington to postulate that the assimilated phenotype had a complex genetic basis, which involved numerous loci distributed over the entire genome. He was thus able

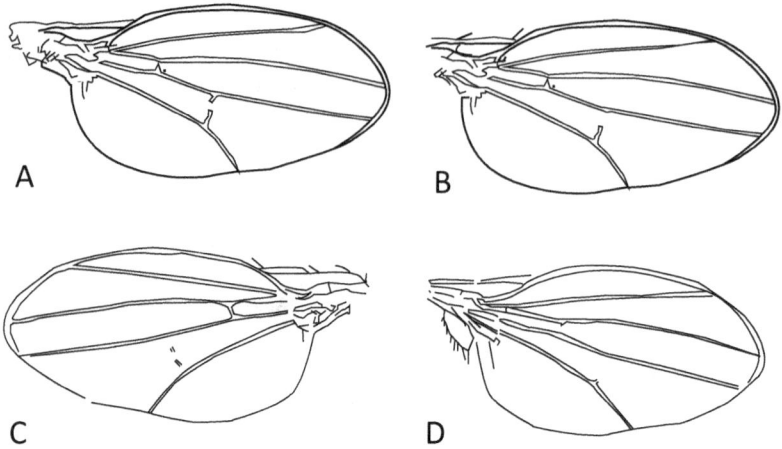

Figure 4.2 *Four crossveinless wings experimentally produced by Waddington.* A:
grade 4, B: grade 3, C: grade 2 and D: grade 1. Reproduced from: C.H.
Waddington, "Genetic assimilation of an acquired character", *Evolution*,
1953, 7, pp. 118–126. With permission from Oxford University Press.

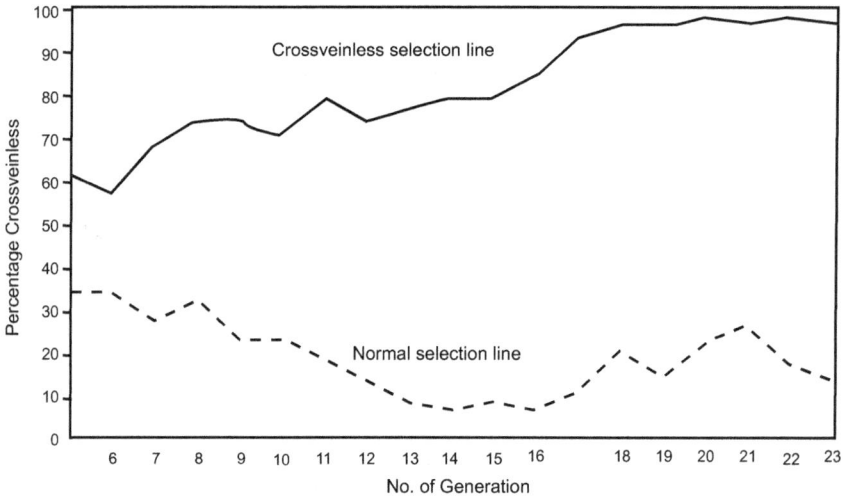

Figure 4.3 *The response to selection obtained by Waddington,* from generation 5 onward,
for "crossveinless" wings ("upward" selection) and normal wings ("downward" selection). Reproduced from: C.H. Waddington, "Genetic assimilation
of an acquired character", *Evolution*, 1953, 7, pp. 118–126. With permission
from Oxford University Press.

to challenge the relevance of other explanations that required the occurrence
of new mutations, such as organic selection:

> [Genetic assimilation] has certainly not happened solely through the
> selection of a chance mutation which happens to mimic the original

acquired character [i.e. a typological understanding of organic selection]. The genetic basis which is eventually built up for crossveinlessness differs from the genotype of the foundation stock in several genes.[76]

In the second series of experiments, Waddington wanted to show that genetic assimilation could be a rather powerful evolutionary mechanism, powerful enough to start what should be considered a macroevolutionary transformation.[77] These experiments were done in 1955 and focused on the "bithorax" phenotype. Bithorax mutants are characterized by a meta-thoracic disk that resembles the normal meso-thorax, that is, that produces wings instead of halters. This time, the stock used was a mass-bred Oregon *K* wild type. Eggs were submitted to an ether vapor for approximately 25 minutes between 2h30 and 3h30 after laying, and then cultures were carried at 25°C until emergence. Ether was known to induce drastic developmental perturbations and significantly altered phenotypes were thus expected. Like in the 1952 experiments, upward and downward selections were applied at each generation (between 2,000 and 3,000 flies per generation). But, this time, two replicates were made of this two-way selection experiment, so that there were eventually four lines instead of two (Figure 4.4).

As for crossveinless phenotypes, bithorax-like phenotypes started to appear in untreated populations rather quickly, as early as the eighth

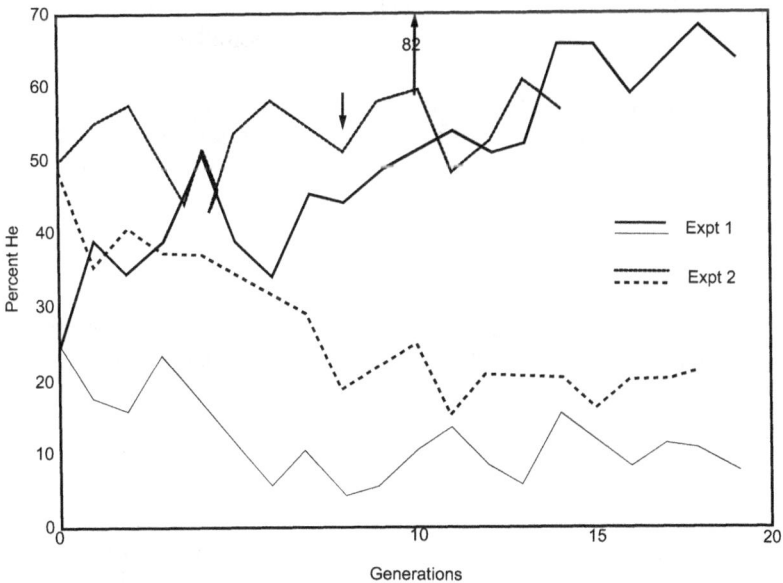

*Figure 4.4 Percentage of* He *(= bithorax-like) phenocopies in successive generations of upward and downward selection, following ether treatment of the eggs.* Reproduced from: C.H. Waddington, "Genetic assimilation of the bithorax phenotype", *Evolution*, 1956, 10, pp. 1–13. With permission from Oxford University Press.

generation for the upward selection line of "experiment 2" (the dark arrow in Figure 4.4.). But this time the genetic determinism of the assimilated phenotype was more complex, and at least two different cases were considered. One of the bithorax-like phenotypes was termed "halter-effect" (*He*). It was this *He* phenotype that was constitutively produced from the eighth generation onward. Different crosses showed that *He* was controlled by a single major gene, "which presumably arose *de novo* by chance".[78] This was a problem for Waddington because such a pattern was exactly what organic selection predicted, as he admitted in the conclusion.[79]

One of the specific features of genetic assimilation, in Waddington's view, was that the assimilated phenotype had to be the product of many minor genes and not of a single allele. Fortunately, a second bithorax-like phenotype, *He\**, seemed to be much in line with his own model. The genetic basis remained complex, to such an extent that Waddington had to continue his work after the initial article[80] but this time the most likely scenario was that assimilation was the consequence of the selective accumulation of minor genes, even if a major gene with a maternal effect was still required. Importantly, it must be emphasized that nowhere in this 1956 article, also published in *Evolution*, was canalization mentioned. This already points to Waddington's own fluid and ambiguous relationship to the causal role of canalization in the process of genetic assimilation. In any case, even if the genetic determinism was complex and remained to be fully elucidated, Waddington was confident in the fact that he had provided substantial evidence in favor of the evolutionary "powers" of genetic assimilation:

> The attempt to assimilate the bithorax character was originally undertaken with the idea that from the evolutionary point of view it was a very bizarre aberration and would provide a good test of the powers of the mechanism of assimilation. The fact that the assimilation has been successfully carried out in a number of generations, which although long in the laboratory are very short in the time scale of nature, suggests that the mechanism can in fact be an extremely powerful one. It seems likely that any modification produced by the environment could, if it were favourable to the animal, be genetically assimilated in a relatively short time.[81]

In his final series of experiments, Waddington wanted to get closer to usual natural conditions, so, this time, he used the same external agent as the main parameter of both the developmental and the selective environments. What was at stake then was the progressive reinforcement through natural (and not artificial) selection of an adaptive character, the size of the anal papillae in *Drosophila* larvae that were supposed to play a key role in osmoregulation. The selected trait was therefore no longer a discontinuously varying trait but rather a quantitative trait. This constituted a second significant difference from his previous work. Could there be genetic assimilation of such quantitative traits?

Like in 1955, Waddington used three Oregon *K* wild-type strains and submitted them to high quantities of sodium chloride that led to considerable mortality (>60%). Like in Gause's experiments on ciliates, what was maintained constant in the course of time was not the salt concentration but the mortality, that is, the intensity of selection during 21 successive generations.[82] In the end, for each of the three strains, he observed that the selected lines had larger anal papillae and better survived high salt concentrations.

Of all the results produced to support genetic assimilation, this work is usually regarded as the weakest.[83] First, it is necessary to emphasize that Waddington made a significant error in his measurements. In the 1980s, Scharloo and his colleagues showed that what he measured as the size of the anal papillae was in fact the size of the inactive epidermal zone *between* the anal papillae.[84] This meant that what he thought was an increase in the size of this organ was in fact a decrease.

But even if we disregard this error, it is hard to see how the experimental design and the data obtained would constitute restrictive evidence in favor of the selective reinforcement of canalization that would end in an assimilated phenotype (i.e., the classical Waddingtonian definition of genetic assimilation). At stake here is the standard selection of a plastic quantitative character that, in the end, is even more plastic than at the beginning: indeed, the functions measuring the size of the anal papillae for larvae resulting from such a selection process show significantly higher directing coefficients, which means that the genotypes obtained have greater plastic accommodation capacities.[85] Especially in this case, canalization seems to be a pointless explanatory device. Waddington himself might have been aware of the many weaknesses of this last experimental attempt. He presented and discussed the results in half of a short *Nature* article (very short in comparison to the detailed *Evolution* syntheses for the two previous series of work) and he did not expand much on them in his later writings.

Most importantly, it must be highlighted that his 1959 experiments were close to Gause's 1940s work on Euplotes (see Chapter 3). In both cases, salt concentration was the selective agent. In both cases, the very same parameter was also responsible for adaptive developmental accommodations. In both cases, experimentation showed that evolution (i.e., ecotypes) and accommodation (i.e., ecophenotypes) followed the same phenotypic direction. Gause's conclusion that "natural selection . . . works in the same direction of strengthening the adaptive modification" could be applied perfectly to Waddington's 1959 experiment. This must be emphasized given that Waddington and his collaborators regularly downgraded Gause's work on the grounds that it only served as an illustration of the less interesting mechanism of organic selection.[86] Here, in particular, the underdetermination of theory by facts reigned.

### 4.3   The "standard" interpretation (GA1): decanalization, genes for conditional accommodation, and genes for adaptation

The aim of the present section is to present the "standard" interpretation of the phenomenon of genetic assimilation. As Waddington eventually acknowledged,[87] "genetic assimilation" could designate both a phenomenon and a mechanism, and though the phenomenon was, roughly speaking, a process of inheritance of acquired characters at the population level, as was the case in his several experiments on various *Drosophila* strains, what was in fact the issue was the mechanism involved. By "standard", three separate things are meant. *Genetic assimilation sense 1* (hereafter GA1) was standard first because it was the most common explanation in Waddington's own writings from the early 1950s onward. Second, GA1 remains the standard way to interpret Waddington's model of adaptive evolution (which is thus a historically legitimate reading). Third, because GA1 is not a threat to the Modern Synthesis and can easily be integrated into the standard framework of evolutionary theory, and indeed has been (see Chapter 8).

The experimental procedures detailed in the previous section show that whatever the specific experimental configuration, fitness, and selection were not themselves an issue for Waddington. For the genetic assimilation of phenotypes crossveinless and bithorax, experimental designs extrinsically forced a fix selective regime on *Drosophila* populations. Rather than the question of whether plasticity could play a causal role in transforming selective pressures, like in organic selection, what was at stake was the embryological *effects* of selection.[88] Waddington was to some extent aware of this aspect (see the introduction of Part II), but his experimental designs were not thought to address the issue. From the outset, the gain in fitness was taken for granted; that is, the embryological dimension was favored at the expense of the ecological one.

GA1's starting point is that developmental accommodation and evolutionary adaptation are embryologically linked: the same developmental path is supposed to be taken in both cases. To Waddington, a trained developmental biologist, this causal connection was central, and this is why he was especially critical of organic selection, which did not make such a hypothesis. This was his main motive for publishing an addendum to Simpson's 1953 founding article. What Waddington wanted to avoid was a possible conflation between genetic assimilation and the then so-called Baldwin effect. And he especially emphasized that the gap in Simpson's reframed formulation was between the initial plastic abilities and the final genetic adaptation: "Is there supposed to be any *connection* between the developmental adaptations and the genes with similar effect, and if so, what?"[89]

The kind of "connection", Waddington was referring to was embryological/structural, as he would later specify. But it must be pointed out that in so doing, Waddington excluded another type of "connection", an ecological connection involving fitness. In organic selection there is also a causal connection between the plastic response and the genetic building of the "coincident" adaptation. Yet, this connection does not imply embryology but ecology: as we have seen, plasticity plays a causal role in altering the selective regime that eventually leads to the genetic fixation of the adaptive variants. That Waddington was unable to see or at least to seriously consider this other form of legitimate connection shows his embryological focus.

In his experimental settings, what was under selection was the "capacity to react", which "must itself be dependent on genes".[90] In subsequent publications, when he interpreted the results he managed to obtain (and especially in his 1957 *The Strategy of the Genes*), Waddington regularly conflated and mixed two divergent theoretical interpretations that need to be distinguished and that are designated in the present book as GA1 and GA2. For the moment, we will only consider GA1 (see Chapter 6 for a characterization of GA2). In GA1, Waddington translated the embryological connection in genetic terms: he supposed that the same genes were responsible both for the ability to react and, when a certain threshold was reached, for the constitutive production of the adaptive phenotype. Waddington made this point very clearly in several places, for instance:

> These facts [his experimental results] find a simple explanation if we suppose that genetic factors, which facilitate some particular developmental response to an environmental stress, also act in the direction of that type of modification even in the absence of the stress.[91]

Given such formulations, it makes sense that he would understand genetic assimilation in terms of the progressive selection of specific alleles, the ones that were originally involved in the ability to react, that is, in adaptive plasticity. For instance, Peter Godfrey-Smith made this very point in 2003 when he argued that in "Waddington's mechanism",

> the genotypes associated with the optimal state – reliable genetic control of the trait – are more genetically accessible from a state in which the population is mostly comprised of good learners than they are from a prior state in which the population is not mostly comprised of good learners.[92]

Such a mechanism is compatible with the standard Darwinian framework in the form of usual threshold selection, an interpretation that would be promoted by most of the architects of the Modern Synthesis (like Mayr) and that grounded the fact that "genetic assimilation" was an unnecessary term for a mechanism that was essentially classical Darwinian selection, and nothing more (see Chapter 8).

When Waddington used GA1 as the main interpretative framework of his experiments, he attributed two other characteristics to the form of threshold selection he was interested in. His specific understanding is especially transparent in the following passage:[93]

> It seems certain, then, that all the genes which enter into the genotype of the assimilated race were already present in the initial population, though in such low frequencies that it was only in a very few individuals that sufficient of them occurred together to give a crossveinless phenotype. What the environmental treatment has done in this case is to reveal sub-threshold concentrations of these genes, and thus made it possible for selection to get a hold.[94]

From this passage, which perfectly fits Waddington's most common formulations,[95] it appears clearly that:

(a) The threshold was reached by progressive concentration of already existing alleles, and not because of a few new mutations that would have arisen during the microevolutionary process. This was important to Waddington because, as we have seen, he thought that this constituted another difference between genetic assimilation and the Baldwin effect or organic selection.

(b) Canalization was reduced to a developmental property that was able to *mask* the effects of alleles on the phenotype. It means that in GA1, individual alleles were still thought of as having specific phenotypic outcomes. In other terms, phenotypic discrete traits remain intrinsic properties of individual alleles. As Waddington highlighted in the previously quoted passage, the environmental perturbation, in de-canalizing development, "reveals" the good set of alleles, that is, it makes their impact on the phenotype apparent. He called this impact of environmental variation on canalization "disruption of canalization".[96]

   As we will see, a different concept of canalization was at the basis of GA2 (see Chapter 6).

This last point appears, retrospectively, rather surprising given that in several other places, when he criticized what he thought to be oversimplifications of theoretical population genetics, Waddington strongly opposed the very possibility of ascribing selective coefficients to individual alleles, partly on the grounds that a gene was almost deprived of any specific and identifiable effect on the phenotype.[97] In contrast, at least to some extent, in GA1, Waddington was prepared to allow that threshold selection was about the "gradual increase in frequency of *appropriate* genes".[98] In those formulations at least, it makes sense to think that he acknowledged that genes have discrete intrinsic phenotypic properties. We will come back to this difficult and pivotal issue in Chapter 6, because in GA2 Waddington endorsed an opposite account in which individual genes were almost entirely deprived of any intrinsic ability to specifically alter the phenotype.

To sum up, GA1, to Waddington, *was a form of threshold selection acting on existing alleles involved in the plastic response. When these alleles are concentrated enough in a single genome, up to a specific threshold, they are causally responsible for the constitutive production of the new phenotype, even in the absence of environmental treatment. In GA1, canalization is only a secondary dimension of the explanatory framework that was supposed to explain why sub-threshold concentration is phenotypically invisible (in other words, the environmental treatment, in disrupting canalization, made visible (and thus selectable) the phenotypic outcomes of alleles).*

GA1 is a developmental take on the relationship between adaptability and adaptation because it is grounded in the idea that the same developmental path is involved in the initial plastic response and in the final evolutionary adaptation, which is not a given in organic selection. Yet, in such a framework, if development is reduced to gene actions, GA1 was also from the start tailor made for being integrable within the Modern Synthesis.

What about Schmalhausen? For him as for Waddington, the developmental connection between adaptability and adaptation is not an issue: all his published work emphasizes this perspective. But was Schmalhausen also assuming something close to GA1, that is, a genetic and somewhat reductionist account of genetic assimilation? As far as I can tell based on his 1941 article "Stabilizing Selection and Its Place among the Factors of Evolution" and of *Factors of Evolution*, GA1 is less present in his writings. For the most part, Schmalhausen's formulations do not seem as committed to a genetic vocabulary as Waddington's. Where Waddington made substantial room for alleles and the actions of genes, Schmalhausen mostly framed the issue in terms of "morphogenetic substances", "systems of correlations", "autonomization of development", and so on. In terms of genetics, Schmalhausen favored a reaction norm perspective. By and large, for him, genetic assimilation was the rebuilding of a population's adaptive norm.[99] This meant that Schmalhausen was even less likely than Waddington to acknowledge that there are at least some cases in which individual alleles have intrinsic phenotypic properties and that they can be submitted to threshold selection.

It must also be noted that on some occasions Schmalhausen too endorsed the view that decanalization "reveals the already existing mutations".[100] He especially emphasized the fact that under specific environmental conditions, what he called the "mobilized reserve", the unexpressed genetic variability, became phenotypically apparent again and thus submitted to selection:

This manifestation of already existing reserves is much more important than the intensity of mutability itself, since mutations, which have previously accumulated in a hidden form [because of canalization], appear suddenly in large concentrations and increase the effectiveness of natural selection.[101]

Such a theoretical positioning, as we have seen, is consistent with GA1. More surprisingly, some passages show a total conflation of GA1 and GA2. The idea that selection accumulates favorable mutations (GA1) and that the phenotypic outcomes of alleles are evolvable (GA2; see Chapter 6) sometimes came together in his formulations:

> The most *viable combinations* are created in the course of natural selection within cross-breeding populations [GA1]. In these combinations, mutations change their expression [GA2], favorable manifestations are strengthened, and harmful expressions are either changed or completely suppressed.[102]

Thereby, the fact that Schmalhausen's wordings were less transparent (at least in English translations) and the fact that only a small fraction of his work is available in English might also explain why GA1 is not as evident as in Waddington's publications. It must also be noted that, in some parts of *Factors*, Schmalhausen used the standard framework of theoretical population genetics, in which a selective coefficient is ascribed to individual alleles,[103] even though he elsewhere contested the validity of such simplifications.[104] Thus, it is probable that, like Waddington, he would not have been reluctant to interpret some results within the GA1 model. In both cases, we have an unresolved tension between two different theoretical interpretations. In the first one, GA1, alleles have discrete and delineable effects on the phenotype, and they can be accumulated by selection on this basis. In the second, GA2 (that will be defined in Chapter 6), they do not.

## 4.4   Chapter's conclusion

The fact that Waddington (and most probably Schmalhausen) supported and conflated two divergent interpretations of genetic assimilation explains why the question of whether genetic assimilation is a truly new mechanism or only a superficial reformulation of standard threshold selection has remained such a contentious issue. Briefly put, GA1 is a form of threshold selection that makes minimal use of canalization (what remains of canalization in GA1 will be further detailed in Section 8.3). It is indeed striking to see that, regularly, Waddington himself did not even mention canalization when he interpreted his own experimental results on *Drosophila*. Such a reading would later be favored by most of his collaborators in Edinburg (Chapter 8). Despite Waddington's repeated claims, if the phenotype can be expressed in genetic terms (GA1), there is no need to consider the fact that organisms develop. GA1 is a model of adaptive evolution that focuses on genotypes and allelic frequencies, as is classically the case in population genetics. Given that Waddington fully endorsed GA1, it makes sense that Mayr and several others in the 1950s would put in doubt the originality of genetic assimilation and attempt to reduce this process in the usual genetic terms. Thereby, in the context of GA1, both logically and historically, it makes sense to translate the embryological causal connection between adaptability and adaptation in genetic terms. Yet, as already suggested, this does not tell the whole story. Less apparent in both Schamlhausen's and Waddington's writings, the GA2 interpretation, much more unorthodox and radical, still represents a stimulating way to engage with the logic of population genetics and the Modern Synthesis. This neglected theorization will be reconstructed and discussed in Chapter 6.

The necessity of canalization in the explanation of the process of genetic assimilation could be disputed not only from the viewpoint of theory but also within the experimental context that was designed by Waddington to support his claim. As I have argued in the present chapter, his experimental designs were especially equivocal and left room for different interpretations. It is also for this reason that these experiments have remained heuristically productive throughout the history of science. They raised important questions for biologists about what such protocols could show and have almost continuously been revived.[105] Waddington can at least be credited with having had an extremely stimulating effect from this point of view. In Chapter 8, we shall see that this "Waddingtonian impetus" also affected, to some extent, quantitative genetics in the late 1950s and early 1960s.

The "theory" of genetic assimilation (as Waddington sometimes liked to call it) was without a doubt underdetermined by the experimental facts produced in the 1950s. These facts can also be interpreted in several other ways, as we have seen. It must once again be stressed that Waddington's last series of experiments (1959), involving increasing salt concentrations and the size of anal papillae in *Drosophila* larvae, were close to the ones performed by Gause in the early 1940s on Euplotes to experimentally document coincident selection (in this case, OS1). The 1890–1950 literature showed that OS1 might imply only a few genes or "congenital variations", sometimes even a single locus, whereas GA1, at least that is what Waddington repeatedly claimed, necessitates the contribution of multiple alleles each with a small phenotypic effect. However, the difference between OS1 and GA1 did not carry the weight that Waddington attributed to it. The idea that OS1 requires the random occurrence of an advantageous mutation (or of a few advantageous mutations) is indeed in line with the first ideas formulated around 1900, when a typological conception of heredity was still largely dominant. But subsequently, this typological conception gradually faded, particularly with the rise of quantitative genetics. In short, the rise of quantitative genetics in the mid-20th century might have strongly weakened the validity of this possible distinction. Moreover, in several of his experimental cases (like for some bithorax phenocopies), as we have seen, Waddington himself had to acknowledge that sometimes a "major gene" is involved in the final genetic stabilization of the new phenotype.

This is to say that while organic selection and genetic assimilation constitute two different explanations of the same Lamarckian phenomenology, it does not mean that there is a radical shift between them or that they are something like "incommensurable paradigms". Quite the opposite. It seems that there is a continuous line between the two, and if OS2 and GA2 are substantially different (they are at the two extremums), the gap between OS1 and GA1 is much narrower and, in the end, may only depend on the genetic architecture of the phenotype under consideration: if this architecture is simple (i.e., implying a major gene), then OS1 would be the best explanatory model; whereas if this architecture is more complex (i.e., implying the coordination of many alleles with small effects), then most probably GA1 should be favored.

## Notes

1 Waddington in particular made this claim repeatedly. For instance, in his 1961 important review article on genetic assimilation that was published in *Advances in Genetics*, he concluded as follows: "It has been conventional to argue, from the fact that "acquired characters" are not inherited, that the development of adaptive modifications during the lifetime of an organism is irrelevant to the evolution of similar genetically determined phenotypes. The theory of genetic assimilation, and the practical demonstration that the process can occur, shows that this argument is misguided, and provides a new way of accounting for all those evolutionary facts for which, in the past, some authors were tempted to advance a 'Lamarckian' explanation" (Waddington, 1961a, p. 290).

2 Waddington (1969c, p. 118).

3 Waddington always stressed this aspect: "The main conclusions to be drawn are, not that any new fundamental genetic principles have been disclosed but rather that it has been shown that well-accepted principles lead to evolutionary consequences quite other than those which have usually been supposed to follow from them" (Waddington, 1961a, p. 290).

4 Waddington (1975a, pp. v–vi).

5 For instance, in 1958, he published in the *Proceedings of the Linnean Society* (London) an article entitled "Inheritance of Acquired Characters" (Waddington, 1958a).

6 Waddington (1959a, p. 382).

7 Gans (1968); Wake (1986).

8 Mark B. Adams published a 32-page book chapter entitled "Severtsov and Schmalhausen: Russian Morphology and the Evolutionary Synthesis". Though this text focuses more on Severtsov, it remains helpful to better situate Schmalhausen within Russian biology (Adams, 1980a). More recently, in 2006, Georgy S. Levit, Uwe Hossfeld, and Lennart Olsson published an 18-page scientific biography. This very informative article is based on many Russian publications, including 18 of Schmalhausen's articles and books (Levit et al., 2006).

9 Adams (1980a, p. 196). It should be added that Ivan Schmalhausen's father, in his botany thesis on plant hybrids (1874, University of Saint-Petersburg), was one of the very few scientists, before 1900, to cite the work of Gregor Mendel (Weinstein, 1977, pp. 343–344).

10 Adams (1980a, p. 195).

11 He published in 1912 a book entitled *Studies on the Theory of Evolution: Individual Development and Evolution* (Adams, 1980a, p. 195).

12 Adams (1980a).

13 Levit et al. (2006, p. 91).
14 Adams (1980a, p. 218).
15 Adams (1988, pp. 281–282); Levit et al. (2006, p. 91).
16 Adams (1980a, p. 220).
17 Levit et al. (2006, p. 93).
18 Historiography of "Organicism" has been renewed this past ten years. See especially: Nicholson (2014); Esposito (2016); Peterson (2016).
19 According to Adams, Severtsov, "at no place in his writings [made] any attempt whatever to integrate genetic explanation into his morphological theories: morphology as he practiced it had not as yet become part of an evolutionary synthesis" (Adams, 1980a, p. 218).
20 Levit et al. (2006, p. 95).
21 Gulick never pushed the idea beyond a rather schematic description, where he briefly contrasted these two forms of selection: "The selection of average forms for propagation tends to produce stability of type; and the selection of extreme, but of opposite, and therefore, balanced deviation from the type produces fluctuating variation; but unbalanced selection, that is, propagation from forms whose average character differs from the average character of the race, changes in some degree the average character of the race in the next generation" (Gulick, 1905, p. 16). "Balanced selection is usually secured by selecting individuals of the average form, and tends to produce increasing stability" (Gulick, 1905, p. 150).
22 Schmalhausen (1949, pp. 100, 106).
23 Robertson (1977); Lewin (1998); Slack (2002); Peterson (2016); Loison (2019).
24 Waddington (1969b, p. 76).
25 Waddington (1969b, p. 79).
26 Robertson (1977, p. 578).
27 Slack (2002, p. 890).
28 Peterson (2016, p. 93).
29 Robertson (1977, p. 590).
30 This count does not take into account books. During the late 1930s and early 1940s, several papers on *Drosophila* were dealing with both genetics and embryology.
31 For instance, he was awarded in 1936 the "Brachet prize" by the Academy of Belgium.
32 Bard (2008, p. 190).
33 Waddington (1940b, p. 4).
34 Waddington (1940b, pp. vii, 2).
35 Morgan (1934a); Robertson (1977, pp. 592–593).
36 Waddington (1940a).
37 Waddington (1940b, p. 54).
38 Waddington (1940b, p. 59).
39 Alfred North Whitehead's so-called influence has been repeatedly stressed (e.g., Peterson, 2011), first by Waddington himself (Waddington, 1969a, 1975b). Yet, it remains to be understood how important Whitehead's processual philosophy truly was for Waddington, and, reversely, if Waddington had a good understanding of Whitehead's difficult notions and wording. To my knowledge, Robertson was the only one who did not take this influence for granted (Robertson, 1977, p. 603).
40 For example, Waddington (1940b, p. 49).
41 Waddington (1942, p. 563).
42 Robertson (1977, p. 580).
43 For instance, Charlotte Auerbach (1899–1994), Douglas Scott Falconer (1913–2004), Geoffrey Herbert Beale (1913–2009) and Alan Robertson (1920–1989).

44 Robertson (1977, p. 583).
45 Waddington (1961a).
46 Wilkins (2008).
47 Waddington (1972b, pp. 283–284).
48 Waddington (1968a).
49 Waddington (1975a).
50 Lewin (1998, p. 157, p. 285); Hall (2001, p. 216); Griffiths (2006, p. 92); Levit et al. (2006, p. 100).
51 With a few rare exceptions, such as Masel (2004).
52 Mayr (1961).
53 Schmalhausen (1941, p. 309).
54 Schmalhausen (1941, p. 318).
55 Schmalhausen (1941, pp. 318, 323).
56 Schmalhausen (1949, p. xxii). Waddington shared the same reading 15 years later: Waddington (1961a, p. 273).
57 Waddington (1961a, p. 273).
58 Schmalhausen (1941, p. 333); Waddington (1957a, pp. 159–162).
59 Schmalhausen (1941, pp. 334–335); Waddington (1942, p. 563).
60 Schmalhausen (1941, p. 335, 1949, p. 25, p. 80); Waddington (1942, pp. 563–564).
61 Schmalhausen (1941, p. 342). Waddington, in his will to distance himself from the ancient tradition of organic selection, minimized the significance of Gause's work: "Gause showed that in several cases natural selection in a particular environment operates to produce changes, in characters such as body size, which are similar to the direct adaptations to that environment exhibited by vegetatively propagated clones. This he considers to be "organic selection" in the sense of Baldwin, since it involves the selection of genes which act in the same direction as the environment. But he does not seem to conceive of the genes as controlling the response of the organism to the external circumstances; he phrases his description always as though the action of the gene was quite independent of the environment" (Waddington, 1957a, p. 165).
62 Schmalhausen (1941, p. 309). My emphasis. See also Schmalhausen (1949, pp. 44–45).
63 "It is inadequate to think of natural selection and variation as being no more essentially connected with one another than would be a heap of pebbles and the gravel-sorter onto which it is thrown. On the contrary, we have to think in terms of circular and not merely unidirectional causal sequences" (Waddington, 1959a, p. 399).
64 Waddington (1942, p. 564).
65 Compare, for example, Masel (2004) to Crispo (2007).
66 See the chapter's conclusion.
67 Schmitt (2000).
68 Goldschmidt (1935).
69 Goldschmidt (1938, pp. 7–8).
70 In a number of publications, Waddington deliberately (and certainly provocatively) used formulations involving the term "acquired characters", which had obvious Lamarckian connotations. At the height of the Lysenko affair, this Lamarckian phraseology, which further complicated the issues of his work, was sometimes criticized (Begg, 1952).
71 For example, in the experiments on the assimilation of the bithorax phenotype, no less than 150,000 flies were described for the lines subject to positive selection (Waddington, 1956, p. 9).
72 Waddington (1953a, p. 118).
73 Waddington (1953a, p. 124).

74  Waddington (1952).
75  Waddington (1953a, p. 120).
76  Waddington (1953a, pp. 123–124).
77  Waddington (1956, p. 1).
78  Waddington (1956, p. 5).
79  Waddington (1956, pp. 10–11).
80  Waddington (1957b).
81  Waddington (1956, p. 10).
82  Waddington (1959b).
83  Robertson (1977, p. 600); Crispo (2007, pp. 2474–2475).
84  Te Velde et al. (1988, p. 52); Scharloo (1991, p. 71).
85  Scharloo (1991, p. 71).
86  Waddington (1957a, pp. 164–166). See also Bateman (1959b, p. 443).
87  Waddington (1961a, p. 259).
88  Except for the work on anal papillae, which was, as we have just seen, very close to other attempts of supporting organic selection.
89  Waddington (1953b, p. 386). My emphasis.
90  Waddington (1953c, "The evolution of adaptation", p. 137).
91  Waddington (1958a, p. 58). In his 1961 developed review, in relation with the famous example of the production of callosities, he put it as follows: "The same gene has the two properties of having a slight tendency to produce callosities in a normal environment [adaptation] and increasing the tendency to react to pressure by the formation of callosities [adaptability]" (Waddington, 1961a, p. 273).
92  Godfrey-Smith (2003, p. 56).
93  This type of explanation can be found in multiple places in Waddington's writings. For example: "The environmental stress acts to reveal, and to expose to natural selection, genes which were already present in the population, but whose action is normally only sub-threshold in intensity and therefore imperceptible" (Waddington, 1958a, p. 58).
94  Waddington (1957a, pp. 177–178).
95  See, for instance, Waddington (1961b, p. 94).
96  Waddington (1961a, p. 280).
97  Waddington (1957a, pp. 65, 110); Waddington (1969c, p. 121). For a general treatment, see Loison (2022).
98  Waddington (1961a, p. 264). My emphasis.
99  Loison (2020a).
100  Schmalhausen (1949, p. 43).
101  Schmalhausen (1949, p. 127).
102  Schmalhausen (1949, p. 94). Emphasis in the original.
103  Schmalhausen (1949, pp. 101–104).
104  Schmalhausen (1949, pp. 218–219).
105  Ho et al. (1983); Gibson & Hognes (1996); Raju et al. (2023). See also the recent work of Ian Dworkin and his colleagues: www.biorxiv.org/content/10.110 1/2022.01.09.475581v1

# 5 A problematic issue

## The creativity of the developmental process

How could genetic assimilation become a creative process at the evolutionary level? This question is decisive in assessing how significant Waddington's and Schmalhausen's views could have been in the history of evolutionary theory. In the Modern Synthesis, the creative force is mostly natural selection: natural selection is the driving causal factor in the exploration of the phenotypic space (which is more or less directly derivable from the genotypic space). It is natural selection that accumulates advantageous mutations with small effects over the generations, producing biological adaptations. If genetic assimilation is a creative process in its own right, then it cannot be reduced to sorting out by the selection of discrete alleles based on their selective values; otherwise, we would be back to square one and the "allmacht" of natural selection. This means, as we saw in the previous chapter, that GA1 is here fully in accordance with the standard view of Modern Synthesis. This account of genetic assimilation (regularly endorsed by Waddington himself) is all about the cumulative selection of alleles that, in the end, results in a threshold effect: the constitutive production of a phenotype that was first elicited by unusual environmental conditions.

For genetic assimilation to be more than just threshold selection, we can at least think of two theoretical perspectives. One of these concerns the evolutionary and genetic construction of a new reaction norm and will be dealt with in detail in the following chapter (GA2). This is where Waddington and Schmalhausen were the most original regarding the orthodoxy of the Modern Synthesis (though surprisingly, this issue went completely unnoticed for decades). The other, most immediate, option is the topic of the present chapter: it focuses on the creativity of the developmental process itself. It assumes that development can produce phenotypes that are "innovative" as regards the genotype and the various environmental challenges that can be faced. In short, it assumes that the phenotype is an emergent property of a developmental system in a given milieu, and this would explain why development should be brought back into evolutionary theory. This is how Waddington's and Schmalhausen's contribution is usually understood and evaluated, especially by the proponents of the Developmental Systems Theory. It is indeed

DOI: 10.4324/9781003422990-10

true that Waddington always emphasized that adaptive evolution was not about "random search",[1] that is, mutation and selection but rather about "adapting and improvising",[2] that is, creatively responding to the environment. But what if the environmentally induced phenotype is entirely reducible to the genotype? If plastic responses are themselves encoded in the genome, this would nullify this part of Waddington's argument. In such a case, plasticity and its developmental effects could be easily integrated within fitness, and adaptive evolution would remain the work of selection, sorting out alleles with different fitness.

As a matter of fact, this represents a core problem for the logic of genetic assimilation, which Waddington never managed to solve. The main reason he failed to do so was that he was especially committed to a vision of strong genetic determinism that was reified in his famous diagram of the epigenetic landscape. He thought of plastic outcomes as hereditary determined. Thus, on the one hand, Waddington needed development to be creative, that is, non-reducible to genetic information. But on the other hand, his account of the epigenetic landscape made development entirely reducible to genetic information. It follows that there is a significant tension in Waddington's writings between his claims as an evolutionary biologist dissatisfied with several aspects of the Modern Synthesis and his claims as a developmental biologist which fully endorsed genetic determinism. In a previous study, I have already addressed the question of the ambiguous role of the concept of environment within Waddington's framework.[3] In the present chapter, Section 5.1 aims to recap and expand on some of these preliminary findings.

A related problem is the question of the origin of the epigenetic landscape: where does this developmental phase-space come from? As early as 1966, George Williams clearly formulated this problem. If genetic assimilation consisted of reconfiguring the topography of this landscape, it required that such topography pre-exists. However, if the landscape's topography must itself be conceived as an adaptation, and thus a product of natural selection – which seems to have been Waddington's conception – then natural selection is still guiding evolutionary change. Genetic assimilation is merely a means of bringing about this change on the ecological scale of microevolution. In the second section, I will examine in detail the validity of Williams' criticisms.

It seems that Waddington never met Williams' challenge, and his constant attempts to demarcate genetic assimilation from the Baldwin effect forced him to put the emphasis on the genetic determination of plasticity, which ended in a strong version of the genetic determination of the whole developmental process. In contrast to Waddington, Schmalhausen appears to have been less committed to a gene-centered view of development. It seems that he had a more open understanding of embryogenesis, which made his concept of genetic assimilation less sensitive to the type of criticisms Williams addressed to Waddington (Section 5.3). In particular, as we shall see, Schmalhausen's explanatory framework, based on the distinction between what he called "modifications" and "morphoses", might have been a more suitable interpretation of Waddington's own experimental results on *Drosophila*.

## 5.1   Problem n°1: the conceptual meaning of the epigenetic landscape's topography

Waddington's theoretical framework was strongly influenced by his epigenetic landscape diagram. This diagram has become an iconic representation in biology, known well beyond embryology. It is mostly valued for its fertile and heuristic qualities and for the ripple effect it produced in science from the late 1940s to the present day.[4] Without denying this very real part of its impact, the following sections look more closely at the implicit assumptions and possible limits of such a representation of developmental dynamics. I will successively focus on the significance of the paths traced in it (Section 5.1.1), its fixed character on the scale of individual development (Section 5.1.2), and finally on Waddington's conception of the origin of such a topography (Section 5.2).

### 5.1.1   *Representing developmental noise or phenotypic plasticity?*

In the biological literature, from the mid-20th century to the present day, the conceptual lines between the notions of plasticity and developmental noise have recurringly been blurred. This ambiguity arises in characterizations of the part of the phenotype that is not under strict genetic control. Does it result from the modifying action of the environment (plasticity), or is it the result of a random component in the succession of differentiation processes (developmental noise)? Can (should?) we draw a line between these two concepts? This question is still far from settled today.

The history of this conceptual distinction begins with Waddington's early speculations on canalization and the way his conceptions stimulated further experimental research in the 1950s. Indeed, while the concept of developmental canalization offered an engaging perspective, it clearly lacked an empirical anchor. How to demonstrate canalization? Could its intensity be measured? The first attempt to address these difficult questions was based on the phenomenon that would later be known as "fluctuating asymmetry" (FA). FA refers to the fact that, in bilateral animals, there are minor differences between the left and right sides of the body, despite identical genomes. From the outset, FA interested biologists, not in itself but rather for what it allowed them to assess and quantify.[5] FA has often been understood as a good indicator of the degree of developmental stability of a phenotypic trait: the more the measure[6] of FA tends toward 0 the more the embryonic development is able to resist "developmental noise", and the higher, it was thought (and often still is), its level of canalization. In what is seen as a seminal work[7] that was published in 1953, it was geneticist Kenneth Mather who apparently first undertook to use FA as a direct measure of developmental canalization.

Kenneth Mather (1911–1990) was one of the most important population geneticists among those who came after Fisher, Haldane, and Wright. Himself a student of Fisher during a postdoc,[8] the whole of his work can be seen as an extension of the synthesis between biometry and genetics largely built up by Fisher since 1918.[9] It is under this lens that we must understand what was most likely his most important book, *Biometrical Genetics* (1949a), one of the founding texts of quantitative genetics and which intended to participate in the development of the "genetic theory of continuous variation".[10] We will see in Chapter 8 that his conceptual distinction between "oligogenes" and "polygenes", so important for the structure of 1950s genetics, was itself derived from this theoretical intention.

In 1953, when he published "Genetical Control of Stability in Development", Mather was already a leading figure in the field of quantitative genetics and had been appointed in 1948 to the Chair of Genetics at the University of Birmingham. Due to the contested reception of his "balanced theory" (see Chapter 8) at a symposium held in Edinburgh in 1950, the relationship between Birmingham and the Institute of Animal Genetics was strained.[11] However, Mather never hesitated to make explicit that part of his own work drew upon Waddington's. This was the case in 1953, when he borrowed Waddington's concept of canalization, as well as its empirical basis (the high phenotypic variability of mutants) and wished to build upon this set of questions by showing how "the study of asymmetry affords . . . a means of investigating the genotypic stabilization of developmental processes".[12] To do this, Mather chose to focus on a quantitative trait that presented an asymmetry and that also appeared to be insensitive to environmental conditions: the number of sternopleural chaetae (chitinous bristles) in *Drosophila melanogaster*. There are usually between 8 and 12 of these chaetae, lying between the fore and mid legs, which can be readily counted. Mather used genotypically comparable strains and made thousands of counts. For each fly, the number of chaetae on the right side was subtracted from that of the left side (R-L). Different variances were obtained for different strains ("Oregon", "Samarkand", etc.), suggesting that developmental canalization differed between them for this character.

To assess the degree of canalization that could be measured by FA quantification, it was necessary to subject these strains to different forms of selection and observe the response to selection. Mather therefore undertook several series of artificial selection experiments which were all based on the same principle: to select both the more asymmetrical parents (high variance line) and those less asymmetrical ones (low variance line). Significant results were obtained: variances both increased and decreased respectively within only a few generations of selection.

According to Mather, the fact that there was a response to selection left "no doubt of the existence of heritable variation"[13] (but this is not the only possible interpretation[14]). In other words, developmental stability was under genetic control. Mather interpreted his results in terms of his polygenic theory of inheritance,[15] which will be detailed in Chapter 8. What needs to be stressed here is that he was able to develop an experimental design capable of quantifying the degree of developmental stability that was explicitly identified with canalization and, on this basis, to modify this developmental property. Mather's work thus inaugurated a double research path. On the one hand, he sought to make the degree of canalization accessible by means of different forms of selection applied to experimental populations. On the other hand, and more specifically, Mather was the first to show how FA could play an important role in understanding developmental stability and in particular the antagonistic consequences of developmental noise and buffering mechanisms. His 1953 paper was thus the founding text of this line of research, which has since remained a continuous route for theoretical and experimental investigation.

But there was also a serious difficulty in Mather's interpretation (which quickly became standard[16]). Were the developmental stability that was supposed to be measured by the degree of FA on the one hand, and the concept of developmental canalization as forged by Waddington on the other, equivalent? Waddington doubted that these two concepts explained the same phenomena and he made an explicit and relatively detailed critique of this possible conflation in *The Strategy of the Genes*. Taking up his pictorial metaphor of the epigenetic landscape, he disputed that the property of resisting developmental noise (which was genetic, as Mather had just demonstrated) could be identifiable with canalization. If the latter could be represented by the depth of the developmental pathways, the former consisted in the degree of roughness of the ball ("the imperfection of the sphericalness") that runs down the network of valleys: a ball that is all the rougher shows a greater sensitivity to developmental noise, and vice versa.[17]

The posterity of this conceptual distinction in the following decades was complex. Often ignored or forgotten,[18] it was also, nonetheless on occasion taken up and reworked. For example, in 1962, Leigh Van Valen adopted an intermediate position between that of Mather and Waddington. For him, resistance to developmental noise and canalization were two non-independent components of the same general process of developmental robustness[19] ("developmental robustness" was here used as a generic category). Decades later, Vladimir Zakharov resurrected a genuine Waddingtonian perspective on these issues and posited the importance of not confusing developmental stability with canalization.[20]

As far as Waddington was concerned, this new conceptual distinction forced him to clarify what the concept of stabilizing selection could cover. We will see in Chapters 6 and 8 that it was essential for Waddington to distinguish the *form* of selection applied to an experimental population from the *effects* it was supposed to induce. If "normalizing selection" consists in a selective process where the advantaged phenotype is the average phenotype of the population, "canalizing selection", in Waddington's account at least, is a selective regime that produces developmental canalization. In 1957, in reaction to Mather's interpretations, Waddington took care to add a third category, "selection for repeatability".[21] For him, there was no doubt that Mather's results concerned this third form of stabilizing selection:

> It is therefore probably not justifiable to take such studies as providing information about the general "developmental stability" or "developmental homeostasis" of the system, as these authors [Mather, Tebb and Thoday] do, at least if these terms are being used as equivalent to developmental canalisation. It seems desirable, either to restrict the term "developmental stability" to mean "lack of noisiness in development", in which case one could speak of selection for similarity between two homologous organs as "stabilising selection", *or to coin another word for such selection, calling it perhaps "selection for repeatability"*.[22]

Thus, the difficult distinction between developmental noise and plasticity (or between developmental stability and canalization) was a central issue for Waddington as early as the 1950s. Against most of his contemporaries (and in particular in opposition to Mather), Waddington was very clear about the absolute necessity of maintaining such a distinction. He believed that developmental stability was simply not achieved by canalizing selection (see Chapter 8). This tells us something about the way he conceived plasticity: for him, it was the ability of a genome to *actively* respond to stimuli from *outside* the developmental system. This capacity materialized in the topography of the epigenetic landscape, which showed multiple and divergent pathways. This topography alone could initiate the process of genetic assimilation. In contrast, developmental noise was conceived as the *passive* result of the action of physico-chemical parameters on biological processes. As such, it was not a genuinely biological property of living systems but simply the inescapable consequence of their materiality.

### 5.1.2 Developmental creativity and the constructive role of the environment: a long-lasting misunderstanding

At this point, another clarification is necessary. Waddington is usually praised for the elaborate notion of environment he opposed the Modern Synthesis view in which the environment is reduced to a fixed and extrinsic selective filter. In a previous article,[23] I examined this issue in detail and here I will limit myself to summarizing its main points in the present section.

It is first necessary to distinguish between two "Waddingtons", so to speak, which only partially overlap. As a critic of theoretical population genetics,[24] Waddington repeatedly emphasized that it made no sense to ascribe selective coefficients to individual alleles, partly because the phenotype is the outcome of both the genome and the environment. In those texts and passages, Waddington considered genes and the environment as symmetrical causal agents in the epigenetic system. The phenotype cannot be reduced to the genotype, which implies that, at least to some extent, he thought of embryogenesis as innovative or creative because the environment actively participated in building the final phenotype. In short, he saw fitness as a strict phenotypic property and he thought that population genetics had failed to acknowledge this crucial aspect as he still argued in his 1969 programmatic "Paradigm for an Evolutionary Process":

> The fitness even of a single phenotype cannot, therefore, be represented by a single-valued coefficient, but only by a matrix, or a continuous distribution of values, which specifies also the variety of environments, in which selection may occur. If we wish to attach "fitness" to a single, multigenic *genotype*, we should have to increase the dimensionality of the matrix so that it could take account also of the epigenetic-programming aspects of the environments; and if we wished to emulate the classical neo-Darwinists and speak of the fitness of single gene, we should have to increase the matrix again to incorporate all the various genetic combinations in which it might occur in the population in question (it seems highly dubious to me that any process of averaging over all these combinations, as advocated, e.g. by Fisher, has any biological validity).[25]

It is indeed true that Waddington tried to develop a richer concept of the environment than the one deployed within the Modern Synthesis. From at least 1954 onward,[26] and in *The Strategy of the Genes*[27] in particular, he paid more and more attention to the complex system of interactions between genes, environment, development, and selection. This theoretical path led him to individualize four main evolutionary sub-systems, as we have seen (the genetic system, the natural selective system, the epigenetic system, and the exploitive system). The environment was a pivotal dimension in all of these except in the genetic system.

Waddington's concept of the "exploitive system", which stresses the active causal role played by organisms in choosing and modifying their environment, has a legacy among the promoters of Niche Construction Theory who praised it as an important first step in the direction of their more refined model.[28] The problem is that Waddington, as a developmental biologist, endorsed a view of embryogenesis that appears to some extent inconsistent with his evolutionary claims: in his view, embryogenesis was anchored in the genome alone (Figure 5.1):

This means that for Waddington-the-developmental-biologist, the environment was not a dimension of the developmental phase-space but an extrinsic perturbation, and, as rightly emphasized by West-Eberhard, the

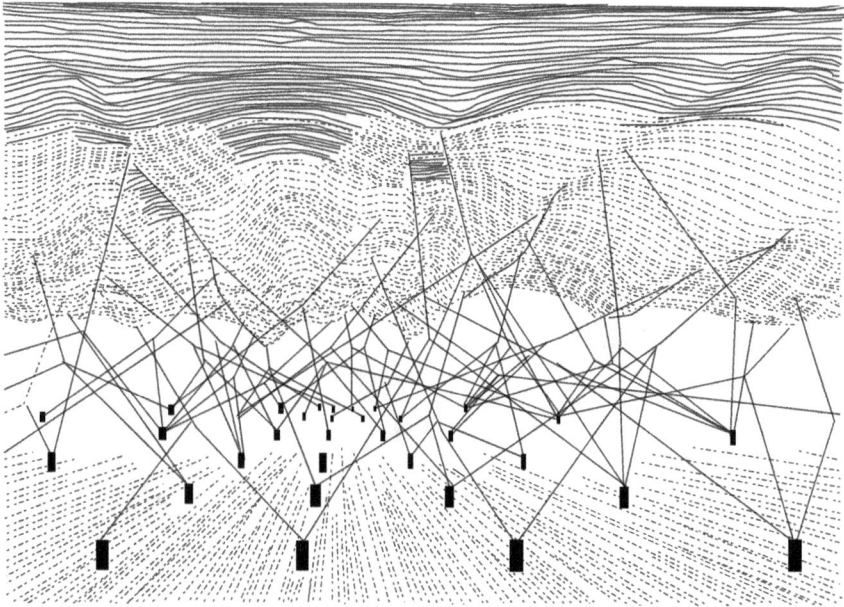

*Figure 5.1 Diagrammatic representation of "the complex system of interactions underlying the epigenetic landscape".* Each peg represents a gene and each string its contribution to the network that causally produces the epigenetic landscape. The developmental environment is not involved in the production of the landscape's topography. Reproduced from: C.H. Waddington, *The Strategy of the Genes*, London, Allen and Unwin, 1957. With permission from Taylor & Francis.

landscape was thus supposed to be a fixed entity at the developmental time scale (Figure 5.2):

> All that the environment can do, in Waddington's scheme, is *deflect development* into a new genetically specified path. . . . The epigenetic landscape is limited and even inconsistent as a model of development in its failure to deal with the dynamic effects of environmental inputs on the epigenetic pathways. Although the landscape has a flexible-looking surface, in fact it is a rigid and static representation of development. Change in the contours requires evolutionary change.[29]

Thereby, it seems that Waddington supported a strong form of genetic determinism as regards the developmental process, even if, as we shall see, he also regularly opposed drawing a causal link between individual alleles with specific phenotypic characters (Section 6.1). Waddington's commitment to a strong view of genetic determinism has already been emphasized, especially by Brian K. Hall.[30] My point here is to stress that such a gene-centered view of embryogenesis immediately paved the way for George William's critical

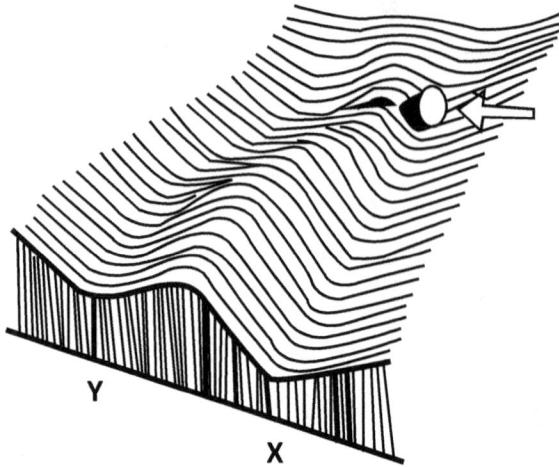

*Figure 5.2 The total independence between the environment and the epigenetic landscape.* In this configuration of the epigenetic landscape, the usual adult character is X. There is a side branch leading to Y, but the developing tissue does not get into the Y path "unless an environmental stimulus (hollow arrow) pushes it over the threshold". This diagrammatic representation suggests that Waddington thought of the epigenetic landscape and the developmental environment as two independent entities. Reproduced from: C.H. Waddington, *The Strategy of the Genes*, London, Allen and Unwin, 1957. With permission from Taylor & Francis.

take on genetic assimilation (Section 5.2). In such a preformationist account of development, what is "new" is the selective and evolutionary building of the landscape, while the subsequent genetic reinforcement of the adaptive path in a given environment seems to be only of secondary importance.

Why was Waddington so committed to a genetic understanding of the epigenetic system that weakened the evolutionary significance of genetic assimilation? At least three reasons led him to promote a strong genetic view of the epigenetic landscape. The first is empirical, the second is linked to his opposition to the Baldwin effect and the third is a consequence of his taste for pictorial representations of scientific concepts.

## (a) The genetic determinism of wing patterns in *Drosophila*

In contrast to other heterodox biologists of the mid-20th century, for instance, Goldschmidt, Waddington never opposed the existence of discrete genes, nor did he deny their causal importance during embryogenesis, in fact, it was quite the contrary. In the late 1930s, Waddington, among others,[31] wanted to tackle the issue of the relation between genetics and embryology on experimental grounds. He was inspired by Thomas Morgan's challenge[32] to bridge the gap between embryology and genetics. His 1940 book *Organisers and Genes* was an attempt in that direction.[33]

Waddington visited Morgan's Caltech Laboratory between January and April 1939 and performed many experiments devoted to the elucidation of the genetic control of wing's development in *Drosophila melanogaster*. From this descriptive effort Waddington produced a 65-page monograph in which he aimed "to analyze the morphogenetic process into its constituent phases, and secondly to determine the ways in which these developmental processes can be modified by gene substitutions".[34] Based on the numerous results obtained (Waddington studied the phenotypic effect of no less than 38 genes during 16 developmental stages in a wide range of genetic backgrounds), he concluded that the deterministic effect of these genes on wing phenotype was inescapable, even if complex.

This in-depth study seems to have had a lasting impact on Waddington and was instrumental in his genetic approach to development. From 1940 onward, he constantly referred to it. In *Organisers and Genes*, Waddington had already been confident that "unless strong evidence to the contrary can be produced, it is safest to assume that all the properties of cells are ultimately determined by genes".[35] Decades later, for instance, in his 1966 *Principles of Development and Differentiation*, he still referred to the "fundamental postulate"[36] of the causal roles of genes in the development of an organ.

## (b) The demarcation between genetic assimilation and the Baldwin effect

As I emphasized in a previous work,[37] in the early 1940s Waddington initially developed a less gene-centric view of embryonic development and the epigenetic landscape than the one he later adopted, in the 1950s and 1960s. In his earlier writings, at least in a few passages, Waddington was prepared to give the environment a more constructive role in embryogenesis:

> It is first necessary to point out the ways in which the environment can influence the developmental system. If we conceptually rigidify such a system into a definite formal scheme, we can think of it as a set of alternative canalized paths; and the environment can act either as a switch [the option later favored], or *as a factor involved in the system of mutually interacting processes to which the buffering of the paths is due.*[38]

The later "hardening" of Waddington's position might be explained by his will to demarcate, as firmly as possible, his mechanism of genetic assimilation from the Baldwin effect. Waddington was convinced that a reliable criterion to distinguish these two evolutionary mechanisms was the genetic basis and evolutionary origin of phenotypic plasticity. In Waddington's understanding of the Baldwin effect, plasticity was somehow taken for granted and had no genetic component: "the initial adaptation [is] a nongenetic phenomenon on which selection has no effect".[39] Whereas in genetic assimilation, plasticity was a genetic property that could evolve because of natural selection. He highlighted the gap between the two models by emphasizing the genetic component of plasticity: very early in the 1950s, he conceived genetic assimilation more and more explicitly as a process based on differences in hereditary (thus genetic) responsiveness to environmental variations. This repeated emphasis led him to minimize the role of the environment, and conversely, to anchor the epigenetic landscape in the genome.

### (c)  The diagram of the epigenetic landscape as an epistemological obstacle

A third reason for Waddington's commitment to a strong genetic determinism might indirectly stem from his ability to picture abstract scientific concepts in suggestive schematic forms. This original and central dimension of his theoretical work is usually positively acknowledged.[40] Its heuristic value is undeniable, and the epigenetic landscape remains a helpful resource for biologists today (Baedke, 2013). However, representing a temporal process – in this case, embryological development – is neither cognitively nor epistemologically neutral. It is indeed very difficult to represent an *open* dynamic process in a schema which is by definition static, and, in the case of development, demands that all possible outcomes be drawn in advance.

In a way, this amounts to the kind of spatialization of temporality that philosopher Henri Bergson criticized in the early 20th century. Bergson's philosophy of "*durée*" might be helpful here to better characterize why Waddington's epigenetic landscape represented an "epistemological obstacle",[41] in a Bachelardian sense. For Bergson, modern physics, from Descartes onward, spatialized the passing of time in a linear and quantitative form that failed to encapsulate the whole essence of what he calls "*la durée*", that is, the emergent and irreducible creation of novelties within time.[42] This is relevant to what happened to Waddington with embryogenesis: his developmental landscape forced him to "rigidify" into "a definite formal scheme"[43] its possible paths which were therefore pre-programmed and genetically anchored from the outset. The creativity of development as a temporal process was thus greatly reduced if not completely negated.

### 5.2    Problem n°2: the origin of the epigenetic landscape

In 1966, American zoologist and evolutionary biologist George C. Williams (1926–2010) published a book titled *Adaptation and Natural Selection, A Critique of Some Current Evolutionary Thought* that would become a landmark in the field of evolutionary theory and that was instrumental in the history of the philosophy of biology, especially regarding the seminal problem of the levels of selection. He wrote this rather short book (around 275 pages), to "purge biology" of "unnecessary distractions that impede the progress of evolutionary theory and the development of a disciplined science for analyzing adaptation".[44] As has been well documented, Williams' central claim is that the concept of adaptation is theoretically "onerous" and "should be only used where it is really necessary".[45]

Group selection was Williams' main target. During the 1950s and early 1960s, group selection was used much too loosely by several evolutionary ecologists to account for what they saw as group adaptation. English zoologist Vero Copner Wynne-Edwards (1906–1997) was at that time one of the main advocates of group selection which, according to him, operated especially at the species level. This was the central argument of his 1962 *Animal Dispersion in Relation to Social Behaviour*, a book that Williams always opposed. For Williams, natural selection, to be efficient, required entities endowed with specific characteristics that were shared by genes and individuals but not by populations or species. Williams' plea constituted a crucial step in the rise of the gene-centered view of evolution later promoted by John Maynard Smith, Richard Dawkins, and others.

Another target of Williams' criticism was Waddington and his concept of genetic assimilation: "The most prominent recent challenge to the adequacy of natural selection for morphogenetic phenomena is that propounded by Waddington (1956 et seq.), who argued that natural selection must be supplemented by another process, which he calls genetic assimilation".[46] He devoted no less than 11 pages to a detailed critical examination of Waddington's mechanism (it is interesting to note that Schmalhausen was absent from his picture). Williams took Waddington very seriously because he had managed to produce experimental results that seemed to ground his mechanism. Williams did not contest Waddington's results,[47] and he took for granted that the "*phenomenon* of genetic assimilation",[48] was "a real one".[49] But according to him, it did not necessarily follow that Waddington's explanation was correct. As we have seen in the previous chapter, Waddington's experimental designs were not restrictive enough to force the acceptance of his model.

Williams contested the usefulness of genetic assimilation as an alternative to standard natural selection and stressed that this evolutionary mechanism faced critical difficulties. His main point was that, in genetic assimilation, the explanandum (what is to be explained) could not be reduced to the final genetic fixation of a plastic response but had to include the previous selective history that was causally responsible for such a plastic ability. Genetic fixation was easy to explain, and even if Waddington had been right, the main point remained unexplained: how did plasticity evolve in the first place? In Williams' reading, Waddington took for granted that natural selection had been efficient enough to account for the evolution of plastic abilities, that is, the evolution of the landscape's topography, and that further explanation was therefore not required:

It seems to me that the sort of process pictured here [the genetic assimilation of a specific phenotypic response] must have occurred many times, but I would question its importance as an explanation of adaptive evolution. To explain adaptation by starting with a facultative response and ending with an obligate response is to beg the question entirely. The process starts with a germ plasm that says: "Thicken the sole if it is mechanically stimulated; do not thicken it if this stimulus is absent," and ends with one that says: "Thicken the soles". I fail to see how anyone could regard this as the origin of an evolutionary adaptation. It represents merely a degeneration of a part of an original adaptation. If the origin of the sole thickening as a fixed response is hard to explain, surely its origin as a facultative response is much more so. It must, as a general rule, take more information to specify a facultative adaptation than a fixed one.[50]

In short, for Williams, Waddington focused on the least interesting aspect of the issue and left the most complicated aspect unexplained:

Waddington gives very little attention to the origin of the facultative responses with which he starts his arguments. He summarized his attitude in one discussion (1958, p. 17) by postulating that natural selection "would, in fact, build into the developmental system a tendency to be easily modified in directions which are useful in dealing with environmental stresses and to be more difficult to divert into useless or harmful paths". It would appear that he finds the theory of natural selection entirely adequate to explain facultative adaptations, but feels that this theory has a "major gap" in its application to fixed adaptations.[51]

There are two steps in Williams' argument against Waddington and genetic assimilation. (a) First, Williams assumed that the epigenetic landscape was an evolutionary construct. (b) Second, because of (a), genetic assimilation was not about adaptation but only about "degeneration", that is, the degenerative simplification of the landscape. Williams pointed toward a genuine weak point in Waddington's reasoning, which he considered to be the weakest point in his theoretical framework. From a logical point of view, if (a) is correct it seems that (b) is unavoidable, thus the point at issue is to discuss (a). In my opinion, Williams' characterization is the correct one. Waddington remained remarkably discreet on the issue, and did not seem to fully understand how crucial it was. In a 2008 article, despite being much more favorable to Waddington than Williams had been, developmental biologist Adam Wilkins also pointed out that Waddington's argument against the Modern Synthesis would remain incomplete as long as the origin of the epigenetic landscape was not properly addressed:

[Waddington] pushed the idea of alternative capacities for morphological trait development and how they could be selected – his idea of "genetic assimilation" – yet he never seems to have asked himself how these alternative capacities [i.e. the epigenetic landscape] might themselves have arisen.[52]

Waddington never directly faced this issue and never identified it as a major problem, so his precise position on this matter can only be reconstructed retrospectively. In the passage cited previously, Williams quotes one of the rare times

Waddington is clear about the fact that he thought of the epigenetic landscape as an adaptation designed by natural selection. This interpretation is also verified by the fact that, from the 1940s onward, Waddington regularly emphasized that the level of canalization of a developmental path was fine-tuned by natural selection.[53] On the whole, he viewed today's organisms as highly evolved products endowed with complex innate properties, such as adaptively reacting to environmental change.[54] There is therefore little to no doubt that he thought of the epigenetic landscape as a whole as historically designed by natural selection.

He could have instead conceived the adaptiveness of embryogenesis as a result of it being a dynamical process sensitive to its environment, that is, a phase-space in a more standard physical account, inherently structured by attractors. This would have been more in line with Waddington's argument that a processual view is key to improve biology as a theoretical science. It is what would have been expected from a developmental biologist so obviously concerned in carving out a theoretical space for development within evolutionary theory. But this was not the path he took. Very rarely indeed, at least in his published writings, did Waddington even consider this alternative. It was only in 1968, in the first volume of the proceedings of the Serbelloni meetings, that he wrote – perhaps as a consequence of his recent interaction with French mathematician René Thom – that "it is an interesting question to discuss how far the existence of chreods [alternative developmental pathways] is necessary and how far it is merely an empirical result of the operation of natural selection".[55] "Interesting" as the idea may have been, he never pushed it further, and what he meant by "necessary" here remained unexplained.

## 5.3 Genetic assimilation with no landscape at all: Schmalhausen

The issue of developmental creativity was not as critical in Schmalhausen's own framework as it was in Waddington's. On the whole, Schmalhausen was not as committed to a gene-centered view of embryogenesis nor was he as concerned by the renewal of the Baldwin effect and he never rigidified the developmental phase-space into a schematic form anchored in the genome. In "Stabilizing Selection" (1941) and in *Factors of Evolution* (1949), there is nothing like Waddington's epigenetic landscape.

Moreover, Schmalhausen elaborated an explicit emergentist view of embryogenesis in which the final plastic accommodations were not reduced to pre-programmed possibilities. In contrast to Waddington, he used terms like "new forms",[56] "new changes",[57] "new modifications arising for the first time",[58] "new reactions",[59] and so on, which were remarkably absent from Waddington's published writings.[60] How did Schmalhausen explain that embryogenesis, in unusual environmental conditions, could produce "new forms"? What did "new" mean in this theoretical context?

This constitutes a substantial difference between Waddington's and Schmalhausen's conceptions of genetic assimilation. To some extent, Waddington was aware of this difference, but he thought that it was detrimental to Schmalhausen. But it may in fact have been the other way around.

During the summer of 1959, Waddington and Dobzhansky exchanged letters about Schmalhausen[61] and Waddington complained that, in his publications, Dobzhansky always favored Schmalhausen's ideas over his.[62] As he explained in the first letter,[63] Waddington thought his model of adaptive evolution was superior to Schmalhausen's for three reasons: (1) he had succeeded in producing what he believed to be an experimental basis; (2) he made a clearer distinction between "stabilizing selection type 1" (i.e., standard normalizing selection) and "stabilizing selection type 2" (canalizing selection);[64] (3) finally, he did not need Schmalhausen's "almost metaphysical"[65] distinction between "morphoses" and "adaptive modifications". Dobzhansky replied that Schmalhausen needed explicit support because of the impact of Lysenkoism on his career and that he did not "see that Schmalhausen implied "any essential physiological difference" between [morphoses and adaptive modifications]".[66] He then gave a telling example:

> Sun tanning is a modification, but blistering is a morphose – the first is a reaction directed by natural selection in a channel which is adaptive, the latter is adaptively ambiguous, at least as it stands at present, but with further selection it might, conceivably, become a modification also.[67]

Dobzhansky's presentation is faithful to Schmalhausen's theorization. It is true that Schmalhausen distinguished between two forms of plastic responses: morphoses and (adaptive) modifications. In short, *his* concept of genetic assimilation was about the evolutionary transformation of morphoses into modifications, under the creative action of stabilizing selection (see Chapter 6). This distinction was not secondary or incidental. Already in "Stabilizing Selection" (1941), and probably before then, Schmalhausen provided a rather detailed definition of the concept of "morphose":

> The presence of unusual conditions, i.e. extreme deviations in environmental factors (temperature, salt levels in water, etc.), or the appearance of an entirely new factor (unusual chemical substances, x-rays etc.) therefore often find the organism vulnerable and, of course, call for changes in the formative process, however, these changes then no longer have an adaptive character. Such non-adaptive modifications can be referred to as morphoses (in literature the term "morphosis" is habitually used more broadly – along with modification).[68]

Schmalhausen was to some extent aware of the significance of the issue of evolutionary/developmental creativity, and he perfectly understood that in his theory of stabilizing selection, evolutionary novelty required true developmental novelty. This was why he promoted such a distinction in the first place. Morphoses were phenotypic "indeterminate effects"[69] that were produced by drastic environmental changes (such as X-rays and unusual chemical substances[70]) and that most of the time were "non-adaptive".[71] In brief, exactly what Waddington had managed to obtain in the first two series of experiments on *Drosophila* in the 1950s. But as we have seen, Waddington interpreted his results in the framework of a stable and pre-existent epigenetic landscape, that is, without thinking that the environment could in any way destabilize the landscape and then causally produce new, not yet canalized paths ("indeterminate effects") that might become the first steps in an evolutionary process. Schmalhausen did differentiate between "modifications" that were designed by natural selection (and as such stored through heredity) and morphoses:

> Nonadaptive modifications or morphoses are of an entirely different character. They arise as new reactions which have not yet attained a historical basis. Either the organism encounters new environmental factors with which it never had to deal before or its norm of reaction is changed (disturbed) as a result of mutation. . . . At best, all these reactions are indifferent; more often they are either pronounced disturbances of an existing organization or obvious deformities which are not very viable under normal conditions. Usually such morphoses are extremely unstable; they easily change their expression when intensity of environmental factors varies.[72]

For him, there was no doubt that only morphoses, which were outside the canalized reaction norm (see Figure 5.3), were able to bring novelty within the evolutionary process: "All really new reactions of the organism are never adaptive".[73] If some of them proved to be advantageous, then stabilizing selection would progressively build "a historical basis" that would make them more robust. They would become part of the hereditary repertoire, that is, outcomes of the new epigenetic landscape. For Schmalhausen, in Waddington's terms, the environment could developmentally destabilize the landscape strongly enough to produce new paths, most of them unstable and pathological (developmental abnormalities).

In a few cases, some morphoses could be the starting points of new evolutionary paths. Note that on some occasions, Schmalhausen thought exactly like Waddington, in terms of a pre-existing set of paths.[74] But his distinction between morphoses and modifications prevented him from rigidifying the developmental phase-space (at the developmental time span) too much and thus made room for a minimal degree of developmental creativity. Today, his framework might provide a more suitable interpretation of Waddington's experimental results on *Drosophila*:

> This specificity of reaction is determined by the historically developed nature of the organism, by its evolution in a certain environment, and by the constant interaction of the latter with internal factors of the organism. On the other hand, environmental factors that do not usually occur in the normal environments of the organisms can, to the extent that they influence the organism, produce only indeterminate effects, manifested in a more or less profound disturbance of its normal structure and function.[75]

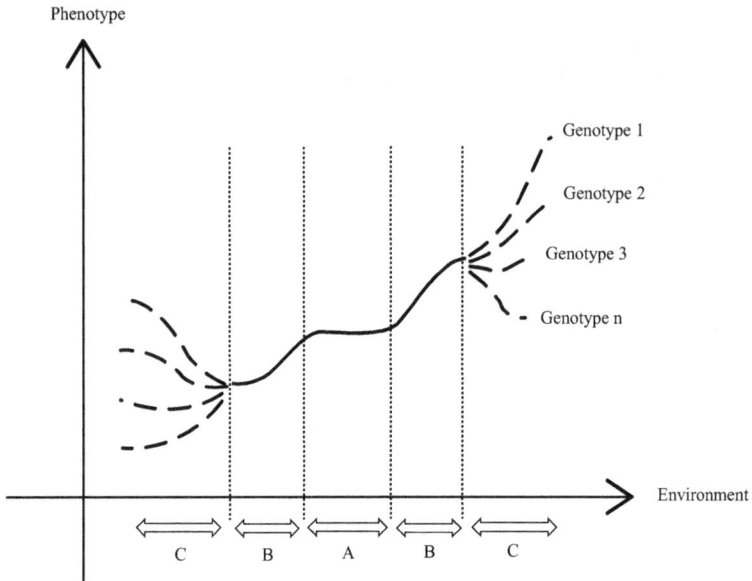

*Figure 5.3 Graphic representation of Schmalhausen's distinction between "morphoses" and "modifications".* The reaction norm of a population (what Schmalhausen sometimes refers to as the "adaptive norm") comprises three types of segments. For certain values of environmental parameters (A), canalization is maximal and phenotypes are identical and constant whatever the genotype considered. For more extreme environmental values, phenotypes show a homogeneous and adaptive variation within the population (B): this is what Schmalhausen called "modifications". At even more extreme environmental values, developmental systems are too strongly destabilized (C): there is strong intra-population phenotypic variability ("morphoses"), revealing previously cryptic genetic diversity.

At this point, again in contrast with Waddington, it must also be emphasized that Schmalhausen regularly favored the cytoplasm and not the genes in the causal production of embryogenesis. In his view, development was the complex consequence of a refined system of mutual interactions between various sub-systems where the cytoplasm had the leading role:

> The *cytoplasm* is the specific *substrate of ontogeny* in which occur all interactions that control determination and differentiation. All differentiations that depend upon the external environment, including all the modifications, are brought about in the cytoplasm.[76]

Finally, it seems that Schmalhausen not only supported a less gene-centered view of the developmental landscape than Waddington but also proposed that the more an organism is evolved, that is, the more it is endowed with "regulative stability", the less we need to think of development within such a predetermined view. During progressive evolution, the epigenetic landscape, or something close to it, progressively vanishes because adaptability becomes the central property of complex organizations:[77]

> The biologic importance of regulative stability is that a stable organism, protected during ontogeny from external accidental changes, can actively alter the environment of its habitat and adapt itself to totally new situations. In this case, typical adaptive norms are not provided as in autoregulated development because the organism is by nature monomorphic, except for possible seasonal variations. The same species may occur in an endless number of variants. Most important is the fact that a particular variation, produced by a functional adaptation of a specific organ to a new situation, immediately adjusts itself by means of functional correlations to a series of other changes in the entire organization. The results of individual adaptation are as complete as in autoregulated development; they are not limited by the existence of two or three historically established typical norms but are developed anew for each individual in a fully harmonious form. In a stable organism of the regulative type, individual development is highly stable and integrated and is independent of any rigidly fixed standard norms.[78]

Even if Waddington also stressed that evolution had favored the "adaptable",[79] such a claim was immediately weakened by his genetic understanding of a developmentally fixed epigenetic landscape. It seems that this was not the case for Schmalhausen, even though further work is still needed given how little of his work is available in English.

## 5.4    Chapter's conclusion

Waddington thought that to introduce developmental biology within the Modern Synthesis would have truly revolutionary consequences, in the Kuhnian sense of the concept. In the preface of his 1975 *The Evolution of an Evolutionist*, he characterized three main "paradigm shifts" in the history of the evolutionary theory since the rise of Mendelian genetics in 1900.[80] The third shift, between the third (the Modern Synthesis) and fourth paradigm, started with a new version of evolutionary theory to which he hoped to have contributed. If Jacques Monod's "Chance and Necessity" offered him a good slogan to characterize the Modern Synthesis, Waddington came up with different ones for the new paradigm that was still in the process of being completed, such as "Learning and Innovation", "Adapting and Improvising", or "Recompiling and Heuristic Search".[81] In the same preface, he acknowledged that throughout his entire career, he "was looking at evolution very definitely from the point of view of a developmental biologist".[82] This was an appropriate assessment of his career and involvement in the field. Whether his work had revolutionary consequences for the Evolutionary Synthesis, on the other hand, is highly debatable.

On the one hand, Waddington's conception of embryonic development was very genocentric. In his view, the topography of the epigenetic landscape was developmentally fixed and genetically determined. From this point of view, it was difficult to understand how the developmental process could produce innovative phenotypes. On the other hand, he probably conceived the topography of the epigenetic landscape as an adaptation resulting from natural selection. So, in the end, natural selection would have to account for the landscape genesis: natural selection remained the main explanatory factor.

To some extent, and based on the literature available in English, it seems that Schmalhausen favored a view of embryogenesis that was less committed to genetic determinism. This might have made his theory a more promising starting point to include embryology within a renewed Evolutionary Synthesis. Unfortunately, Schmalhausen did not go very far in that direction. He never formalized his hypotheses in a workable model and never performed any research program in that direction. And, because of Lysenkoism, he had to quit theoretical biology around the time he finished *Factors of Evolution*. Therefore, while his framework might be seen as a more suitable interpretation of Waddington's own experimental work, these ideas have remained on the drawing board.

To this day, from a theoretical point of view, any model that aims to reintroduce the active role of embryonic development within evolutionary causality faces one crucial problem: the heredity of phenotypes. Indeed, the seemingly serious limitations encountered by Waddington could end up being easily solved by relaxing the genocentric basis of his theory making more room for the constructive role of the environment, as many biologists (often on the side of evolutionary ecology), such as Mary-Jane West-Eberhard, have recently called for. We could imagine developmental potentialities that are not rigidly predetermined: they would emerge from the establishment of a complex (and quasi-symmetrical) relationship between the developmental system (not reduced to the genetic information of the nucleus) and the surrounding environment. In other words, we would end up with a theoretical model closer to the one envisaged by Schmalhausen.

If so, would such a conception have "revolutionary" consequences for evolutionary theory? If heredity is essentially conveyed by genes, then, probably not. Even if the phenotype were to have a lot of freedom in relation to the genotype, this would not alter the causal dynamics of the evolutionary process, since natural selection can only act on heritable variation, as Richard Dawkins has regularly reminded us:

Jablonka and the school of thought dubbed "Developmental Systems Theorists" think that the complexity of embryonic development somehow detracts from the validity of the gene's eye view of Darwinism. But we must not allow complexity to become a euphemism for muddle. [. . .] You may be sick of hearing my *leitmotif* but we are just going to have to play it one more time as a finale. It doesn't matter how complicated the developmental support structure, nor how utterly dependent

DNA may be upon it, the central question remains: which elements of the Great Batesonian Nexus of development have the property that *variations* in them are replicated, with the type of fidelity that potentially carries them through an indefinitely large number of evolutionary generations? Genes certainly meet the criterion.[83]

## Notes

1 Waddington (1968c).
2 Waddington (1975a, p. vi).
3 Loison (2022).
4 Baedke (2013).
5 Leamy & Klingenberg (2005, p. 2).
6 Since the 1950s, many FA measurement modalities have been developed. All of them are based on a difference between the value of the character on one side of the body (usually the right side, R) and that of the same character on the other side (usually the left side, L), but, in detail, there are many different indexes (some are based on the variance of R-L, others on the mean of the absolute value of R-L, etc.). In their seminal 1986 paper, Palmer and Strobeck count no less than 22 different indexes, which makes meta-analyses difficult (Palmer & Strobeck, 1986).
7 Leamy & Klingenberg (2005, p. 1).
8 For biographical details, see Lewis (1992).
9 Fisher (1918).
10 Mather (1949a, p. viii).
11 Falconer (1993, p. 140). Still in 1955, in a letter to Lerner, Mather reported on this difficult relationship between the Birmingham and Edinburgh groups: "Both Falconer and Alan Robertson have visited us in Birmingham, which means, I hope, that relations will lose some of the strain noticeable in the past (although I have never regarded them as mamalekites like some others I could name). I am not suggesting that the geneticists of this country will soon be like the heavenly choir, but every little helps" (Mather to Lerner, 6th October, 1955, Lerner's archives).
12 Mather (1953, p. 299).
13 Mather (1953, p. 318).
14 As early as 1962, Leigh Van Valen proposed another explanation: it could be that what Mather had measured was not in fact the fluctuating asymmetry of the trait under consideration but rather another form of asymmetry called "antisymmetry" (Van Valen, 1962, p. 126).
15 Mather (1953, p. 328).
16 Tebb & Thoday (1954).
17 Waddington (1957a, p. 40).
18 Palmer & Strobeck (1986).
19 Van Valen (1962, p. 138).
20 Zakharov (1992); Zakharov et al. (2020). For a very enlightening definitional clarification, see Debat & David (2001).
21 Waddington (1957a, pp. 41, 72). This term seems to have had no legacy.
22 Waddington (1957a, pp. 40–41). My emphasis.
23 Loison (2022).
24 Wilkins (2008).
25 Waddington (1969c, p. 121). Emphasis in the original.
26 Waddington et al. (1954).
27 Waddington (1957a, p. 107).
28 See, for instance, Odling-Smee (2010).
29 West-Eberhard (2003, pp. 13–14).

30 Hall (1992, 2003).
31 For instance, Boris Ephrussi: Sapp (1987, pp. 123–162); Burian & Gayon (1999).
32 See Morgan's 1934 book, *Embryology and Genetics* (Morgan, 1934a). See especially his 1933 Nobel Lecture "The Relation of Genetics to Physiology and Medicine" (Morgan, 1934b).
33 Robertson (1977, pp. 592–593); Peterson (2016, p. 125).
34 Waddington (1940a, p. 94).
35 Waddington (1940b, p. 52).
36 Waddington (1966a, p. 44).
37 Loison (2022).
38 Waddington (1942, p. 564), my emphasis.
39 Waddington (1961a, p. 287).
40 Nicoglou (2018).
41 Bachelard's concept of an epistemological obstacle was theorized in *The Formation of the Scientific Mind*, first published in 1938 (Bachelard, 1938).
42 See especially Bergson's famous book *Creative Evolution* (Bergson, 1907).
43 Waddington (1942, p. 564).
44 Williams (1966, p. 4).
45 Williams (1966, p. 4).
46 Williams (1966, p. 71).
47 Williams (1966, p. 73).
48 Williams (1966, p. 72). My emphasis.
49 Williams (1966, p. 72).
50 Williams (1966, p. 80).
51 Williams (1966, p. 82).
52 Wilkins (2008, pp. 229–230).
53 Waddington (1942, p. 564).
54 Waddington (1959a, pp. 399–401).
55 Waddington (1968b, p. 13).
56 Schmalhausen (1941, pp. 332, 343, 346; 1949, p. 83).
57 Schmalhausen (1949, p. 3).
58 Schmalhausen (1949, p. 3).
59 Schmalhausen (1941, pp. 309, 317; 1949, pp. 8, 34, 242–243).
60 It is telling I think that in the few places where Waddington was considering "novelty" and "new forms" in this sense, he developed ideas that were *in contradiction* with his standard account of the epigenetic landscape. See especially Waddington (1969b, p. 120).
61 Waddington published part of this correspondence in *The Evolution of an Evolutionist* (1975a).
62 Waddington (1975a, p. 96).
63 Written the 25th of July.
64 See also Waddington (1957a, p. 73).
65 Waddington (1975a, p. 96).
66 Waddington (1975a, p. 98).
67 Waddington (1975a, p. 98). Dobzhansky replied to Waddington the 15th of August.
68 Schmalhausen (1941, p. 317).
69 Schmalhausen (1949, p. 3).
70 Schmalhausen (1949, p. 3).
71 Schmalhausen (1949, p. 3).
72 Schmalhausen (1949, p. 8).
73 Schmalhausen (1949, p. 8).
74 For instance: "The species as a whole is adapted not in one form, but in two or more forms; not to one season or biotope with its characteristic set of conditions, but to several. Accordingly, the species does not become extinct when

environmental variations cause the disappearance of one particular aggregate of conditions which characterized one of the seasons or biotopes to which the species was adapted. There remains another set of factors to which the species has already become adapted. . . . The adaptive norm is altered and the former principal reaction loses its value and gives way to a new reaction, which previously had only secondary importance. All these reactions had acquired earlier an integrated character so that the entire organism is transformed immediately when the physical factors of the environment vary rapidly" (Schmalhausen, 1949, p. 196).

75 Schmalhausen (1949, p. 3).
76 Schmalhausen (1949, p. 28). Emphasis in the original.
77 See also Schmalhausen (1949, p. 275).
78 Schmalhausen (1949, pp. 42–43).
79 See especially chapter 5 of his *The Strategy of the Genes*, "The Survival of the Adaptable" (Waddington, 1957a).
80 Waddington (1975a, pp. v–vi).
81 Waddington (1975a, p. vi).
82 Waddington (1975a, p. viii).
83 Dawkins (2004, pp. 391–392). Emphasis in the original.

# 6 An overlooked revolution? Creativity in the evolutionary building of a new reaction norm

The conclusion reached in the previous chapter can be expressed in mathematical language: in Waddington's framework at least, it appears that the phenotype remains a function of the genotype. This function may be a very complex one, as denotes the epigenetic landscape itself, but it remains a function, that is, a pre-existing rule that gives the value of a dependent variable (the phenotype) that corresponds to specified values of one or more independent variables (the genotype). This means that Waddington's originality and heterodoxy was most probably not on this side of things; that is, his account was still gene-centered at least in this sense. But Waddington can still be considered original and thought-provoking in other, more unexpected places.

The present chapter's argument starts as follows: in Waddington's and Schmalhausen's views, the phenotype is still a function of the hereditary material but more precisely, and crucially, a function of the *genome as a whole* and not of the genotype in its usual additive understanding. In short, their gene-centered view was in fact a genome-centered view, and to both of them, the genome was much more than its constitutive parts, that is, individual genes (Section 6.1). Such a holistic and interactionist account could have significant and nontrivial consequences for the way embryogenesis and evolution can be reframed.

Its most immediate consequence is that the mathematical function that links the genome to the phenotype is necessarily highly non-injective. In biological terms, this means that the same phenotype can be built on very different genetic bases. A weak understanding of this idea would be to assume that, even if the genotype and phenotype do not map up perfectly one-to-one, there exist several sets of individual alleles that can bring into being the same phenotype. In such a case, the action of genes is additive and the genome remains the sum of its parts. On some occasions, Waddington's and Schmalhausen's writings could be interpreted along these lines, which do not challenge the Modern Synthesis. This view simply implies that, given a specific phenotype, the genotype cannot be precisely deduced, which is something obvious for most evolutionary geneticists.

DOI: 10.4324/9781003422990-11

But there is more to be found in Waddington's and Schmalhausen's ideas, and this "more" is what is termed GA2 in the present book, in other words, a more radical and original concept of genetic assimilation than GA1. The present chapter is thus devoted to the historical and philosophical elucidation of this heterodox and neglected concept of genetic assimilation. In the chapter's conclusion, a comparative table summarizes the main findings of the next three sections to help the reader visualize how related concepts, such as canalization, stabilizing selection, and phenotypic variation, do not have the same meaning in GA1 and GA2, as detailed in Sections 6.2 and 6.3. It must be stressed from the outset that GA1 and GA2 were usually conflated by Waddington and Schmalhausen themselves, which did not help the reception and discussion of their evolutionary models.

## 6.1 The genome as a canalized system: against genetic reductionism

Instead of the standard atomistic view, where each gene has one or a few discernable causal roles in the building of the phenotype, both Schmalhausen and Waddington shared a more systemic understanding of the genome that paved the way to a systemic account of development. Other biologists, for instance, Richard Goldschmidt, presented holistic accounts in the mid-20th century. Goldschmidt always put the emphasis on pleiotropy rather than linearity.[1] In the late 1930s, he also came close to the idea of canalization of the wild type,[2] an idea that would be a central argument for Waddington and Schmalhausen a few years later. Overall, Goldschmidt favored a systemic view of the genome that he labeled "system of genic action"[3] or "system of gene-controlled reaction",[4] in which genetic causality was distributed throughout the whole genome and not individualized into discrete alleles. This led him to contest that the "gene as a hereditary unit" was still a workable hypothesis.[5] It is well known that, in the end, and especially in *The Material Basis of Evolution* (1940), Goldschmidt opposed the gradualism of the Modern Synthesis.[6] The main evolutionary steps were the results of "systemic mutations" that introduce a gap between micro- and macroevolution.

Even if Waddington regularly praised Goldschmidt's insights,[7] neither Waddington nor Schmalhausen went as far as him in their theorizations: their systemic account of the working of the genome and of embryogenesis did not fuel a saltational type of evolution which explicitly went against the main tenets of the Modern Synthesis. Waddington did not advocate for a complete upheaval of the standard account; rather, he wanted to improve it.[8] He still thought of microevolution and macroevolution as processes of the same nature. In his view, the genome was still central in setting up the phenotype, as we saw in the previous chapter. But the systemic organization of the genome introduced what Waddington thought was "an inescapable indeterminism", since "identical phenotypes may have different genotypes, and identical genotypes may give rise to different phenotypes".[9]

In Waddington's and Schmalhausen's conception, canalization was, in the end, responsible for this indeterminism. This constituted the final step in their reasoning, the one with the most radical consequences for evolutionary theory, as we shall see in the next two sections. But again, this challenging view was not the only one they assumed when they thought of the genetic basis of a phenotype. In their published writings, Waddington and Schmalhausen endorsed at least three related views in their critique of the idea that a phenotypic trait could be the consequence of a simple (additive) genetic determinism. There were different levels in their understanding of how the genome accommodates phenotypic change:

(a) A first level assumes that a single phenotypic character is the consequence of several independent and discrete genes. Waddington favored this interpretation in the experiments that showed the genetic assimilation of the *crossveinless* phenotype. In contrast to what was supposed to be the case in the Baldwin effect, the assimilated phenotype was not the consequence of the selective fixation of a single point mutation, but of at least several of them. These mutations could be very numerous indeed, and already in *Organisers and Genes* (1940), in the wake of his experimental work on wing development at Caltech, he assumed that "all the genes are concerned in all developmental processes".[10] But even then, the interactions between the genes involved could remain additive.

(b) A second level assumes that *different* sets of genes can be responsible for the same phenotype. This second level was envisioned by Waddington as early as 1953 for the genetic assimilation of the *crossveinless* phenotype. At the end of the *Evolution* paper, he wrote, "It seems quite possible that if a similar selection for crossveinless formation was repeated, perhaps using a different foundation stock, *the same phenotypic effect might be produced with quite a different genetic basis*".[11] Later, he would assume that "the differences between the assimilated strains and their foundation stocks always involve all the chromosomes".[12] For both Waddington and Schmalhausen, the high frequency in natural populations of the "wild type" was a textbook example of such a multi-realizability of the same phenotype.[13] Yet, even then, it was still possible to evaluate the quantitative contribution of each gene to the phenotype produced. The genome remained the sum of its parts, even if the same sum could be reached in several different ways.

(c) The third and last level is the most challenging one and will be given full attention in the following paragraphs and in the next section. In this view, the genome is a highly integrated system: what matters is not the individual genes themselves but the topology of their causal relationships. As Schmalhausen put it: "The normal development of the entire organism depends upon the genotype as a whole if the system of correlations is complex. When the parts of the organism are connected, the role of the individual gene is reduced to the level of modifiers".[14] In this case, a new property emerges, namely canalization, that is responsible for the partial decoupling of genotype and phenotype.

Levels (a), (b), and (c) were not always well distinguished by Waddington and Schmalhausen and they most usually presented a mix of the three. Levels (a) and (b) do not pose any theoretical difficulty and are fully compatible with the Modern Synthesis. Level (c) is of course the most heterodox, original, and controversial and is the focus of the next pages. The first thing to emphasize is that canalization was always thought of as a property of the developmental system or even the genotype rather than of the phenotype. For instance, already in his 1942 "Canalization of Development and the Inheritance of Acquired Characters", Waddington started by writing that "the main thesis is that developmental reactions [. . .] are in general canalized", and specified, a few lines later, that canalization must be understood as the "buffering of the genotype".[15] Exactly like a reaction norm, canalization was a genomic property, and as such could be directly selected as Waddington wanted to demonstrate in the late 1950s.[16] Waddington, probably more than Schmalhausen, always stressed that natural selection was responsible for the degree of canalization achieved. As he put it, canalization is "built up by natural selection".[17]

But the fact that canalization can be subjected to natural selection does not imply that, for Waddington and Schmalhausen, there exist "genes for canalization" (see Section 8.2 for a more detailed analysis). Rather the opposite: both seem to understand canalization as an emergent and systemic property of the genome, that is, the most important consequence of its integrative functioning. As Waddington summarized, "The 'canalization of development', must be regarded as a result of the complex interactions between the numerous gene-controlled processes by which development is brought about".[18] A form of generalized epistasis, under the guidance of natural selection, creates canalization. Therefore, there is no need to oppose these two aspects:[19] canalization as an adaptation designed by natural (canalizing) selection and canalization as a systemic property of the genome. Both were endorsed by Waddington[20] (and most probably by Schmalhausen) but represented different aspects of the same property: canalization was thought of as an evolutionary adaptation brought about by epistatic genetic interactions.

Schmalhausen[21] and Waddington assumed that a high degree of pleiotropy and epistasis was the general rule. This meant that development could not be analyzed in standard genetic terms, at least for evolved organisms. Developmental canalization emerged as the result of the systemic and interrelated working of the genome and the whole developmental system.[22] In several places, Schmalhausen expanded on this crucial aspect:

> As the systems of correlations become more complex, they lose their genetic character; in other words, the effects on individual genes can no longer be distinguished (more precisely their disruption by mutations has lethal consequences). Morphophysiologic interrelationships and their systems may be regarded as entities that are not capable of analysis in genetic terms. These correlation systems consist fundamentally of the mutually related vital functions of the organism, which are responsible for the harmonious course of the general physiologic processes.[23]

As canalization evolves, "the decisive role may pass from the individual genes to the genotype as a whole".[24] Waddington reified the exact same idea in one of his diagrammatic representations of the epigenetic landscape that we discussed in the previous chapter (Figure 6.2): he aimed to picture "the complex system of interactions underlying the epigenetic landscape".[25] As we have seen, each peg represents a gene and each string its contribution to the network that causally produces the epigenetic landscape. Later in the book, Waddington made crystal clear that canalization was a systemic property of the genome because of a dense network of epistatic interactions:

> The epigenetic feed-back mechanisms on which canalisation depends can, of course, be regarded as examples of gene interaction. Interaction between two allelomorphs is referred to by such terms as dominance, recessiveness, over-dominance, etc. Interactions between different loci come under the heading of epistasis. This is perhaps most usually thought of in terms of interactions between only two or three loci. *We know, however, that in the development of any one organ very many genes may be involved, and in canalized epigenetic systems we are probably confronted with interactions between comparatively large numbers of genes.*[26]

This "generalized epistasis" implied that all the genes were involved in every single character individualized in the phenotype. It therefore seemed unrealistic to ascribe a definite selective coefficient to individual alleles. From the viewpoint of embryogenesis, the genome could not be atomized. Waddington and Schmalhausen shared this critical appreciation of what they saw as a necessary assumption of theoretical population genetics.[27] In *Factors of Evolution*, Schmalhausen repeatedly emphasized that "an individual mutation does not necessarily have a constant effect".[28] The exact same criticism can be found in many places in Waddington's numerous publications: "to attach coefficients of selective value to genotypes" was opposed as a "basic error".[29] In *The Strategy of the Genes*, he argued, like Schmalhausen, that "fitness is a quality of the organism as a whole".[30] Because he interpreted the genome as a system, he saw standard genetic reductionism as a dead end that might be workable in some rare situations but that failed to give a realistic picture of usual cases. The "inescapable indeterminism" Waddington put forth and that he explained with canalization, was, according to him, the main reason why theoretical population genetics should be seen as a very approximate and inadequate framework to model adaptive evolution. In his view, genetic assimilation not only acknowledged but was based on this alleged systemic working of the genome. This interactionist view of the genome was the basis of GA2 that was supposed to oppose the standard blind mutation/selection process of the Modern Synthesis.

## 6.2   Phenotypic variation as a property of the developmental system (GA2)

Waddington, probably without realizing it, used the concept of canalization in at least two quite significantly different senses. Most of the time, he saw canalization as a property of the genome simply capable of *masking* the phenotypic effects of genetic mutations. This is the meaning found in most of his publications which also underpins his standard conception of genetic assimilation (GA1). In other, rarer cases, however, Waddington conceived of developmental canalization not merely as shielding the phenotypic effect of mutations but as *producing* it. This more heterodox conception, that is the subject of this section, underpins a more speculative and original view of genetic assimilation (GA2).

### 6.2.1   *Toward a different concept of canalization and genetic assimilation*

Waddington clearly expressed these more heterodox conceptions of canalization and genetic assimilation in at least six different publications, between 1953 and 1968. This means that it represented a significant aspect of Waddington's theoretical activity regarding adaptive evolution. In GA2, the systemic understanding of both the genome and development was at its highest. As a result, canalization was conceived as an emergent property of these interrelated systems with significant powers. Canalization was no longer simply a screen or a filter that could prevent the phenotypic expression of specific mutations but an intermediary level of biological organization that was powerful enough to "*guide* the phenotypic effects of the mutations available".[31] In other places, Waddington used different terms – such as "conditioned",[32] "influenced",[33] "characteristic of the epigenetic system",[34] and so on – to characterize how the effects of mutations were transformed by canalization. In the epilogue of *The Strategy of the Genes* he was fully explicit about this more radical view of genetic assimilation:

> We have been led to conclude that natural selection for the ability to develop adaptively in relation to the environment will build up *an epigenetic landscape which in its turn guides the phenotypic effects of the*

*mutations available.* In the light of this, the conventional statement that the raw materials of evolution are provided by random mutation appears hollow. The changes which occur in the nucleoproteins of the chromosomes may well be indeterminate, but the phenotypic effects of the alleles which have not yet been utilized in evolution cannot adequately be characterized as "random": they are *conditioned by the modelling of the epigenetic landscape* into a form which favours those paths of development which lead to end-states adapted to the environment.[35]

This crucial passage exemplifies the central argument of GA2 and as such requires several comments. The first is that what Waddington contested here (and in other similar paragraphs) was the so-called blindness or randomness of genetic mutations. This was a core dimension of Darwinism, from Darwin and Weismann to the Modern Synthesis: hereditary variation is blind as regards adaptation; that is, the probability of a mutation is independent of its phenotypic effect.[36] In general, critics of the MS speculated that genes were somehow able to mutate in the adaptive direction, which made phenotypic variation inherently adaptive and then natural selection of secondary importance. For example, in the late 1980s and 1990s, the discovery of possible mechanisms for directed mutations in bacteria fueled such speculations. Several "mutator systems" were then characterized and were sometimes understood as concrete arguments against the logic of Modern Synthesis.[37]

Already in the 1950s and 1960s, Waddington thought that such a possibility was mostly improbable.[38] This was not what he had in mind when, in the framework of GA2, he contested the randomness of genetic mutation. His criticism was of a different, more complex, and subtle nature. As the previous quotation makes clear, he did not deny that mutations, as molecular events, were random. He acknowledged that "the changes which occur in the nucleoproteins of the chromosomes may well be indeterminate".[39] What Waddington opposed was that randomness at the molecular level implied randomness at the phenotypic level. DNA alterations could be random, but it did not follow that hereditary variation was random at the level of the developmental system. Between the genotype and the phenotype, Waddington always put the emphasis on an intermediate level of biological organization that he reified into the epigenetic landscape. At this level, genetic mutations, whatever their intrinsic nature, were only the "raw material" of phenotypic variation but in a sense substantially different from the one embedded in the Modern Synthesis. These genetic (molecular) variations were almost deprived of any constant phenotypic properties. He conceived phenotypic variation as a systemic property that depended on the canalized genome and almost insensitive to what happened at specific loci. This was what Waddington meant when he wrote that mutations "are conditioned by the modelling of the epigenetic landscape".[40] In 1961, he rephrased the same idea as follows:

> As the genotype of an evolving population becomes moulded by selec-tion for its capacity to react in a satisfactory manner with environ-mental stresses, it will have built into it certain chreods with particular homeorhetic characteristics. When new gene mutations now occur, by what we may consider random changes in the nucleo-proteins of the chromosomes, *the effects these mutations produce on the phenotype of the animals containing them will not be entirely random, but will be, to some extent at least, characteristic of the epigenetic system in which the gene is operating.*[41]

In more philosophical terms, this meant that the effects of mutations on the phenotype switched from "intrinsic" to "extrinsic property". In metaphysics there exists a long tradition regarding conceptual distinctions between sev-eral kinds of properties and the most fundamental is between two main kinds that can be ascribed to entities: extrinsic *versus* intrinsic. This distinction is usually introduced as follows: intrinsic properties are those that character-ize an entity by virtue of what it is (mass is the textbook example of this), whereas extrinsic properties result from the interactions between this entity and the world (such as weight).

David Lewis's clarificatory work, in particular from his 1983 article "Extrinsic Properties", is of particular interest to my argument. As Lewis argued, "extrinsic properties" do not only "depend on that thing" (here a mutation or an allele) but may also depend, "wholly or partly, on something else". More importantly for our purposes, he adds that "if something has an intrinsic property, then so does any duplicate of that thing; whereas dupli-cates situated in different surroundings will differ in their extrinsic proper-ties".[42] In GA2, different "surroundings" are different developmental systems and different evolved genomes: what a mutation can do is highly constrained or even guided by the whole epigenetic system, as Waddington repeatedly stressed.

In several publications,[43] philosopher of biology Lucie Laplane has elabo-rated a more precise typology distinguishing four kinds of properties: cat-egorical, dispositional, relational, and systemic. A categorical property is an intrinsic property that is independent of any interaction with surrounding entities (e.g., the atomic mass of an element). A dispositional property is also an intrinsic feature but one that only manifests upon interaction with exter-nal stimuli (e.g., fragility, which is apparent only on impact). A relational property is an extrinsic property that relies on the interaction between enti-ties (e.g., body weight, which depends on gravity). Lastly, a systemic prop-erty is a characteristic that is provided and maintained by the system, that is, one in which the intrinsic properties of the entity considered are almost irrelevant. In my view, in GA2, phenotypic variation can be considered as a systemic property – the system being here the developmental system, and not any more as a dispositional property of discrete alleles. It is especially remarkable that, as early as 1941 (and most probably even before), Schmalhausen also perfectly expressed the idea that phenotypic traits were not properties of genes but of the developing organism:

> However, unlike the [ideas put forward by] foreign geneticists, we see dominance, robustness, trait expression etc. *not as properties of genes* (see M. Kamshilov), not just as a result of their coordinate actions or their higher or lower [levels of] activity, but rather as an expression of interdependence between parts *in the correlational system of the developing organism.*[44]

Waddington tried to explain this refined view of the relationships between the genome, the developmental system, and the phenotype by emphasizing that DNA molecules and organisms' traits are two different levels in biological organization. For him, the phenotype was much more complex, and there exists between the phenotype and the genotype a distance of several orders of magnitude that makes it impossible for randomness at the molecular level to produce randomness at the phenotypic level.[45] At least on one occasion, Waddington emphasized this aspect using a metaphor that resembles one of Darwin's.[46] In Darwin's very famous example, the shape stones are random and the architect (natural selection) purposefully assembles them. In Waddington's example, gravel (DNA mutations) might be random but not the concrete (phenotypic variation) that allows the architect (natural selection) to build a bridge (evolutionary adaptation). Natural selection, in this view, does not act on random phenotypic variation:

> For present purposes we do not need to look any further into the generally accepted principle that these primary genetic changes can be regarded as random mutations. Does it follow that the variations on which natural selection acts are also random? The pebbles forming the gravel on a river bed have their form determined by random processes, i.e. are the results of random search; it does *not* follow that random search plays any important part in the erection of a bridge built of concrete made out of this gravel. The factors that have to be considered in the engineering of the bridge belong, we may say, to a different order of complexity to those involved in the formation of the aggregate of which the concrete is constituted. We need to ask whether the attribution of evolution to random search does not depend on a similar confusion of different orders of complexity – an error very similar in logical type to Whitehead's well-known "Fallacy of Misplaced Concreteness".[47]

Overall, Waddington (and most probably Schmalhausen) supported, along-side GA1 (and often mixed with GA1), a more radical account of genetic assimilation, namely GA2, that was based on a different concept of canaliza-tion. *In GA2, canalization is not about masking the effects of mutations on the phenotype but about producing the adapted phenotypic outcome regard-less of the precise genetic variation actually present in the population. What is required, in this view, is enough genetic variation for the epigenetic system to be able to recruit it to stabilize the new phenotype.* Standing variation can most probably do the job, and does not require waiting for the spontaneous and chance occurrence of the "good" adaptive allele, as Waddington thought it was the case in the Simpsonian form of the Baldwin effect.[48] Or, to put it more briefly, in GA2, *the response to selection that leads to adaptive evolu-tion requires two things only: a sufficiently high degree of canalization and standing genetic variation.*

Waddington did not conceive orientation of the phenotypic effect of muta-tions only in terms of scaling laws between the molecular level and the cel-lular and macroscopic levels. For him, current developmental systems were above all the result of a long evolutionary process directed by the relentless action of natural selection. He thus imagined that the adaptive character of the effects of developmental canalization was also a form of adaptation which had been gradually elaborated by natural selection.[49] While biologi-cal organizations had not learned to mutate adaptively, they had acquired the ability to partly direct the phenotypic effect of random mutations, at least to some extent.[50] In this adaptationist viewpoint, genomes had become *evolved* genomes with maximized evolvability. Such an idea was embedded in his complex concept of the "tuning" of canalization, which Waddington believed to be one of the cornerstones of his model of genetic assimilation.

### 6.2.2   *The "tuning" of canalization*

Another important aspect of genetic assimilation must be emphasized and made as explicit as possible. It is related to the causal process that ultimately leads to the genetic fixation ("assimilation") of the adapted phenotype. As we have seen, this theoretical issue was a serious problem within the frame-work of organic selection. It was already a problem for Baldwin back in the early 1900s, when contemporaries, like Delage, opposed the efficiency of organic selection on the grounds that plasticity masks hereditary variation and thus prevents the operation of natural selection.

In comparison, the hereditary stabilization of the new phenotype is quite easy to understand within the framework of GA1. Given that the same genes are responsible for adaptability and adaptation, natural selection will tend to improve their frequencies in genotypes as long as the selective regime is maintained, up to a point where their combined action will be sufficient to produce the adaptive phenotype without plasticity: "Assimilation would then require no more than that the gene-dosage was raised above the threshold which protects the initially normal developmental path".[51] In such cases, genetic fixation is a simple threshold effect of the standard cumulative action of natural selection: "The remodeling of the epigenetic landscape goes so far that what was initially the side-valley [plastic phenotype], reached over a threshold, becomes the most easy path of change".[52]

Nonetheless, this rather straightforward explanation was consistent only in GA1's framework when genes are endowed with specific phenotypic properties. This might be why Waddington tried to develop an additional concept that he termed the "tuning" of canalization. He too struggled to understand why, at least in the Baldwin effect, selection "should genetically fixed a character which, it might seem, could have been directly produced in each generation".[53] His solution rested here on canalization and its so-called tuning. Unfortunately, even though he thought that "this process of tuning is, from the theoretical point of view, perhaps the most important aspect of genetic assimilation",[54] he never expanded much on the subject and barely touched it on a couple of occasions.[55] Waddington's idea was basically the same as the one Baldwin sketched in the early years of the 20th century: plasticity was not enough to reach the adaptive optimum, which creates a fitness deficit that will allow the final genetic building of the adaptation. These final selective steps were what he called the "tuning":

> Finally, canalizing selection would, in many if not most cases in Nature, have still another task to perform, namely to guide the new path of development so that it reaches exactly the most valuable end. We might call this the "tuning" of the adaptive phenotype.[56]

In his 1959 experiments on the size of anal papillae in the larva, Waddington stressed that adaptability is limited, "and it is clearly *because* of this limitation that some degree of genetic assimilation has occurred".[57] He represented the same exact idea in several of his diagrams of genetic assimilation from the late 1950s onward (Figure 6.1).

Waddington produced another reason to explain why natural selection would favor the tuning of canalization. In his understanding, environments were not fixed entities, and most usually there existed a difference, usually slight, between the developmental and the selective environment. If the most common case is that there is a difference between the developmental and the selective environment, then selection will favor genetic fixation over plastic accommodation.[58]

In the short epilogue of *The Strategy of the Genes*, Waddington emphasized that the tuning of canalization "acquires a much wider importance than as merely providing an explanation for the fixation of ecotypes".[59] In his view – but, unfortunately, he remained very elusive on this crucial aspect – it was the tuning of the epigenetic landscape, under the guidance of canalizing selection, that allowed canalization to do more than simply mask the effects of mutations, and to "guide the phenotypic effect of the mutations available",[60] as we saw in the previous section. It remains difficult to ascertain exactly how he envisioned the relation between the tuning of canalization

A          B          C

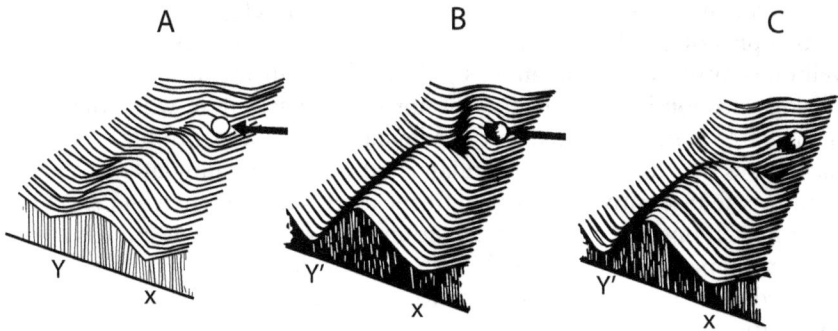

*Figure 6.1 Stages in the genetic assimilation process showing the tuning of the "acquired character".* From left to right, under the action of canalizing selection, the topography of the epigenetic landscape is progressively modified over the generations. In A, when environmental variation (black arrow) is necessary for the genesis of the new phenotype, the developmental path taken leads to point Y. In C, when the evolutionary transformation is complete, Y' corresponds to a "valley" dug deeper into the epigenetic landscape (its degree of canalization is greater). It is the gap between Y' and Y that Waddington called "tuning". In his view, this gap in the degree of canalization corresponded to a fitness gap, which accounted for the definitive genetic stabilization of the initially plastic phenotype. Reproduced from: C.H. Waddington, *The Strategy of the Genes*, London, Allen and Unwin, 1957. With permission from Taylor & Francis.

and GA2, but it is clear that he thought of this relation as a cornerstone of his theory of genetic assimilation. He most probably thought of this tuning step as a process that strongly constraining the part of the phenotypic space that would be reachable in the next generations. In short, tuning was about setting the direction of subsequent evolutionary change: mutations would necessarily follow given that their phenotypic outcome was constrained by the modeling of the epigenetic landscape itself.

## 6.3   Stabilizing selection as a creative process

Exactly as for "genetic assimilation" and "canalization", the term "stabilizing selection" went through different definitions and conceptual contents from the early 1940s onward. In the mid-20th century, biologists coined multiple terms to designate processes that might overlap but only partially: "canalizing selection", "normalizing selection", "stabilizing selection", "centrifugal selection", and so on. The present section does not aim to provide an exhaustive mapping of this rich and equivocal vocabulary, and focuses instead on one specific concept: Waddington's "canalizing selection", which was close to Schmalhausen's previous concept of "stabilizing selection".[61] For the sake of clarity, I will use Schmalhausen's wording: what was stabilizing selection supposed to be?

At first glance, the answer appears straightforward: stabilizing selection is a particular form of normalizing selection in which repeated selection for the intermediate phenotype eventually produces developmental canalization for that phenotype. This first approximation is not erroneous and maps quite well onto most of Schmalhausen's published statements. It implies that stabilizing selection is a subcategory of normalizing selection, when the action of the latter is not restricted to the elimination of alleles producing deviant phenotypes but consists more fundamentally in the accumulation of alleles ensuring the insensitivity of development to genetic and environmental perturbations. If one accepts such a definition, then stabilizing selection is a form of normalizing selection that is characterized by the *effects* it produces: unlike standard normalizing selection, stabilizing selection produces developmental canalization.

However, such a definition is problematic because the distinction between normalizing and stabilizing selection only becomes possible after the fact: both have the same selective target (the average phenotype) but only stabilizing selection produces canalization. In other words, stabilizing selection cannot be defined by its mode of operation. Such a limitation has presented a problem for experimenters since the late 1950s, as the geneticist John M. Thoday, for example, very explicitly stated:

> Waddington (1953, 1958) has classified selection according to the *effects* it may be expected to produce, rather than according to the measurable characters of the individuals selected. While of value for theoretical discussion, such a classification cannot be used in designing selection experiments and is not therefore used here.[62]

Waddington was aware of this possible limitation (Schmalhausen probably much less so), which led him, in the late 1950s and early 1960s, to conceive experiments with extremely complex designs to show that it was possible to produce developmental canalization (these experiments, which were Waddington's last experimental work, will be studied in detail in Chapter 8). However, to do this, Waddington developed procedures that did not rely on selection of the average phenotype at all. In other words, stabilizing selection was no longer, at that time, a form of normalizing selection and indeed could *only* be defined in terms of its effects.

The relative success of Waddington's results in these later works eventually led one of the few "Waddingtonian" biologists of the 1960s and 1970s, Jim Rendel, to argue that the distinction between stabilizing and normalizing selection (contested by many geneticists, including Thoday) made operational sense and, as such, remained a useful distinction:

> [Thoday] comments in reference to Waddington's terms normalizing and canalizing selection that although it is valuable to separate the two ideas – achievement of the optimum mean and achievement of minimal

variation about it – Waddington's terms should not be applied to selection since the two cannot be distinguished operationally, and selection is nothing if not operational. However, Waddington has designed and carried out an experiment which does select for absence of variation about a mean expression without in any way tending to pick on a specific mean [Waddington 1960], and since several people have done the converse, Waddington's terms must be held to have more force operationally than at first appears.[63]

Therefore, in artificial conditions at least, stabilizing selection could not necessarily be identified by the selective regime it required but by its alleged consequences on canalization, and, ultimately, on the whole embryological process. Waddington and Schmalhausen claimed that, under certain circumstances, stabilizing selection produces developmental canalization and, eventually, genetic assimilation. This aspect is not included in the standard concept of normalizing selection. Waddington stressed this distinction as follows:

It is a platitude to point out that natural selection is not only responsible for the elaboration of evolutionary novelties, but that it has an even more universal function in the preservation of normality by the elimination of aberrant individuals. This second type of action has been referred to as stabilizing selection, and may operate *in two rather different ways*. In the first place, natural selection will tend to eliminate those alleles which in the normal environment cause the development of abnormal phenotypes; this may be called normalizing selection. A rather different type of natural selection will tend to remove those alleles which render the developing animal sensitive to the potentially disturbing effects of environmental stresses, and will build up genotypes which produce the optimum phenotype even under sub-optimal or unusual environmental situations. This type of selection has been referred to as "canalizing selection", since it brings it about that the epigenetic systems of the animals in the population are canalized, in the sense that they have a more or less strong tendency to develop into adults of the favoured type.[64]

What stabilizing selection was and how it could be a creative process in evolution was the main topic of *Factors of Evolution* whose subtitle was *The Theory of Stabilizing Selection*.[65] Already in 1941, in "Stabilizing Selection and Its Place among the Factors of Evolution", Schmalhausen distinguished between stabilizing selection and what he then termed "ordinary direct selection".[66] In the years that followed, he elaborated the distinction and eventually contrasted two "types" of natural selection, "dynamic" and "stabilizing", as he explained in section 2.D of *Factors*. What he calls the "dynamic role" of natural selection is the standard understanding: when the environment and selective pressures change, natural selection alters the reaction norm to adapt the phenotype to the new set of conditions. Stabilizing selection is about the long-term consequences of a fixed (or at least stable) selective regime. What happens in the long run, over hundreds or thousands of generations? For Schmalhausen (and Waddington), stabilizing selection proceeds by steps, from the partial stabilization of transient phenotypic responses to a complete reorganization of the interactions between the genome and the developmental system to optimize the reliability of the new reaction norm. In the end, the evolutionary stabilized reaction norm becomes resistant to both genetic and environmental perturbations, which would later be called "genetic canalization" and "environmental canalization".[67]

In contrast to the usual concept of normalizing selection (see Waddington's previous quote), stabilizing selection is all about canalization, and Waddington and Schmalhausen assumed that there was a reciprocal causation between the two phenomena: stabilizing selection produces developmental canalization,[68] and developmental canalization, in turn, as we have seen, constrains the phenotypic consequences of genetic mutations. Thus, as soon as a developmental system is endowed with some degree of canalization, a positive feed-back loop is initiated between canalization and stabilizing selection and this dimension of the evolutionary process is self-reinforcing.[69] This is why Schmalhausen and, to a lesser extent, Waddington, always emphasized that over time, as more complex organizations evolve, stabilizing selection tends to become more important and more "creative" than standard "dynamic" selection.[70]

What does "creative" mean as regards stabilizing selection? The issue of the so-called creativity of natural selection is a difficult question within the philosophy of biology. John Beatty recently gave a helpful overview of a century and a half of debates on the topic.[71] "Darwinism", from Darwin to the Modern Synthesis, requires natural selection to be the main creative factor, that is, the force responsible for the building of evolutionary adaptations. In this sense, natural selection is creative because it gradually accumulates genetic variations in the direction of maximized fitness that will end up producing organic adaptations, such as the eye of the vertebrate (which did not pre-exist the long-term cumulative action of natural selection).

It must first be emphasized that Schmalhausen, in contrast to Waddington, made regular use of the word "creative" when he characterized the evolutionary role of stabilizing selection. Second, he was aware that the meaning of "creative" was not obvious, nor equivalent to what the adjective usually meant regarding natural selection. He made this point clear as early as 1941:

> Stabilizing selection does not lead to establishing new traits of organization and new physiological reactions, and is not tied to an increase in

individual adaptability in the morpho-physiological traits of the adult organism. To the contrary, it leads to the destruction of many reactions and lowers the capacity for adaptive modifications. The results of the actions of stabilizing selection leave an impression of general specialization. However, in that case it would be inaccurate to assign only negative characteristics to the role of stabilizing selection. Rather, we are obligated to also note its *creative significance*, even if it has not become obvious yet.[72]

For both Schmalhausen and Waddington, stabilizing selection was creative because this form of selection progressively reorganized the genetic architecture of a phenotypic trait; and, as we have seen in previous sections, this could be done in many ways given their systemic understanding of the functioning of the genome. Creativity was thus an attribute of the way a new reaction norm was hereditarily stabilized under the guidance of stabilizing selection: it created new biological organizations.[73] The way the new adaptive norm became canalized was termed the "positive" and "creative role" of stabilizing selection in Schmalhausen's writings.[74] One of its direct theoretical consequences was that (developmental) heredity was not thought of as an independent factor from natural selection but as the long-term consequence of the working out of stabilizing selection:

> The creation of such an internal apparatus of development, i.e. a system of morphogenetic correlations (broadly construed), appears to be the main expression of stabilizing selection's creative role. Adaptive reactions can be established only following this form of stabilization, i.e. [only] following the creation of the internal mechanism of their development can one talk about their hereditary fixation.[75]

In the preface of *Factors*, Schmalhausen summarizes the whole idea by pointing out that stabilizing selection creates "new forms of ontogenesis which are more protected against the disturbing influence of the environment".[76] What allows stabilizing selection to produce much more significant effects than standard normalizing selection? The issue of how canalization evolved in biological systems remains a difficult one. As we will see in Chapter 8, this required a great deal of experimental ingenuity on Waddington's part to achieve any kind of demonstrable results. Within the standard theoretical framework of quantitative genetics, some of Waddington's collaborators were able to show as early as the 1950s that normalizing selection cannot (in general) produce an increase in developmental canalization (see Section 8.1.1). Contemporary models show that normalizing selection only marginally produces canalization and that nontrivial and additional assumptions are required.[77] Neither Schmalhausen nor Waddington avoided the problem, but they were of course limited in the possible answers they could envision.

Schmalhausen was the first to tackle the issue and, as early as 1941, he tried to show which ecological conditions would result in an enhanced developmental canalization.[78] He came back to the issue in *Factors* and supported the hypothesis that fluctuating environmental "conditions" might favor the evolutionary reinforcement of canalization. According to him, this hypothesis had already been experimentally addressed by Kamshilov in 1939, who used *Drosophila* as an experimental system.[79]

A decade later, Waddington was also concerned with the conditions that made the evolution of canalization possible. In Section 8.1.2, I will recount Waddington's experimental work on the sensitivity of the Bar phenotype to variable temperatures. In their general designs, these experiments, published in 1960, may have been rather close to Kamshilov's.[80] Waddington succeeded in reducing the environmental sensitivity of the Bar stocks, that is, he proved that what he called "family selection", that is, stabilizing selection operating in fluctuating environmental conditions, might improve canalization.[81] Waddington, like Schmalhausen, was especially confident about the fact that stabilizing selection would produce even more profound effects in more usual conditions:

> Since canalizing selection, operating through families, has been so extremely effective on this abnormal developmental system after only five generations, one is left with a feeling of considerable confidence as to its powers when acting on more normal populations.[82]

To conclude, it is worth discussing a final theoretical dimension of stabilizing selection which has given rise to a series of misunderstandings, namely that the evolution of canalization does not imply a loss in the average plasticity of a given organism. Usually, in contemporary scientific literature, canalization and plasticity are understood as two opposite properties of genotypes.[83] The more a genotype is canalized, the less it is assumed to be plastic, and vice versa. This is not how Schmalhausen and Waddington conceptualized the issue and their originality on this point has been mostly ignored. Genetic assimilation is regularly dismissed because it is thought to lead to a loss in plasticity.[84] For Schmalhausen and Waddington, this was not true: genetic assimilation was about the transformation of the whole developmental phase-space. This is especially transparent in *Factors*, in which Schmalhausen repeatedly argued that stabilizing selection did not lead to a decrease in plasticity (an argument that other Soviet biologists already opposed to him back in the 1940s) but exactly to the opposite:

> In stressing the importance of the stabilization of morphogenesis and the role of stabilizing selection in this process, in order to avoid misunderstanding it must be pointed out that, in the opinion of the author, this stabilization of form does not exclude the converse process of labilization. . . . Simultaneously with the continuous stabilization of special

types of reaction, in particular, of functional differentiations which have attained permanent value in the given conditions of existence, there are also acquired entirely new reactions forms. And the capacity for individual adaptability, including the capacity for adaptive modifications, is based upon these new reaction forms. The capacity for more extensive adaptation is an important acquisition which leads the organism into new paths of progressive evolution.[85]

This led him to conclude *Factors* by emphasizing that the main evolutionary trend, on a geological time scale, was an increase in adaptability that was proportional to the developmental complexity produced by stabilizing selection.[86] He explained this counter-intuitive consequence of stabilizing selection and canalization at the macroevolutionary level as follows:

The creative role of natural selection is not manifested merely as continuous adaptation and a general increase in the complexity of organization. It is also expressed in capacity for new adaptive reactions to variations in the external environment, that is, as a heightened individual adaptability of the organism. In turn, this increased adaptability aids the further differentiation of such reactions and the incorporation of their effects (if they acquire permanent value) into the normal organization. As individual adaptability develops, the regulating processes become more important and the general stability of the organism increases. The apparatus of individual development, as well as the entire system of mechanisms protecting normal morphogenesis, becomes more complex.[87]

Waddington's theoretical positioning was more or less the same. Like Schmalhausen, he never restricted genetic assimilation to only an ineluctable loss in plasticity, even if the developmental rigidity of his epigenetic landscape was a major conceptual obstacle, as we have seen in the previous chapter. Like *Factors*, *The Strategy of the Genes* was a plea for the significance of adaptability in evolution. In chapter 5, "The Survival of the Adaptable", Waddington repeatedly argued that adaptability was an emerging property of organisms that tended to become more important in evolution than standard genetic adaptation.[88] Yet, on this aspect, it must be noted that Waddington was much more equivocal than Schmalhausen. For instance, he too sometimes opposed canalization and plasticity in a rather simplistic manner.[89]

## 6.4    Chapter's conclusion

The aim of the present chapter was to make explicit the most original aspects of Waddington's and Schmalhausen's theoretical work. In contrast to the standard idea – especially at work in Simpson's reformulation of the Baldwin effect (see Chapter 7) – that phenotypic stabilization requires specific adaptive alleles able to mimic the initial plastic accommodation, both supported, at least on some occasions, a systemic understanding of the genome that substantially altered the form of the explanation.

Robustness and canalization became properties of the developmental system, and genetic assimilation, whatever the trait under scrutiny, was supposed to involve the genome as a whole. In short, phenotypic stabilization and, ultimately, heredity, was a genomic property gradually produced by the cumulative effect of a specific and underappreciated form of selection, stabilizing selection. In such a framework, evolution is a creative process not only in a standard Darwinian sense (because natural selection creates new forms and new adaptations) but also because the biological organization can stabilize new phenotypes in many ways, almost irrespective of the alleles present in the gene pool. This implies another difference between Waddington's and Schmalhausen's developmental view and standard population genetics: because of canalization evolution became endowed with a form of inertia that is somehow stored in the genome's architecture.

In such a framework, the phenotype is a function of the genome as a whole; that is, it is a function of a system that, to some extent, captures its own history within its internal architecture. Because of epistasis, they saw canalization itself as an emergent property of the genomic architecture. This means, at least for evolved organisms which experience a complex developmental process, that the topology of the causal interactions among loci becomes more important than the specific alleles present in the genome. This explains why they believed this function to be highly injective, as I proposed in this chapter's introduction. In my reading, this is how it might be possible to retrospectively interpret the controversial idea that "the phenotypic effects of the alleles . . . are conditioned by the modelling of the epigenetic landscape".[90]

Certainly, Schmalhausen and Waddington were vague and regularly conflated two different concepts of genetic assimilation (GA1 and GA2). Save a few minor occasions,[91] they did not try to develop their speculative intuitions into formal workable models. But it remains that their effort to devise new concepts and new terms might have been, to a certain degree, productive because some of these concepts (GA2) are not as easily integrable into the Modern Synthesis' framework as others (GA1). As a historian of science, I do not have the skills required to determine to what extent it would be scientifically sound and interesting to resurrect and push some of their ideas further. My aim was to make them as explicit as possible and to contrast their theoretical content with what Bachelard called the "sanctioned science".

To recap the main conclusions reached in the three previous chapters I summarize some of the main characteristics of GA1 and GA2 (Table 6.1) to stress the key differences between the two concepts of genetic assimilation. As we will see in Part III, GA2 has almost entirely disappeared from the scientific literature since the 1960s.

*Table 6.1* Comparative table summarizing the main differences between the two
concepts of genetic assimilation (GA1 and GA2).

|  | GA1 | GA2 |
|---|---|---|
| Mutations' effects on the phenotype | Dispositional (intrinsic) properties. Mutations can be expressed or not depending on the context (genetic background, developmental environment), but how they impact the phenotype is mostly a consequence of their molecular nature. | Systemic (extrinsic) properties. The developmental system is mostly responsible for how mutations impact the phenotype. Mutations are the "raw material" of evolution in a strict sense. |
| Target of selection | Individual alleles | Epigenetic system |
| Causal mechanism | Threshold selection | Stabilizing selection |
| Canalization's function | Masks mutations' effects | Guides mutations' effects |

# Notes

1  Goldschmidt (1938, pp. 78–81).
2  Goldschmidt (1938, pp. 122–123).
3  Goldschmidt (1938, p. 172).
4  Goldschmidt (1938, p. 173).
5  Goldschmidt (1938, p. 303). For a global treatment, see Allen (1974).
6  See especially Dietrich (1995).
7  For instance, in 1941, he did not hesitate to write that "Goldschmidt's book [*The Material Basis of Evolution*] is one of the most important of recent contributions to the theory of evolution" (Waddington, 1941, p. 109).
8  See especially the preface of his 1975 *The Evolution of an Evolutionist* (Waddington, 1975a).
9  Waddington (1975a, p. vi).
10  Waddington (1940b p. 59).
11  Waddington (1953a, p. 124), my emphasis.
12  Waddington (1961a, p. 264).
13  Schmalhausen (1949, p. 80).
14  Schmalhausen (1949, p. 219).
15  Waddington (1942, p. 563).
16  Waddington (1960).
17  Waddington (1942, p. 564).
18  Waddington (1959b, p. 1654).
19  For example, Siegal and Bergman recently did (Siegal & Bergman, 2002).
20  See, for instance, Waddington (1966a, pp. 47–48). See also Waddington (1968b, p. 13).
21  Schmalhausen (1949, pp. 14–15).
22  Schmalhausen (1949, pp. 233–234).
23  Schmalhausen (1949, p. 219).
24  Schmalhausen (1949, p. 229).

25  Waddington (1957a, p. 36).
26  Waddington (1957a, p. 131), my emphasis. See also Waddington (1966a, p. 22).
27  Schmalhausen (1941, pp. 312–313).
28  Schmalhausen (1949, p. 142). See also p. 154.
29  Waddington (1969a, p. 7). See also Waddington (1974).
30  Waddington (1957a, p. 110).
31  Waddington (1957a, p. 188).
32  Waddington (1957a, p. 188).
33  Waddington (1957a, p. 190); Waddington (1959a, p. 396).
34  Waddington (1961a, p. 288).
35  Waddington (1957a, p. 190). My emphasis.
36  Dobzhansky (1970, p. 92); Gayon (1998); Beatty (2008); Merlin (2013). As is well known, Darwin already expanded on this crucial theoretical aspect: Darwin (1868, p. 430).
37  See especially Jablonka & Lamb, 1999, chapter 3 ("Induced genetic variations")
38  Waddington (1957a, p. 180; 1958b, p. 15).
39  Waddington (1957a, p. 190). See also Waddington (1959a, p. 396).
40  Waddington (1957a, p. 190).
41  Waddington (1961a, p. 288). My emphasis.
42  Lewis (1983, p. 197).
43  Laplane (2016); Laplane & Solary (2019).
44  Schmalhausen (1941, pp. 348–349). My emphasis.
45  "However, neo-Darwinism seems to me to be wrong in so far as it usually tacitly assumes that randomness of genetic mutation implies randomness of phenotypic variation" (Waddington, 1968b, p. 18).
46  Darwin (1868, p. 430).
47  Waddington (1968c, p. 112). Emphasis in the original.
48  Waddington (1961a, pp. 287–288).
49  Waddington (1957a, p. 188).
50  "The characteristics built into the epigenetic landscape influence, *but do not completely determine*, the phenotypic effects of mutations" (Waddington, 1957a, p. 190, my emphasis).
51  Waddington (1957a, p. 172).
52  Waddington (1957a, p. 166).
53  Waddington (1957a, p. 163).
54  Waddington (1957a, p. 172).
55  In a few paragraphs in *The Strategy of the Genes* (1957a) and in his 1961 review.
56  Waddington (1957a, p. 172).
57  Waddington (1959b, p. 1655). My emphasis.
58  Waddington (1957a, p. 168).
59  Waddington (1957a, p. 188).
60  Waddington (1957a, p. 188).
61  Waddington (1953b, p. 387).
62  Thoday (1959, p. 187).
63  Rendel (1967, pp. 111–112).
64  Waddington (1960, pp. 140). My emphasis. See also Waddington (1958a, pp. 58–59).
65  In the preface, Schmalhausen pointed out: "This book contains a more specialized and complete discussion of the role of the stabilizing form of natural selection" (Schmalhausen, 1949, p. xxi).
66  Schmalhausen (1941, pp. 320–321).
67  These precise terms are absent from Waddington's and Schmalhausen's writings, even if the idea was already there (e.g., Schmalhausen, 1941, p. 318; Waddington, 1957a, p. 123; 1966a, p. 48).

68 Schmalhausen (1949, p. 79); Waddington (1957a, p. 72).

69 Schmalhausen (1949, p. 244).

70 Schmalhausen (1949, pp. 238–245).

71 He published a two-part essay in the *Journal of the History of Biology* untitled "The Creativity of Natural Selection" (Beatty, 2016, 2019).

72 Schmalhausen (1941, p. 324). My emphasis.

73 Schmalhausen (1941, p. 325); Schmalhausen (1949, pp. 92–94, 242); Waddington (1957a, p. 79).

74 Schmalhausen (1941, p. 329).

75 Schmalhausen (1941, p. 329). In *Factors*, Schmalhausen specified this aspect as follows: "Stability of the organism, as expressed in the hereditary transmission of its properties, is by no means due to the nucleus, whose structure can be altered relatively easily, but rather to the cellular system as a whole together with the mechanisms controlling it, and to the correlation system of the developing organism" (pp. 44–45).

76 Schmalhausen (1949, p. xx).

77 G. Wagner et al. (1997); Rice (1998); Rünneburger & Le Rouzic (2016).

78 Schmalhausen (1941, p. 321).

79 Schmalhausen (1949, pp. 82–83). As far as I know, Kamshilov's work on *Drosophila* was published only in Russian, and for the moment I had no direct access to it.

80 But I have not been able to verify this directly.

81 Waddington (1960).

82 Waddington (1960, p. 148).

83 For a critical evaluation of this classic opposition, see Pigliucci (2010, pp. 357–358).

84 See especially West-Eberhard (2003) and Crispo (2007).

85 Schmalhausen (1941, pp. 242–243).

86 Schmalhausen (1949, pp. 270, 272).

87 Schmalhausen (1949, p. 275). See also pp. 282–283.

88 Waddington (1957a, pp. 145–146).

89 Waddington (1957a, p. 149, p. 158).

90 Waddington (1957a, p. 190).

91 Waddington (1957a, pp. 73–75) (this did not escape the notice of Günter Wagner and his collaborators back in the 1990s when they developed the first population genetic model of canalization: G. Wagner et al., 1997, p. 340); Waddington (1960, pp. 148–149).

# Conclusion of Part II

This part was devoted to the elucidation of Schmalhausen's and Waddington's understanding of genetic assimilation. As already pointed out, my aim was not to give a full account of their scientific work: here the focus was limited to their views about adaptive evolution. Schmalhausen's work as a vertebrate paleo-morphologist and Waddington's achievements in causal embryology, that is, what constitutes, at least quantitatively, their most significant scientific contribution, were almost entirely set aside.

The first thesis defended here is that genetic assimilation, in contrast to organic selection, looked at adaptability from an embryological perspective instead of an ecological one. In contrast to organic selection, genetic assimilation is a mechanism-based argument which is primarily concerned with the progressive embryological reorganization that starts with a plastic and transient path. In brief, in organic selection, the causal focus is outside the organism: the impact of adaptability is ecological. Organic selection is concerned with the survival of a population in a given ecological context and how plasticity interacts with both. Whereas in genetic assimilation, the causal focus is inside the organism: the impact of adaptability is embryological. Genetic assimilation takes survival for granted and focuses on the long-term embryological consequence of the new selective regime. Organic selection and genetic assimilation are thus two concepts of the same family but they do not share the same theoretical perspective. The Lamarckian phenomenology was not explained in the same way in these two frameworks. Plasticity is a concept at the intersection of three main fields: ecology, developmental biology, and genetics. Organic selection favored ecology, genetic assimilation favored developmental biology, and, without surprise, the Modern Synthesis's reinterpretations of these concepts favored genetics. An important follow up question was to understand whether one concept could be reduced to another. The ambition of the Modern Synthesis was indeed to promote genetic models and interpretations of both. To what extent this reductive program was fruitful and legitimate is the central focus of the next part.

The second thesis supported in Part II is that Schmalhausen's and Waddington's developed (but also sometimes unclear) theoretical elaboration

DOI: 10.4324/9781003422990-12

contained two different concepts of genetic assimilation, termed GA1 and GA2. These concepts were not well distinguished at the time, and in most of their writings, they were closely intertwined and mixed which complicates the analysis of what was really at stake with genetic assimilation. This can explain why most of their contemporaries were skeptics about the need for their new terms, given that GA1 at least could be easily translated within a standard genetic population framework, as we will see in Chapter 8. In a sense, GA2, a more radical concept with respect to the Modern Synthesis, was most of the time hidden by GA1. To make it fully apparent requires a significant work of retrospective reconstruction. This was the main ambition of the present part. While it was truly challenging, GA2 could have had transformative consequences on the evolutionary theory back in the 1950s and 1960s. However, it did not. With no rigorous characterization, no formal development, and an equivocal experimental basis, GA2 disappeared from the scientific literature in the 1970s.

## Part III

# (Dis?)Integration into the Modern Synthesis

## Thinking plasticity genetically

# Introduction of Part III

Since evolution is a change in the genetic composition of populations, the mechanisms of evolution constitute problems of population genetics.

– Dobzhansky, 1937[1]

I have thus far attempted to unfold the different layers of meaning of the concepts of organic selection and genetic assimilation that have accumulated over time. While organic selection has privileged the ecological dimension (that of the "why?" in terms of natural selection), genetic assimilation has focused first on the embryological dimension (that of the "how?" in terms of developmental reconstruction). In both cases, there were at least two different conceptual orientations, which I have tried to characterize in the simplest and historically accurate way possible. If my reconstruction is correct, we can now consider how these concepts were understood and used during the period 1940–1970, that is, at the time of the rise of Modern Synthesis.

The theoretical identity and history of the Modern Synthesis[2] is still a hotly contested topic.[3] It is therefore necessary to first clarify the position I will adopt in the following chapters. First, I consider that the Synthetic theory of evolution was indeed the name of a "theoretical event". Without denying the strong social and institutional dimension of what took place in the United States in the 1940s at the time of the creation of the *Society for the Study of Evolution* (1946) and the journal *Evolution* (1947), and without minimizing the problems that have arisen from the involvement of some of the actors themselves in the writing of history – in particular Ernst Mayr – the fact remains that something significant happened from the point of view of biological theory. My commitment is therefore to historical realism. This is not to say that I consider the "MS" category to be fully satisfactory and unproblematic. It covers very different positions, tends to minimize certain internal tensions, and also masks important transformations which took place during the period 1930–1970. However, despite all these limitations, there is a relatively consensual and robust conceptual characterization of its content, which I adopt, at least as a working hypothesis, in this final part.

DOI: 10.4324/9781003422990-14

In a classical sense, I consider the Modern Synthesis to be the theoretical framework that resulted on the one hand from the founding work in population genetics of Fisher, Wright, and Haldane and on the other hand from the subsequent syntheses of which the books of Dobzhansky (1937), Mayr (1942), and Simpson (1944) drew the main lines.[4] This amounts to considering that the Modern Synthesis was based on the idea that evolution (whatever the level considered) ultimately consists in a change of allelic frequencies within Mendelian populations and that the deterministic force driving this change is natural selection, the only creative force at the scale of micro- as well as macroevolution. As this minimal definition implies, the evolutionary synthesis has thus mainly represented a drastic *constriction*[5] of the possible hypotheses to explain the mechanism of evolution. Nothing was left of the very speculative "factors" of the preceding period (1880–1930), such as the orthogenesis of the paleontologists, the mutationist theory of the first Mendelians, or the heredity of acquired characters of a good number of Lamarckian naturalists:[6] none of these theories could be translated within the parametric space of the new population genetics.[7] In other words, none of these factors could be characterized in terms of their biasing effects on gene frequencies.[8] In essence, the Modern Synthesis is therefore a theoretical framework that seeks to evaluate a phenomenon from the point of view of its genetic significance.

This is exactly what happened to organic selection and genetic assimilation in the 1950s: the focus was on how these hypothetical processes could alter allelic frequencies and thus find evolutionary translations. This is why they were both reframed as the "Baldwin effect" and "threshold selection", respectively. The aim of this part is thus twofold. On the one hand, I will describe the modalities of these redefinition attempts. In other words, I will reconstruct the theoretical and experimental frameworks that motivated both the emergence of the Baldwin effect as Simpson, and later Mayr, wished it to be, and threshold selection as this model took shape within the quantitative genetics of the 1950s (and in particular in Waddington's own laboratory in Edinburgh). On the other hand, I will evaluate these re-conceptualizations, or attempt to understand to what extent they did justice – or not – to the wealth of meanings that accumulated over the course of a history of over half a century long. Did these specific "constrictions" succeed in preserving the integrity of the concepts while making them more operative, or did they result in significant displacements, even impoverishments? Not surprisingly, as we shall see, there were major disparities depending on the concept. OS1, OS2, GA1, and GA2 were not "received" identically, simply because they were not given the same degree of understanding or the same level of interest. We therefore do not have a homogeneous history but rather a juxtaposition and interweaving of multiple histories, each of which has its own temporality, and which all reveal unexpected events.

# Notes

1 Dobzhansky (1937, pp. 11–12).
2 In the present part, I use "Modern Synthesis" and "synthetic theory of evolution" as synonymous. For a discussion on their possible divergent meanings, see Gayon & Huneman (2019, p. 524).
3 There is already a large body of work on these issues: Mayr & Provine (1980); Smocovitis (1996); Gould (2002); Cain (2009); Gayon & Huneman (2019).
4 Beyond these three seminal books, Provine provided a list of 15 texts that he thought were "the major works of the synthesis" (Provine, 1980, p. 400).
5 According to Provine (1992, p. 177): "What was new in this conception of evolution [Modern Synthesis] was not the individual variables, most of which had long been recognized, but the idea that evolution depended on relatively so few of them. So has gone much of science. Phenomena that appear so complex as to baffle the imagination, whether the motion of the heart, action of the tides, or periodicity of the comets, have been shown by science to depend essentially upon surprisingly few variables. Instead of a synthesis, evolutionary biologists during the 1930s and 1940s came to agree resoundingly upon a relatively small set of variables as crucial for understanding the evolution in nature. This I will now call the 'evolutionary constriction', which seems to me to be a more accurate description of what actually happened to evolutionary biology".
6 For an overview of these alternative hypotheses see Bowler (1983).
7 Gayon (1992, pp. 334–336).
8 The fact that a genuine evolutionary cause should be assessable by its impact on allelic frequencies within a population grounds the very framework of the Modern Synthesis. For a philosophical treatment, see Sober (1984), especially chapter 1: "Evolutionary Theory as a Theory of Forces".

# 7 The Baldwin effect
## "De-ecologizing" organic selection

Like any process in which a conceptual form has a reception within a different theoretical context, the reception of organic selection by the scientists who founded the synthetic theory of evolution was complex. This complexity can be measured on at least three levels:

(a) First, different scientists considered this evolutionary mechanism in different ways. I cannot provide an exhaustive study of these in this short chapter. I have therefore limited myself to the most central figures of the Modern Synthesis: Huxley, Mayr, Simpson, and, to a lesser extent (because he did not have much to say about organic selection), Dobzhansky.

(b) Second, as we shall see, during the period 1940–1970, these scientists' positions often shifted and sometimes went through considerable and unexpected changes. The case of Mayr will be particularly enlightening here, first through his rich dialogue with Thorpe's ideas, and then in the way he further hardened the concept of the Baldwin effect as recast by Simpson in 1953.

(c) Finally, the two causal mechanisms, OS1 and OS2, often conflated under the term "organic selection", ended up knowing very different fates and it is important to individualize the specific course followed by each one. This history is further complicated by the fact that each mechanism was at some point considered in this reception, but not necessarily in relation to Baldwin's (or others') formulation. In other words, both versions of organic selection were at some point reinvented from scratch in total ignorance of the rich debates and literature of the first half of the 20th century. This disassociation between the vocabulary used and the causal mechanisms targeted makes the reconstruction of this history even more difficult. Words and things had their own parallel histories, which only occasionally overlapped.

However, the general tone of this reception was as follows: as time went by, there was a considerable *impoverishment* of the conceptual content attached to this family of mechanisms, the major step of which was undoubtedly the

DOI: 10.4324/9781003422990-15

theoretical reconfiguration carried out by Simpson in 1953 when he substituted the term "Baldwin effect" for the set of terms in use until then. The history we are dealing with here is therefore above all that of a misunderstanding which led to the oblivion and erasure of the several ecological dimensions of organic selection.

## 7.1  A promising start? Huxley on organic selection in *Evolution, The Modern Synthesis*

Sir Julian Huxley (1887–1975) was a zoologist whose interests covered vast areas of knowledge, including the history of science. This is perhaps not surprising considering he was the grandson of Thomas Henry Huxley (1825–1895), Darwin's famous so-called bulldog and one of the founders of evolutionary theory. As we have seen, it was Julian Huxley who made Thorpe aware of the full significance of his experimental work on olfactory conditioning in insects, showing him how it related to the old question of organic selection and the work of Baldwin and Lloyd Morgan (Chapter 3).

It was from this perspective that Huxley dealt with organic selection in his seminal 1942 book, *Evolution, The Modern Synthesis*, which was a cornerstone in the development of modern evolutionary theory. Yet, despite Huxley's obvious interest in what he then called "the Baldwin and Lloyd-Morgan principle of Organic Selection",[1] he was unable to encapsulate the whole conceptual content of half a century of rich history in the few pages that he devoted to it. While his characterization of OS1 (touched upon in Chapter 3), undoubtedly constituted an important milestone, he missed the idea that plastic response can direct selection pressures by creating a new ecology. Huxley's definition of OS1 has the advantage of being very clear and constituted a reference in the years that followed. For example, Gause used it word for word in his important review "Problems of Evolution"[2] (1947). This definition remained unchanged in the successive editions of Huxley's book (1963, 1974):

> We have here [Thorpe's results] a beautiful case of the principle of organic selection (p. 523), as enunciated by Baldwin (1896, 1902) and Lloyd Morgan (1900), according to which modifications repeated for a number of generations may serve as the first step in evolutionary change, not by becoming impressed upon the germ-plasm, but by *holding the strain* in an environment where mutations tending in the same direction will be selected and incorporated into the constitution.[3]

The idea that plasticity plays a causal role because it prevents extinction is transparent here. However, and unlike Thorpe, Huxley apparently never understood that the genetic fixation of the phenotype required that the attainment of the adaptive peak could not be completely achieved in a plastic manner. He did however see that the "principle of organic selection" was also a possible mechanism for sympatric speciation. As an ethologist, Huxley was particularly sensitive to the ethological isolation associated with behavioral change. It was in fact this aspect of organic selection that he developed the most, relying heavily on Thorpe's work. The problem of "biological differentiation" or "ecobiotic divergence" gave rise to important controversies, as we shall see in the next section. By reproductively isolating certain populations, phenotypic plasticity, could, according to Huxley, play the role of a causal factor in sympatric speciation.

Huxley, in his book, tries to accumulate as many examples as possible of animals in which speciation can evidently be linked to this form of organic selection. A particularly demonstrative case for him of this mode of speciation is the reproductive isolation taking place in certain populations of birds (incipient species) due to ethological isolation linked to plasticity in the mastery of a certain type of song.[4] In many species, the song produced in the adult state is not genetically fixed and there is a lot of room for learning, that is, for behavioral plasticity. For Huxley, there was no doubt that this could initiate sympatric speciation.[5]

Based on a long list of examples, Huxley, still relying heavily on Thorpe, could conclude that "a survey of any group will reveal many cases in which physiological and ecological divergence must have been primary, morphological distinctions having been added in the course of later evolution".[6] For him, and in opposition to others (such as Mayr), there was little doubt that behavioral plasticity in certain animal species could have a major ecological and evolutionary role in speciation, similarly to a geographical barrier. His conception of sympatric speciation and of the establishment of genetic isolating mechanisms is in fact very close to that of Mayr, except that the geographical barrier (i.e., a physical discontinuity) is replaced by an ethological barrier (i.e., an ethological discontinuity resulting from the behavioral plasticity of individuals). Based on Thorpe's results, Huxley summarized his conception as follows:

> Once genetic adaptation to a particular host has begun, selection will step in to prevent the biological waste which would be caused by the desposition of eggs on other hosts. The mechanism of olfactory conditioning provides a certain reserve of plasticity; but this plasticity will become hedged about by genetic safeguards. Genetically-determined attractions to the normal host will become established, and also specific assortative mating reactions to prevent cross-mating. Thus ecobiotic isolation here has the same general effects as geographical or ecotopic isolation, but operates by rather a different mechanism, and follows a somewhat different course as regards the degree of divergence in morphological, physiological, and reproductive characters respectively.[7]

Thus, among the founding works of the Modern Synthesis, Huxley's book was unquestionably the one that gave the most space to organic selection. This was a topic Huxley had a real interest in, especially because of Thorpe's results and ideas. Out of nearly 600 pages, a good 20 are mainly devoted to the discussion of the evolutionary significance of organic selection. This certainly does not make it the central theme of the book, but it is not marginal either, especially when compared to the non-reception of organic selection at play in contemporary books by Dobzhansky and Mayr.[8] Huxley paid special attention to the causal link between plasticity and speciation and most of the 20 pages are concerned with this issue, that is, a possible mechanism for sympatric speciation. However, it was his definition of organic selection (OS1) that had the longest legacy. Despite his interest, Huxley did not identify the other mechanism of the orientation of selection pressures (OS2). Thus, despite the relative richness of Huxley's discussion, this marks the beginning of a conceptual impoverishment of organic selection. Most importantly, Huxley failed to convince the other "architects" of the MS of the possible efficiency and evolutionary scope of this mechanism.

## 7.2 Focusing on gene flow: Mayr's evolving position about ecological factors in speciation

In the middle of the 20th century, Ernst Mayr (1904–2005) was the main advocate of geographical speciation, and in numerous publications he strove to show the weaknesses of sympatric models. This ambition was one of the main perspectives of his magnum opus, *Systematics and the Origin of Species*, published the same year as Huxley's book.[9] Mayr developed three main theses which can be summarized as follows: (1) the new population genetics is not a sufficient theoretical framework to account for the entire evolutionary process; (2) speciation, in particular, escapes this description, even though it is the key process of evolution; (3) in animals at least, speciation occurs for the vast majority of cases in a geographic (or allopatric) manner.

The argumentative power and the clarity of the book are impressive, and in support of his interpretation Mayr was able to make available a considerable amount of biogeographic data, especially related to his own work as an ornithologist in the islands of Southeast Asia (especially New Guinea). It was in this work that Mayr developed his theory of allopatric speciation, based on the necessity of a physical discontinuity to interrupt the genetic flow between two populations, a necessary step to achieve speciation.

A fervent promoter of "population thinking", Mayr based his reasoning on the population heterogeneity of species (what he called, following Huxley, the "polytypic" species concept[10]). Each species consists of an aggregate of populations that differ morphologically, ethologically, and genetically, populating different habitats. But as long as there is spatial continuity between these populations, speciation, that is, reproductive isolation, is made impossible by the homogenizing effect of gene flow, hence the absolute necessity of breaking this continuity by establishing an extrinsic, that is, physical, barrier. It is only when gene flow is thus interrupted that the populations can diverge genetically up to a certain threshold which makes any "secondary intergradation" impossible.

Obviously, within such a causal scheme, there was no room for plasticity. Plasticity could not play any significant causal role, and Mayr strongly opposed, at least in animals,[11] the very possibility of sympatric speciation. It was such exclusivity of geographical speciation that Thorpe challenged, as we have seen. Thorpe did not contest that *Systematics and the Origin of Species* was an "excellent book";[12] nor did he oppose the efficiency of allopatric speciation as modeled by Mayr. But he argued that a place for sympatric speciation should be maintained because of the (behavioral) adaptability of animals and their ability to learn. In short, habitat selection was not only a mechanism that came at the end of the speciation process, to reinforce the genetic isolation between incipient species, but, at least in some cases, it could be the first step in that process, thus giving phenotypic plasticity the leading role. This was the main thesis of his 1945 article "The Evolutionary Significance of Habitat Selection". Thorpe-the-zoologist was then putting to good use some of the most recent results of genetics. Indeed, as we shall see in the next chapter, the biometrician and population geneticist Kenneth Mather was at that time developing a rather speculative and complex model: each chromosome was seen as a highly evolved construct in which genes with opposing phenotypic effects were in a balanced relationship. This precarious equilibrium, tightly controlled by natural selection, could be quite easily disrupted as soon as gene flow was impeded, suggesting how behavioral plasticity could play an effective role in speciation:

> Mather has pointed out that the kinds of character which are likely to distinguish races and species are polygenic and that any small decrease in mating freedom between two populations, whether brought about by natural obstacles or by any other cause, will be sufficient to lessen the intensity of selection for good "relational balance" (i.e., the balance existing in an outbreeding organism between pairs of different homologous combinations). The result of this will be that when members of two such populations or strain cross, the offspring will be heterotic. Mather further brings forward evidence to show that heterotic individuals are less fit than the parental types and that the avoidance of heterosis is probably the most widespread stimulant of isolating devices. According to Mather, such a system, once started, is self-propagating and irreversible.[13]

Therefore, there was absolutely no need for gene flow to be interrupted for genetic divergence between two populations to take place, at least according to this hypothesis. The gene flow only needed to be sufficiently reduced for this genetic balance to be disturbed.

In the 1940s, Thorpe was not the only one who wanted to bring sympatric speciation back to the forefront through a reconsideration of organic selection. On a much more fragile empirical basis, ornithologist John E. Cushing pushed in the same direction. In the first half of the 1940s, he published a series of articles in which, for the most part, organic selection was discussed in connection with the issue of speciation. In 1941, he published a short paper in *The Condor* titled "Non-Genetic Mating Preference as a Factor in Evolution". This first article is not based on any original work and builds on results obtained by others. It is also not very rigorous in its formulation of the ecological and ethological mechanisms that the author envisaged. Nonetheless, as the title makes clear, Cushing was interested in plasticity not as a product of adaptive evolution but as a causal factor. Moreover, Cushing explicitly linked his own conceptions not only to those of Baldwin but also to those of Gulick,[14] a much rarer occurrence.

Encouraged by Thorpe's successes, Cushing also took on experimental work on olfactory conditioning in *Drosophila*. At Caltech, under the patronage of Surtevant, individuals of the species *Drosophila guttifera* were reared in the laboratory on two types of substrates: one corresponding to the fungi usually used by this species, the other being an artificial mixture specially designed for the experiment ("Laboratory Medium"). If the traditional substrate was always widely used as oviposition substrate whatever the conditioning, Cushing observed that the "Laboratory Medium" saw its frequency substantially increase from 9.91% to 21.20% of the cases (in terms of the number of eggs laid) when the flies were reared on an equivalent substrate.

These results reinforced Cushing's belief that olfactory conditioning – and thus developmental plasticity – could play a role as an "isolating mechanism".[15] In a more synthetic article published in 1944, Cushing showed greater conceptual clarity: plastic behaviors could hold a genuine causal role in the evolution of populations insofar as they "seem capable of influencing selection pressure".[16] This time, his argument was based on some empirical material regarding learning and perpetuation of traditions in raptors with respect to food preferences. Based on a few examples, described rather quickly, he concluded very explicitly that genetic change was the consequence of new selective pressures created by behavioral plasticity (OS2):

> With the most characteristic trait of raptors, that of the taking of living prey for food, apparently not directly maintained by genetic factors, it follows that the genetic complex characteristic of raptors owes its existence to the selection pressure exerted by a non-genetic habit. This follows because of the well-known postulate that unless a character (for example, the talons of raptors) is subjected to a constant selection pressure, it will deteriorate as the population accumulates genes that

decrease its efficiency. An obvious corollary to this postulate is that the form of such heritable characters as feet and bills is model by the kind of selection pressure to which it is subjected. *Therefore, the genetic basis of structures relating to the raptorial habit would seem to be in large part determined by the direction in which selection pressure is exerted through the medium of traditional differences in feeding habits.*[17]

What was Mayr's reaction to this revival of the sympatric speciation model linked, on the one hand, to a number of experimental works concerning conditioning and, on the other hand, to a renewed interest in the old theme of organic selection? In 1947, he published a synthetic article in *Evolution* entitled "Ecological Factors in Speciation", which rapidly became a landmark on this set of questions. A few years later, in 1951, he published a second, lesser-known, text soberly titled "Speciation in Birds", which helps us understand, a quite significant – though temporary – evolution in his theoretical positioning.

The 1947 text is conceptually very dense and shows how well Mayr mastered almost the entire literature on these issues. I will therefore not give an exhaustive characterization of its content but only emphasize a few aspects. Mayr wrote this text as a reaction to Thorpe's work, which he considered of great importance (the same was not true about Cushing, whose results he found inconclusive[18]). It was indeed in direct dialogue with Thorpe that Mayr wrote the first version.[19] In its broad outline, and as the summary seems to attest, it is an unambiguous defense of his allopatric model and a detailed critique of the competing sympatric model in the form given by Thorpe. However, in some places Mayr is much more nuanced. He was prepared to concede that, in some cases, ecological accommodation, habitat selection, and organic selection could play a joint role in the first steps of a speciation event:

> The adaptation for life in a given habitat includes the faculty of the individual of the species to select this habitat (from a vast array of other possible ones!) during the dispersal phase. Although genetic factors must to some extent control this ability of habitat selection, nevertheless it is not entirely rigid and undeviating. There is a certain amount of ecological plasticity which is greater in some species, less in others.
>
> The establishment of individuals in a new environment will be assisted by conditioning, as emphasized by Thorpe (1945), as well as by "organic selection," that is, the gradual substitution of modifications by mutations (Baldwin, 1902; Gause, 1947). A new population within the species will thus come into being and will permit the species to spread into areas that were previously outside the breeding range. It is probable that most range expansions of species are caused by the origin of such new intraspecific populations.[20]

Without question, plastic divergent accommodation was not enough to prevent gene flow,[21] thus such mechanisms could not be a sufficient cause in speciation. Consequently, strictly sympatric speciation was doubtful and could only represent a marginal process in animal evolution. However, Mayr also acknowledged that ecological specialization was not only the final refinement of a speciation event but could also, as Thorpe argued, be at the very beginning of the process, even if a subsequent geographic separation remained in his view necessary for speciation to be achieved:

> The first step in the speciation process is the founding of a new intraspecific population. This is very often possible only through a shift in ecological tolerance, and insofar *Thorpe's statement is quite correct that an ecological rather than a geographical event may be the first step in speciation*, or – since the new population will be not only ecologically different but also spatially segregated from the parental one – that the ecological event (adaptation to a new ecological niche) is at least simultaneous with the first geographical event. This first step leads to the establishment of a spatially segregated population that is exposed to different selection pressure owing to the more or less differing ecological conditions under which it lives. The ecological factors here *lead* to evolutionary divergence. These populations will drift apart genetically (probably at an accelerating rate!) until a discontinuity through reproductive isolation develops, provided extrinsic barriers reduce dispersal to such an extent that it can no longer neutralize the effects of the different selection pressures in the two populations. It is still unknown whether this can happen between contiguous populations without the help of extrinsic factors that reduce gene flow.[22]

Mayr's second text carries the same ambiguity as the more compact 1947 synthesis. In his 1951 article "Speciation in Birds", he again seems to uncompromisingly defend the geographic model. Nevertheless, in the detail of his argument, we find the same concession as in 1947: if behavioral plasticity alone cannot lead to speciation, the way it transforms the selective pressures allows the genetic evolution of the population,[23] the first potential step in speciation. Mayr explicitly acknowledged that "this hypothesis is sound as long as it does not include the concept that two such ecologically slightly differing populations could co-exist sympatrically without reproductive isolating mechanisms".[24]

But this time there might have been more at play than a mere revision of his model in the light of Thorpe's work. Indeed, Mayr seemed to have a positive impulse toward the ecological factors of speciation. He conceded that the "geographical approach" had been dominant these last years and that it would be necessary to emphasize "the even more important ecological aspects of species and speciation".[25] He then went as far as stating that the future of systematics, at least as far as birds were concerned, lay with the growing understanding of "the ecological factors that are active during this process of speciation".[26] Mayr was in fact not present at the Tenth International Ornithological Congress, which was held in Uppsala in June 1950, where his text was read before Huxley, among others. It did not escape Huxley that this time, Mayr, left a lot more room to ecological factors and in particular to behavioral plasticity in the process of speciation. It was important to Huxley that this renewed attention should not overlook the founding work of Baldwin and Lloyd Morgan, whose names were absent from Mayr's 1951 text:

> Dr Mayr referred to the importance of modifications of habit as often paving the way for genetic evolutionary change. I would hope that full credit will be given to Mr. Baldwin and, C. Lloyd Morgan, who first drew attention to this phenomenon.[27]

It thus seems that Mayr subtly changed his conception of the ecological dimension of speciation. Around 1942, he saw ecological factors as consequential to a process causally produced by an extrinsic physical separation that genetically isolated two populations. As such, they were integrated with other reproductive isolating mechanisms that were the *product* of this divergent evolution. At least during the period 1947–1951, he considered ecological factors as possible initial *causes* of genetic divergence: phenotypic plasticity allowed the extension of the range of certain populations, which initiated new selection pressures, producing, and accentuating genetic divergence. Of course, without geographic isolation, Mayr felt that this could not be sufficient to form two distinct species, that is, two separate genetic systems. Gene flow could not be interrupted based on this ethological and ecological differentiation alone. Nevertheless, in doing so, Mayr was clearly relaxing the logic of his explanatory model, an adjustment that, to my knowledge, had never been noticed.

The cause of this change is clear: like Huxley but with much less enthusiasm, Mayr had to take into account the significance of Thorpe's experimental results and conceptions. I believe this change in his position went unnoticed, partly because Mayr expressed it rather timidly and it thus requires particular vigilance but also because it was only temporary. There is no trace of Mayr's adjustment, for example, in his refined model of peripatric speciation that he started to elaborate in 1954 (see Section 7.4).

### 7.3    Simpson's take: a clarifying constriction or a misinterpretation?

In his comments about Mayr's text, Huxley also complained about the lack of clarity attached to the vocabulary of organic selection and encouraged his colleagues to develop a more appropriate terminology: "Some brief term is required for the phenomenon, and I hope that evolutionists will find one".[28] This is exactly what Simpson did in 1953 in his seminal text on "The Baldwin Effect" published in *Evolution*.

This article is undoubtedly one of the most important milestones in the history whose meaning my book aims to reactivate. Alongside the books by Schmalhausen (*Factors of Evolution*, 1949) and Waddington (*The Strategy of The Gene*, 1957), Simpson's article was certainly the most significant publication of this second debate about the causal relations between individual plasticity and evolution. Its importance should not be underestimated. Indeed, this text profoundly reconfigured how these questions were approached and also imposed a new terminology to talk about them. Many saw it as a welcome conceptual clarification, as well as a rigorous account of the place of this mechanism within the theoretical framework of the Modern Synthesis.

Paleontologist George Gaylord Simpson (1902–1984) was another founder of the new version of the evolutionary theory, and it is well known that his specific contribution was to show that the data of paleontology were fully compatible with the most current knowledge in population genetics. He did a lot to import quantitative methods into the study of fossils and developed various models to link microevolution to macroevolution. All of this is to be found in his most important book, *Tempo and Mode in Evolution* (1944), the third publication of the famous seminal triad after those of Dobzhansky (1937) and Mayr (1942). Already in *Tempo and Mode* was Simpson interested in the various forms that selection could take. To clarify a fluctuant vocabulary, he proposed the terms "centripetal selection" and "centrifugal selection" that he found more rigorous than, respectively, "stabilizing selection" and "disruptive selection". As we shall see in the next chapter, even though in the long run Simpson failed to impose this new terminology, some geneticists of the 1950s (for instance, Douglas Falconer) would, for a while, take it into account.[29]

Given his personal involvement in the conceptual characterization of the various forms of selection, it is not surprising that Simpson was particularly interested in the American translation of Schmalhausen's book, whose subtitle was "The Theory of Stabilizing Selection". He read the book immediately and published a rather extensive, and very positive, review in *The Journal of Heredity* already in 1949. Simpson's criticism of Schmalhausen mainly concerned the book's form (and it is hard to disagree with him): he viewed the book as poorly constructed and very repetitive, making it tedious to read. Moreover, the use of traditional Russian morphological terms and the poor quality of the translation did not help.[30]

Two more central aspects of his review are particularly enlightening. The first directly concerns what would later be at stake with the Baldwin effect in 1953: already in 1949, Simpson stipulated that (1) this evolutionary mechanism is very probable but also most certainly marginal; and (2) it is in any case fully compatible with the theoretical framework of the Modern Synthesis. Simpson's detailed characterization shows the extreme vigilance and competence with which he endeavored to distinguish between the different concepts of selection discussed at that time in the biological literature. Indeed, Simpson was well aware of the vagueness that accompanied Schmalhausen's use of the term "stabilizing selection". According to him, at least three inter-twined meanings were used by Schmalhausen:

> Baboons may be supposed once to have had a labile reaction norm, with callosities appearing only where and when there was repeated contact with the ground and in proportion to the duration and roughness of such contact. Stabilizing selection then favored a new norm in which callosities appear autonomously, by heredity, even before any contact occurs.
>
> In this sense, "stabilizing selection" is most strictly defined and is narrowly equivalent to the "organic selection" of Baldwin and others, down to Gause. Schmalhausen, however, also uses "stabilizing selection" in at least two other senses, without clearly explicit distinction. In some passages it seems merely to mean selection in favor of wild type, or of established norm – what is sometimes called "centripetal selection". Elsewhere, and more frequently, it means selection favoring and, in a sense, producing "internal regulating mechanisms which counteract the harmful influences of the external environment" or of disadvantageous mutations and other disintegrating factors. Among the mechanisms of this sort discussed at some length are diploidy, dominance of wild type, balanced genetic systems, determination of morphogenesis by the genome as a whole, wide range of normal tissue reactivity, and complexity of morphogenetic correlations.[31]

In my opinion, before Waddington himself clarified the issue in the 1950s (see Chapter 6), Simpson here perfectly identifies the three meanings of the stabilizing selection concept actually present in Schmalhausen's book: (1) stabilizing selection as an equivalent of the Baldwin effect (deemed negligible by Schmalhausen or at least highly unlikely, as we have seen); (2) stabilizing selection as *normalizing selection* acting on allelic frequencies and homogenizing the gene pool (deemed secondary by Schmalhausen); (3) stabilizing selection as *canalizing selection* leading to the establishment of homeostatic devices making embryonic development buffered against genetic and environmental perturbations (the meaning that was most central to Schmalhausen). Thus, the 1953 article did not come out of nowhere for Simpson: it marked a stage (and very probably a stage close to a conclusion) in an in-depth reflection on the nature and scope of centripetal selection, a reflection he had started at least a decade earlier.

Given the rest of the contemporary literature, that was so terminologically confused, conceptually difficult, and empirically fragile, it is easy to see why Simpson's 1953 article quickly became a landmark. It has at least four major qualities.

(a) *Historical depth.* Most of the major scientists we encountered in the first part of this book – from Baldwin, Lloyd Morgan, and Osborn to Schmalhausen, Gause, and Thorpe – are mentioned in the eight pages of the article. Even Hovasse is discussed in depth by Simpson,[32] admittedly mostly as a paradigmatic example of unjustified generalization. Of course, some important figures are absent, such as Gulick, but to the reader in the 1950s this text would nonetheless present an enlightening historical synthesis of the evolution of a particularly tangled scientific problem. It should be noted that Simpson mentioned Waddington much less than Scmalhausen. Waddington is only cited at the end on the question of the genetic determinism of plasticity. This is not necessarily surprising since, on December 19, 1952, when Simpson submitted this text to *Evolution*, the first results obtained by Waddington on what he would call genetic assimilation had not yet appeared in print.[33]

(b) *Experimental information.* Simpson not only recalled the main milestones of this history since 1896 but was also particularly vigilant about the demonstrative significance of the empirical or experimental arguments produced over time. For example, his critique of Gause's numerous works is extensive and highlights some of the limits of this experimental program.[34]

(c) *Terminological precision.* Although the term "Baldwin effect" gradually fell into disuse in the 1970s, it was for two decades widely used, and then preferred terminology compared to the many alternative expressions of the 1930s and 1940s. On this point, Simpson had perfectly met the challenge formulated by Huxley in 1950: to elaborate a concise and epistemologically adequate terminology. If Simpson was directly inspired by Hovasse and his "Baldwin principle", the replacement of "principle" by "effect" is indeed far from anecdotal. It made it possible to highlight that the eventual result (the genetic fixation of an adaptive character) was the end product of a chain of causal factors whose sequence had to be identified.

(d) *Conceptual clarity.* This was undoubtedly what was seen as Simpson's major contribution to the debate. Whereas up to 1953 definitions had remained at best ambiguous, often expressed in an outdated or even pre-genetic vocabulary (as in Baldwin's writings), he succeeded in producing a renewed definition which worked perfectly with the updated content of genetics and evolutionary theory. This definition established the Baldwin effect as a microevolutionary process taking place in three major stages:

(1) Individual organisms interact with the environment in such a way as systematically to produce in them behavioral, physiological, or structural modifications that are not hereditary as such but that are advantageous for survival, that is, are adaptive for the individuals having them.

(2) There occur in the population genetic factors producing hereditary characteristics similar to the individual

modification referred to in (1), or having the same sorts of adaptive advantages.

(3) The genetic factors of (2) are favored by natural selection and tend to spread in the population over the course of generations. The net result is that adaptation originally individual and non-hereditary becomes hereditary.[35]

The crucial question, then, is the extent to which this definition does justice to the conceptual content identified in the first part of this book. In my view, Simpson's definition represents a textbook case of clarification based on considerable conceptual impoverishment. Indeed, not only is OS2 totally absent, but even OS1, that is, the first, more central causal mechanism identified within the banner of "organic selection" is conceptually mutilated. As we have seen, from Baldwin and Lloyd Morgan to Gause and Thorpe, OS1 was progressively constituted as an evolutionary factor based on a precarious balance between the ability to avoid extinction and the inability to remove selection pressures. Plasticity had to be situated between a minimal level below which it did not allow for the survival of the population in the new environmental conditions, and an upper level beyond which it rendered inoperative the selection of genes whose phenotypic effect tended in the same direction. However, in Simpson's definition, the issue of extinction is almost completely erased. It remains only marginally in the formulation "that are advantageous for survival, i.e., are adaptive for the individuals having them". But "adaptive" is not the opposite of a state that leads to extinction. It simply means an increase in fitness.

One should not underestimate the lasting consequences of the transformation that Simpson operated on OS1 with this simple shift. Simpson's interpretation (henceforth OS1') was likely instrumental in changing for decades to come and, until the present day, the conceptual alternative at play. In its previous formulations, from Baldwin onward, the alternative had been between extinction and genetic adaptation. In such a view, OS1 was seen as a mechanism allowing for what is now called "evolutionary rescue".[36] In Simpson's formulation, this dimension, previously so central, simply no longer existed. Thus, even with respect to OS1, without taking into account OS2 or the issue of speciation, the concept of the Baldwin effect as formulated by Simpson underwent a strong deproblematization: its ecological dimension was simply ignored.

The Baldwin effect underwent a prolonged eclipse that started in the early 1970s, but there was renewed interest in this possible mechanism of adaptive evolution from the end of the 1980s onward, when computer simulations made it possible to measure its scope more concretely.[37] However, the concept that was resurrected at that time was not OS1 but was most probably derived from OS1'. The question of extinction, once again, was then replaced by the speed of genetic adaptation.[38] The alternative was therefore not between extinction and genetic adaptation but only between slow and fast genetic adaptation: it was thought that the Baldwin effect could, in some cases, be responsible for faster genetic adaptation. Yet, even without plasticity, the population would most probably not go extinct but would only adapt more slowly, through the standard mutation/selection process. Still in the early 21st century, when the first mathematical models were developed, it was to Simpson that biologists returned. Consequently, in the models designed, population size was no longer a parameter to be considered.[39]

## 7.4 Mayr on behavioral plasticity after the "Baldwin effect"

Simpson's three-stage reconceptualization was perceived positively by his contemporaries: he had established that this adaptive mechanism, which at best played a minimal role in the evolution of species, could now be rightly overlooked. Mayr immediately seized upon this and adopted a definition closely modeled on Simpson's:

> Simpson (1953b), who gives an excellent critical discussion of the Baldwin effect, points out that it involves three steps (rephrased by me):
>
> (1) The genetically determined reaction norms of individual organisms permit the development of behavioral, physiological, or structural modifications of the phenotype which are not hereditary as such but which are advantageous for survival and permit the descendants of the organisms that have them to continue in the given environment;
> (2) Mutations (and gene combinations) occur in this population that produce the favored phenotype obligatory and rigidly rather than as a facultative modification;
> (3) The genetic factors under (2) are favored by natural selection and therefore spread in the population over the course of generations until the facultative character becomes obligatory and fixed.[40]

Let us note two modifications between Simpson's account and Mayr's ten years later (1963). On the one hand, the issue of extinction is less absent than in Simpson's text, since Mayr, in addition to emphasizing the gain in fitness ("advantageous for survival"), adds this time "permit the descendants of the organisms that have them to *continue* in the given environment". On the other hand, unlike Simpson, Mayr was careful to specify that plasticity was itself a genetic property of organisms ("the *genetically determined* reaction norms of individual organisms"). This clarification was most probably a result of Waddington's positioning on this subject. Indeed, as we have seen, an important argument for Waddington in his quest to distinguish genetic assimilation from the Baldwin effect had been to make it clear that in his own model, phenotypic plasticity was a genetic property over which natural selection had control. Very often in his publications, to illustrate what he thought was a significant difference, Waddington referred to Mayr,[41] who indeed usually used terms like "nongenetic plasticity", "nongenetic variation", or "nongenetic modification".[42] Nevertheless, this was a simple misunderstanding rather than a conceptual difference. What Mayr meant with these formulations was that the final phenotype was not determined by the genotype alone (i.e., that the environment had a causal impact) but not that plasticity was not a genetic property. He made sure to make this point crystal clear in *Animal Species and Evolution*: "It must not be overlooked that the capacity of a genotype to produce several phenotypes is very much under genetic control. 'Nongenetic' in the present discussion means simply that the differences as such between the modified phenotypes are not caused by genetic differences".[43]

Building directly on Simpson's take, Mayr was therefore prepared to discard the Baldwin effect, and to some extent this is what happened. In *Animal Species and Evolution*, as in other articles and books, Mayr made it clear that individual plasticity, instead of paving the way for genetic evolution, "has a retarding effect on evolution, contrary to the claims of the adherents of the 'Baldwin effect'".[44] He was globally even more severe than Simpson:

> Yet even Simpson leaves the door open for a role of the Baldwin effect in the evolutionary acquisition of certain characters, for instance, calluses. It seems to me, however, that a more detailed analysis shows that the conceptual assumptions underlying the hypothesis of the Baldwin effect make it *desirable to discard this concept altogether*.[45]

It is therefore not surprising that in the causality implemented in his new peripatric speciation model, to which we shall return in detail (see Chapter 9), ecological factors such as behavioral plasticity no longer played any causal role, contrary to what was the case during the 1947–1951 period. All this is consistent: the almost complete rejection of the effectiveness of the Baldwin effect, coupled, at exactly the same time (i.e., after 1953) with the reconfiguration of the allopatric model into a scheme excluding any causal role for phenotypic plasticity. However, this is not the whole of the story. Indeed, in the same movement, and sometimes in the same publications, Mayr was apparently giving more and more significance to behavioral changes, not in speciation but more generally in adaptive evolution and especially at the level of macroevolution. Some pages of *Animal Species* are here especially telling, for instance:

> A shift into a new niche or adaptive zone is, almost without exception, initiated by a change in behavior. The other adaptations to the new niche, particularly the structural ones, are acquired secondarily.[46]

Thus, as far as transpecific evolution is concerned, "evolutionary shifts" could be the long-term consequence "initiated" by a change in behavior. In the abridged 1970 version of *Animal Species*, Mayr went even further in that direction, writing that "as far as animals are concerned, behavior is the most important evolutionary *determinant*, particularly in the initiation of new evolutionary trends" and concluding the section in emphasizing that "the *enormous role* played by behavior in initiating transpecific evolution is being increasingly appreciated by evolutionists".[47]

While these passages show that Mayr was willing to argue that in animals, morphological evolution is often the consequence and not the cause of behavioral evolution, there remained substantial ambiguity about the determinism of behavioral changes. Indeed, Mayr could be read as arguing that behavioral change is itself generated by genetic change. Such an interpretation would be perfectly consistent with the general spirit of the Modern Synthesis and with Mayr's own analysis of the Baldwin effect. On this decisive question, Mayr does not tell us much, neither in *Animal Species* nor in the updated 1970 version. He merely states that "how the behavior changes themselves originate", is "a problem still poorly understood".[48]

A clue to a "Thorpian" understanding of the plastic origin of these behavioral changes is, however, given by the example used by Mayr at this point in his reasoning: "the recently developed habit of British titmice, mostly *Parus major* . . . of opening milk bottles and drinking the cream".[49] This example indeed seems difficult to explain in terms of genetic variation but more likely implies the ability to learn and mimic certain behaviors, that is, phenotypic plasticity.

Decisive evidence for such an understanding must be sought in another publication by Mayr, his contribution to the volume edited by Anne Roe and George Gaylord Simpson entitled *Behavior and Evolution* (1958). This book combined the proceedings of two major conferences held in April 1955 (in New York) and in May 1956 (in Princeton). Simpson, in close collaboration with his wife Anne Roe, a well-known experimental psychologist, attempted to *incorporate* the behavioral sciences into the Synthesis[50] and invited among their most eminent representatives, such as Nikolaas Tinbergen (Thorpe was however absent from the list). To Waddington's disappointment, most contributions reduced behaviors to evolutionary products, depicting them as consequences of natural selection. That behavior could play the role of an evolutionary factor which could modify the allelic composition of a population was usually ignored, even if Simpson himself briefly indicated this possibility but did not dwell on it.[51]

Mayr's chapter ("Behavior and Systematics") is indeed the only one that places much more emphasis on behavior as an evolutionary factor. Like in *Animal Species*, Mayr was quite willing to consider the driving role played by behavior in the phylogenetic evolution of structure and morphology. But this time, he distinguished very clearly between the two main possibilities for the genesis of new behaviors on a population scale and contrasted the genetic explanation and the one involving plasticity and learning:

> There are at least two different possibilities for the acquisition of a new behavior pattern by a species.
>
> 1. The new behavior may have a genetic basis right from the beginning. Since much behavioral variability is correlated with the genetic variability of the species, any factor affecting the gene content of the species may also affect behavior. Some of this may happen as an incidental by-product of genes selected for very different properties. Some of the behavioral variability described earlier may have this source.
> 2. A new behavior is at first a nongenetic modification of an existing behavior, as a result of learning, conditioning, or habituation, and is replaced by an unknown process by genetically controlled behavior.
>
> The study of a new behavior "fashion" might be very revealing. When titmice in England acquired the habit of opening milk bottles, it was observed that the technique was highly variable (Fisher & Hinde, 1949).[52]

It seems, therefore, that Mayr was willing to accept not only that "evolutionary novelties" and "shifts" may have been the consequence of behavioral changes but that these behavioral changes may not be the result of initial genetic differences, on the contrary finding their origin in plasticity and learning.

How can we reconcile this theoretical positioning with the way Mayr simultaneously very explicitly stated the negligible character of the Baldwin effect? The solution to this puzzle, in my opinion, is to be found in the preceding quotation, in which Mayr, surprisingly, qualifies as an "unknown process" the mechanism that would be likely to ensure the genetic fixation of plastic behavior. At this point, of course, Mayr was not unaware of the Baldwin effect, and in fact refers to it critically a few lines further on.[53] What Mayr most likely meant was that, for him, the Baldwin effect could not be responsible for this general process of genetic reinforcement of new plastic behaviors.

By the Baldwin effect, we must understand here the extremely simplistic mechanism as defined by Simpson in 1953, that is, OS1'. What mechanism did Mayr intend to substitute for it? Mayr briefly considered this issue in a few places in his work, for example in *Populations, Species and Evolution* when he writes that "changes in habit . . . permit shift in niche occupation that set up new selection pressures".[54] It seems almost certain that what Mayr had in mind here, this "unknown process" capable of genetically stabilizing new forms of behavior and thereby generating major morphological transitions on the scale of macroevolution, was none other than OS2, that is, one of the dimensions of organic selection that Simpson had excluded from the Baldwin effect.

Thus, Mayr's complex trajectory from the late 1930s to the early 1970s can be reconstructed as follows:

(1) In the late 1930s and early 1940s, Mayr was unaware of the mechanism of organic selection and made no room for behavioral plasticity in the first version of his model of geographic speciation.

(2) After reading Thorpe's latest work in the mid-1940s, and in particular his most speculative considerations on the role of conditioning in sympatric speciation, Mayr relaxed the geographical logic of his model and envisaged that behavioral plasticity could play a causal role (not necessary and not sufficient) in the genetic isolation of two populations. He then sometimes,[55] but not systematically,[56] linked his thoughts to the question of organic selection.

(3) Mayr followed Simpson and accepted his redefinition of the Baldwin effect, as well as his general assessment of the very low probability that this mechanism played an important role in adaptive evolution. From 1954 onward,[57] when he transformed his model into what he called the peripatric model of speciation,[58] behavioral plasticity no longer played a causal role.

(4) On the other hand, from the end of the 1950s to at least the beginning of the 1970s, Mayr allotted more and more space to behavioral plasticity insofar as it was capable of polarizing selection pressures and thus of initiating trans-specific morphological evolution.

Ultimately, Mayr's rejection of organic selection was somewhat ironic and rested heavily on Simpson's reading of the history. If Simpson had seen that Baldwin's and especially Gulick's hypothesis were also about behavior directing selection pressures, if he had seen that "social heredity" potentiated the effects of individual plasticity, as illustrated by the case of the new behavioral pattern in some populations of *Parus major*, then most likely Mayr would have been much more enthusiastic about the Baldwin effect.

## 7.5 Chapter's conclusion

By and large, it is easy to characterize the reception of organic selection during the first two decades of the Modern Synthesis: there was a before and after Simpson. Before 1953, there was a real (though limited) interest in how phenotypic plasticity, and in particular behavioral plasticity, might play a causal role in animal evolution. Huxley's overview, while it left out important aspects (OS2), was relatively exhaustive and based on substantial empirical documentation, in which Thorpe's work played the leading role. Mayr himself, as we have seen, was forced to attenuate the geographical logic of his allopatric model to make room for ethological and ecological factors in the speciation process.

Simpson's pivotal article acted as a kind of break, since it emptied the organic selection of almost all conceptual content and did not do justice to the rich history of this notion. OS2 and the issue of speciation were totally ignored in the redefinition produced, and OS1 subsided only in an impoverished form, OS1', in which the question of extinction was almost erased. In short, Simpson removed from organic selection what had been its core since the 1890s, namely its ecological dimension. Both OS1 and OS2 aimed to elucidate the causal relationship between phenotypic plasticity and natural selection in an ecological perspective, although of course formulations were incredibly diverse and sometimes obscure. Simpson's Baldwin effect appears in contrast to be a very poor proposal, emptied of most of this ecological dimension.

But, as we have seen, the clarity of the mechanism proposed by Simpson allowed for a very positive reception and still impacts the framing of problems today. Mayr immediately seized upon this new Baldwin effect and promoted it while insisting on the very low effectiveness of such a mechanism under real conditions, even more than Simpson did. As Mayr reserved an equivalent treatment to Waddington's genetic assimilation (see Chapter 9), he believed that he could definitively put this type of mechanism aside, while reaffirming the ineluctability of the alternative between the mutation/selection logic of the Modern Synthesis and the Lamarckian concept of inheritance of acquired characters. This was to be expected since the logic of the Modern Synthesis was indeed strengthened by this setting aside of the Baldwin effect. It should be stressed here that the context of Lysenkoism and of the "Cold War in biology"[59] was not conducive to the elaboration of middle ground views, and clear-cut distinctions between what was seen as the keystone of the synthetic theory (the absence of a causal link between plastic variation and genetic adaptation) and the mechanism promoted by any Lamarckian theory of evolution (the postulation of a direct causal and physiological link between plastic variation and genetic adaptation) were preferred.

However, as surprising as it may seem, the dismissal of the Baldwin effect in the theoretical framework of the Modern Synthesis did not mean the complete eradication of the question of the causal role of plasticity in evolution. Indeed, Mayr's scientific production *after* 1953 shows that, at least as far as trans-specific evolution was concerned, he made an increasing and not negligible place for OS2. He was simply unaware that the question of the links between behavioral plasticity, learning, and adaptive evolution had been one of the major dimensions of organic selection as early as the work of Baldwin and Gulick. This historical blindness is even more astonishing since, although Mayr probably did not have first-hand knowledge of Baldwin's writings, he was, on the other hand, well acquainted with Gulick's work on *Achatinella* snails, which he cited on many occasions, in particular, to illustrate the role of chance and accidental events in speciation. What Mayr's case testifies, then, is that Simpson's seminal paper once again broke the historical thread, that is, cut biologists in the 1950s and 1960s off from the reality of what the conceptual content of organic selection had been in the first half of the 20th century. Although Simpson called this evolutionary mechanism the Baldwin effect, his redefinition led to the neglect of Baldwin's (and others) own work. And when these questions resurfaced in the late 1980s and early 1990s,[60] it was first in complete ignorance of what organic selection had been around 1900.[61]

## Notes

1  Huxley (1942, p. 296, p. 523).
2  Gause (1947, p. 21).
3  Huxley (1942, p. 304). My emphasis.
4  Huxley (1942, pp. 305–307).
5  Huxley (1942, p. 305).
6  Huxley (1942, p. 316).
7  Huxley (1942, p. 304).
8  In the first edition of his book *Genetics and the Origin of Species* (1937), Dobzhansky makes no mention of organic selection and nowhere discusses the possibility that plasticity might play a causal role in evolution. The same observation can be made with respect to Mayr's *Systematics and the Origin of Species* (1942).
9  Mayr's book was apparently nearly completed by the late 1930s, but its publication was subsequently delayed.
10  Mayr (1942, chapter VI). It might be interesting to note that Cuénot (see Chapter 4) had already developed this concept in the 1930s and that Mayr was well aware of the content of Cuénot's ideas (Mayr, 1942, p. 131), whose work received sustained attention from most of the founders of Modern Synthesis.
11  Mayr admitted that in plants sympatric speciation could result from polyploidy: instantaneous chromosomal accidents could be at the origin of a complete reproductive isolation between individuals (Mayr, 1942, p. 191).
12  Thorpe (1945b, p. 67).
13  Thorpe (1945b, p. 69).
14  Cushing (1941a, p. 235).
15  Cushing (1941b, p. 499).
16  Cushing (1944, p. 265).
17  Cushing (1944, p. 268), my emphasis.
18  Mayr (1947, p. 274).
19  He wrote at the end of the paper: "I am deeply indebted to Dr. W.H. Thorpe for an extensive criticism of the first tentative draft of this paper" (Mayr, 1947, p. 286).
20  Mayr (1947, pp. 266–267).
21  Mayr (1947, p. 275, p. 281).
22  Mayr (1947, p. 286). My emphasis.
23  Mayr (1951, p. 123).
24  Mayr (1951, p. 118).
25  Mayr (1951, p. 116).

26   Mayr (1951, p. 124).
27   Huxley (1951, p. 125).
28   Huxley (1951, p. 125).
29   Falconer (1957).
30   Simpson (1949a, p. 324).
31   Simpson (1949a, p. 323).
32   Simpson was known to be proficient in several languages, including non-European ones (Whittington, 1986, p. 530). Apparently French was among them (Hovasse's work was never translated) but not Russian or Ukrainian.
33   As we have seen, the first set of experimental results on the genetic assimilation of the "crossveinless" phenotype were published in the same year in *Evolution* (Waddington, 1953a). Waddington's paper was submitted on October 9, 1952, that is, approximately two months before Simpson's synthesis.
34   Simpson (1953, p. 114).
35   Simpson (1953, p. 112).
36   The modern concept of evolutionary rescue is broader than OS1: in most cases, phenotypic plasticity plays no role.
37   The paper that rekindled the interest of evolutionary biologists (but also computer scientists) in the Baldwin effect is: S.J. Hinton, G.E. Nolan, 1987, "How learning can guide evolution", *Complex Systems*, 1, pp. 495–502.
38   For example, Mayley (1997).
39   Ancel's work here is particularly representative of this "Simpsonian" way of framing the issue: Ancel (1999, 2000).
40   Mayr (1963, p. 610).
41   For example, Waddington (1961a, p. 287).
42   For example, Mayr (1958, p. 354).
43   Mayr (1963, p. 139).
44   Mayr (1963, p. 147).
45   Mayr (1963, pp. 610–611). My emphasis.
46   Mayr (1963, p. 604).
47   Mayr (1970, pp. 364–365). My emphasis.
48   Mayr (1963, p. 605).
49   Mayr (1963, p. 605).
50   "The generally accepted modern theory of evolution is called 'synthetic', but comparative psychology has been an element not yet fully incorporated in the synthesis. Realization of these shortcomings and a hope to do something about them led to organization of the conferences mentioned in the preface and so eventually to the book now before you" (Roe & Simpson, 1958, p. 1). For a more detailed contextualization of these two conferences and the project behind them, see Grodwohl (2019).
51   Simpson (1958, pp. 9 and 21).
52   Mayr (1958, p. 354).
53   Mayr (1958, p. 354).
54   Mayr (1970, p. 364).
55   Mayr (1947).
56   Mayr (1951).
57   Mayr (1954).
58   Mayr (1982).
59   Lindegren (1966).
60   Schlichting (1986); West-Eberhard (1989); Scheiner (1993).
61   Scheiner (2014).

# 8 Re-working Waddingtonian concepts within quantitative genetics

Waddington was a very intuitive scientist who was good at bringing up new questions and original perspectives. But he was less good at transforming his intuitions into fruitful research programs. This was not so much because he couldn't bridge pictorial speculations and operative concepts, nor because he struggled to imagine experimental designs which could test their relevance (some of his protocols were in fact particularly ingenious; see Section 8.1.2) but rather because he lacked the perseverance that is necessary for rigorous quantitative approach of such difficult questions.[1] As we shall see, it was a kind of fatigue that led Waddington to gradually turn away from experimental science from the middle of the 1960s.[2] But, at the same time, he was an excellent promoter of evolutionary genetics, and he was able to attract the attention of many leading researchers on the disputed issues of stabilizing selection, canalization, and genetic assimilation. In Part II of this book, I deliberately isolated Waddington's views to give as clear a synoptic account as possible. The purpose of the present chapter is to show how, during the 1950s and 1960s, Waddington's ideas stimulated various series of experiments and theoretical reflections performed in Great Britain, Australia, and, to a lesser extent, the United States.

This research took place within the nascent field of quantitative genetics, which was becoming a well-established discipline from both a theoretical and institutional point of view. This expansion phase of quantitative genetics remains ambiguous in its relationship to population genetics and the Modern Synthesis. On the one hand, it can be seen as a period of refinement of the evolutionary models which came out of the pioneering work of Fisher, Haldane, and Wright, in which the phenotypes under selection were metric traits subject to a complex genetic determinism. On the other hand, especially during the 1950s, some evolutionary geneticists believed that their field was close to a state of crisis, and, as we shall see, leading protagonists such as Michael Lerner or Kenneth Mather put forth bold new hypotheses at that time.[3] This moment of theoretical indecision was therefore particularly favorable to the reception of such openly speculative and heterodox conceptions as those of Waddington and Schmalhausen.

DOI: 10.4324/9781003422990-16

This was also the golden age of the famous Institute of Animal Genetics in Edinburgh, directed during this period by Waddington himself. Waddington could be disconcerting in his role as a director: he was able to instigate a powerful research dynamic, but he was rarely present in the day-to-day life of his own institute leaving the researchers he had recruited totally free to choose their own paths.[4] While Waddington's institute was one of the most important institutions devoted to genetics in the world at that time,[5] it did not rear a new generation of Waddingtonians. In fact, it counted among its ranks many researchers, such as Alan Robertson and, to a lesser extent, Douglas Falconer, who were very skeptical about the value of Waddington's concepts. Even Jim Rendel, who was the most faithful to the general orientations of Waddington's speculations, remained very cautious about the transformative scope of this way of linking genetics, evolution, and development. Rendel did not believe a drastic reform of the Darwinian framework that had stabilized during the previous two decades was needed.

As we will see in this chapter, these attempts at reconceptualization within quantitative genetics, in which Waddington himself sometimes participated, were successful on at least two counts. First, they were fruitful: on this occasion, several experiments were performed, and many theoretical models were elaborated, which contributed, albeit modestly, to the development of quantitative genetics as such. In addition, these biologists were sometimes able to find operational refinements of several Waddingtonian concepts. The question of whether these attempts were faithful to Waddington's theoretical intentions, or whether some of the cognitive content was lost or distorted when it became operationalized in the models of quantitative genetics is a difficult one which I address at the end of this chapter. But first, I discuss the most significant work carried out in this context.

This experimental and theoretical research mobilized dozens of biologists for about two decades. Most of them knew each other personally and exchanged informally during colloquia or research visits, or through correspondence. It is not my intention here to reconstruct the complete ecosystem of these exchanges in detail, but I will sketch their main conceptual lines. In these exchanges, many topics were addressed in an often intertwined and complex manner, particularly in the late 1950s and early 1960s. The division of the present chapter into three sections is meant to facilitate the understanding of these debates that were sometimes extremely abstruse.

## 8.1　Experimenting with canalization

Waddington's main and undisputed legacy remains without a doubt the concept of developmental canalization. But this stimulating vision was never a simple one and gave rise to a remarkably wide range of interpretations. A crucial problem that arose from the outset (as early as the 1940s) was that of measuring and quantifying canalization (either of a particular phenotypic trait or of a developmental system in general). As Robertson[6] (who was never fully convinced of the value of Waddington's concepts) was fond of pointing out "the concept is not quantifiable in the sense that we do not know how to measure the canalization of any specific measurement".[7] Canalization thus posed a considerable challenge to quantitative genetics, the science of measurement par excellence: to develop theoretical and experimental paths capable of making this question soluble within this general framework. Several geneticists in the 1950s took up this challenge head on, producing some experimental settings, and using a variety of methodologies that will be outlined in this section.[8]

It must be emphasized that this body of work was not driven solely by theoretical motives. A primary motivation was to make quantitative genetics a tool for improving breeding procedures: if a trait of interest was canalized, directional selection had little chance of allowing its modification. Most of the biologists involved in this research believed it was necessary to determine and quantify the parameters controlling the response to selection. Empirical procedures of artificial selection had indeed shown their limits in the previous decades, which encouraged Lysenkist extravagances.[9] Directional selection seemed to run up against poorly understood barriers, just as it greatly reduced the viability or fertility of farm animals. Waddington's "unit" in the Institute of Animal Genetics in Edinburgh was financed by the Animal Breeding and Genetics Research Organization. During the difficult post-war years, the problem of livestock efficiency had become a major political issue. In this applied science perspective, the genetic understanding of developmental canalization was primordial also because it directly impacted selection and thus breeding procedures.

### 8.1.1  Does normalizing selection produce canalization?

As we saw at the end of Chapter 6, Waddington's proposed distinctions between different forms of selection contained conceptual ambiguities. Remember that one of his main distinctions was between normalizing selection (selection for the average phenotype) and canalizing selection (selection supposed to produce developmental canalization). Were they exclusive categories designating disjoint sets, or were there areas of overlap between them? Could normalizing selection, which was defined by its modus operandi, have canalizing effects, making canalizing selection a special case of normalizing selection? The question of the genetic consequences of normalizing selection quickly became a debated issue in the second half of the 1950s: the aim was to understand whether selecting the average phenotype had an impact on the degree of developmental canalization.

Surprisingly, this question was not so much addressed because of Waddington's speculations but primarily because of the considerable influence of the very heterodox theses that Michael Lerner had gathered in his 1954 book *Genetic Homeostasis*. As we will see, it was in fact more in reference to Lerner that, even in Edinburgh,[10] geneticists wanted to address the issue of the genetic consequences of normalizing selection.

Michael Lerner (1910–1977), like his contemporary Kenneth Mather, was one of the most brilliant of the second generation of population geneticists and his theoretical speculations played a major role in the development of quantitative genetics in the 1950s. Of Russian origin, raised in a very privileged cultural environment, he was forced to emigrate because of the 1917 Revolution, and trained as a biologist in the United States at The University of California, Berkeley, where he spent most of his career.[11] Against the necessary but also problematic atomistic simplifications of the first models of the 1930s, he wished to reorient population genetics toward a much more holistic and integrated understanding of the genome, first on the scale of individuals but especially on the scale of Mendelian populations.[12]

This populational perspective found its most refined expression in his book *Genetic Homeostasis*. We will return to the content of the theory of genetic homeostasis in more detail in the next chapter (Section 9.2). For the moment, it will suffice to say that (a) what Lerner called genetic homeostasis was the ability of a population to restore its gene pool when its equilibrium was upset by external disturbances, and (b) this population property was mostly a consequence of the heterozygosity level of individuals, which itself was responsible for their degree of canalization (developmental homeostasis). Thus, for Lerner, there was a very simple and direct causal link between the developmental homeostasis of individuals (due to the high degree of heterozygosity of their genomes) and the evolutionary homeostasis of a population (due to the allelic diversity within the gene pool resulting from this high rate of heterozygosity). By favoring canalized genotypes, natural selection, at least in its stabilizing form, also favored the genetic cohesion of Mendelian populations.

One of the significant empirical findings of the mid-20th century, also widely used by Lerner, was that homozygous individuals were more sensitive to environmental variations. In Lerner's framework, this means that the individuals that deviate most from the average phenotype, that is, those with the lowest level of heterozygosity, are the most plastic. It therefore became possible, by selecting the most extreme phenotypes over the generations, and by mating individuals with symmetrically inverse characteristics, to quantitatively evaluate the following question: would there be an increase in environmental variance over the generations? This project guided the first work of Falconer and Robertson in Edinburgh, whose results were published in 1956.[13] Douglas Scott Falconer (1913–2004) was at the time one of the leading specialists in the genetics of mice,[14] which was used as the experimental system. As one of the top quantitative geneticists of the time, his main contribution was probably one of the most important textbooks in the field, *Introduction to Quantitative Genetics*, first published in 1960. Alan Roberston

(1920–1989) was one of the main theorists of quantitative genetics. Though they both worked in the department headed by Waddington, they shared a constant skepticism toward Waddingtonian concepts and models.

The principle of their first series of experiments was to evaluate the sensitivity of the body weight of individuals to variations in the environment. Two selection lines were instituted: one in which the heaviest individuals were mated with the leanest individuals (the "extreme" line), and one in which the individuals closest to the average were selected as exclusive breeders (the "central" line). Selection was carried out through 13 generations and did not show significant results, in either of the two lines. These negative results were difficult to interpret and remained "ambiguous",[15] especially because the genetic diversity of the extreme line was problematic to evaluate. It nonetheless suggested that, given that environmental sensitivity in the central line remained unchanged, normalizing selection had no clear effect on canalization.

Nevertheless, these early experimental results encouraged Robertson to produce two theoretical models to try to account for the possible effects of normalizing selection. It should be noted that his work explicitly followed that of Lerner (one of the two models is directly derived from the "genetic homeostasis" hypothesis), whereas neither Waddington nor Schmalhausen was even mentioned in the article.[16] He first showed that in a simple setting with one locus and two alleles, normalizing selection would result in genetic fixation. In this first model, extreme phenotypes are less fit because they are extreme. In such a theoretical configuration, normalizing selection thus cannot be "responsible for the maintenance of genetic variability in the population".[17] The second formal model, termed "homeostatic model", was directly inspired by Lerner and is more complex. This time, extreme phenotypes are less fit not because they are extreme but because they are homozygote. Robertson found that, in such a configuration, normalizing selection would maintain genetic variability, which led him to support Lerner's general thesis that heterozygotes have an enhanced fitness. This example perfectly illustrates Robertson's remarkable ability to build formal models from verbally stated concepts.[18] For Robertson, there was no doubt that the second, less simplistic, model was more interesting because it paved the way for many experimental controls, particularly with regard to the speed with which genetic homeostasis returns the population to equilibrium once the selective forces cancel out.[19]

Robertson's theoretical work in return stimulated Falconer, who produced a large-scale empirical study of the effects of normalizing selection in *Drosophila*. There was no doubt at the time that what was lacking in these highly speculative debates was, above all, solid empirical data that would make it possible to decide between the different interpretations.[20] Falconer, like others, was not convinced by Waddington's distinction between "normalizing selection" and "canalizing selection" because it was based solely on the putative effects of what was thought to be the same selective regime favoring the mediating phenotype. This is why he used Simpson's more neutral "centripetal selection".[21] Given that centripetal selection was supposed to produce a reduction of the phenotypic variance of a metric character, Falconer wanted to be able to decide between three major classes of hypotheses.[22] This possible reduction of variance could be the consequence either of an increased canalization (Waddington's and Schmalhausen's hypothesis) or of a reduced allelic diversity (the standard hypothesis within the MS), or be based on the superiority, in terms of fitness, of heterozygous genotypes (Lerner's hypothesis).

To test these hypotheses, he performed selection experiments by choosing a fitness-neutral trait, the number of bristles on the ventral surface of the fourth and fifth abdominal segments. A second advantage of this trait was that by comparing the number of bristles between the two segments of the same individual, it was possible to distinguish genetic variance from environmental variance. The experimental design was standard: selection consisted of breeding flies as close as possible to the average values. After 13 generations, results were unfortunately negative: "The selection of intermediates produced no detectable effect on the variance".[23] These negative results seemed to go against the possibility that normalizing selection could produce developmental canalization, at least on the time scale of laboratory work. Falconer deduced that, at least for this trait, hypothetical "stabilizing genes" did not show sufficient allelic diversity on which selection could operate.[24]

Other experimental results were produced in the following years in the US and the UK that went against the evolvability of canalization via normalizing selection. For example, Timothy Prout, at the University of California, published a large study in 1962 on the impact of normalizing selection on the length of development from oviposition to emergence of *Drosophila melanogaster*. Using artificial selection procedures similar to the previous ones, this time, he obtained significant results: the variance of lines subjected to normalizing selection decreased over the generations.

However, as we have seen, a decrease in variance could have other causes than an increase in developmental canalization. This is why Prout wanted to evaluate the allelic polymorphism of the different lines by subjecting them to a brief directional selective regime to see if a response to selection could be measured. The results clearly showed that the S line, the one subjected to normalizing selection, could no longer respond to selection which meant that it had become genetically homogenous. Therefore, the decrease in phenotypic variance was not due to an increase in canalization but simply to the "reduction of the additive genetic component".[25] His results were thereby in line with Robertson's first theoretical model.

We can therefore observe that, during the second half of the 1950s, the experimental results and theoretical models available seemed to go directly against what can be understood as Waddington's and Schmalhausen's intuitions: far from allowing the evolution of developmental canalization, normalizing selection seemed to lead instead to a reduction in allelic diversity. Were Waddington and Schmalhausen so wrong? Or, as appears most likely, had the attempts to express some of their ideas within the framework of quantitative genetics missed a key point? For canalization to evolve, the selection pressure must indeed be adequately designed. This was not the case in the previous experiments and models, because the mean phenotype was selected in a constant environment, that is, without developmental canalization being necessary for the effective realization of this phenotype. In short, the selection pressure for canalization was almost zero, as Waddington rapidly stressed.

### 8.1.2  Back to Waddington and canalizing selection

Waddington did not immediately take on the question of the experimental demonstration of canalization and its evolvability. In the 1950s, as we have seen (Chapter 4), he was primarily occupied with the issue of the demonstration of the phenomenon of genetic assimilation. It was only at the end of this decade, after all the work previously mentioned and at the time when Rendel was launching his vast research program (see Section 8.2), that Waddington attempted to clear up some misunderstandings.

All in all – and this is a surprise – Waddington only carried out two series of experiments on this decisive and highly debated question. The first was carried out in 1959 and published in 1960 in *Genetics Research*. In this first article, he opened with a critique of the work done so far and in particular that of his colleagues in Edinburgh. Without the application of a developmental stressor (environmental or genetic), the selection pressure was too weak to allow for the evolution of canalization.[26] Falconer and Robertson's negative results were therefore of little demonstrative significance. The challenge was to develop an experimental design that would produce selective pressures effectively directed toward developmental canalization. In this first work, Waddington made two experimental attempts, of which only the second proved fruitful.

He decided to work with various well-identified mutants in *Drosophila*, which were known to have a lower level of developmental canalization compared to the wild type (*Bar, Dumpy, Cubitus-interruptus*). It was finally the *Bar* mutant that proved to be the most relevant for what was at stake. *Bar* individuals show a specific eye structure, where the number of facets is markedly reduced. Besides, this number is greater at a low temperature than at a higher one. Waddington applied to experimental populations two forms of selective regime named "alternate selection" and "family selection", both designed to increase developmental canalization.

"Alternate selection", as the name indicated, concerned the selection of individuals in a population that was alternately submitted to two distinct temperatures (18°C and 25°C): organisms were reared at 18°C during one or more generations, then at 25°C, then at 18°C, and so on, for a total of 18 generations. At each generation, selection was made against the known effect of the environment on the phenotype: at 18°C, the selected flies were those with the smallest number of eye facets, while at 25°C, it was the opposite. This form of selection yielded some results, with the sensitivity of the *Bar* mutants to temperature variations decreasing. However, Waddington did not find them significant enough, so he developed a second, more complex, selective regime, that of "family selection":

> A very different and, as it turned out, much more effective type of selection was applied by means of family selection. In this, a series of pair-matings were set up, and from each pair of parents a number of offspring were reared at 25° and a number at 18°. Family averages of the difference between the phenotypes at these two temperatures were then ascertained by measuring a certain number (usually twenty) of the offspring at each temperature. Selection was made on the basis of these family averages, lines being carried on by breeding together sibs from the families which exhibited the least environmental sensitivity and in which the difference between the offspring at the two temperatures were smallest.[27]

This time the results were spectacular, and in only a few generations (6) "marked reductions in environmental sensitivity have been achieved".[28] Whereas the control populations showed a significant difference in the number of facets according to the rearing temperature, this difference became very low in the populations subjected to family selection (Table 8.1):

Table 8.1 *Some experimental results obtained by Waddington in using a complex selective procedure termed "family selection" applied to various strains of* Drosophila melanogaster *(Waddington, 1960). The character under selection was the facet numbers in* Bar *females in two different developmental environments (at 18°C and 25°C). The differences in temperature sensibility between the selected (6a, 6b, and 6c) and unselected lines are spectacular.*

|  | Average family mean at 18°C | Average family mean at 25°C | Difference in family means |
|---|---|---|---|
| Unselected | 156.25 | 55.46 | 100.79 |
| Selected 6a | 106.02 | 91.51 | 14.51 |
| Selected 6b | 100.76 | 95.30 | 5.46 |
| Selected 6c | 111.39 | 98.48 | 12.91 |

Also, despite the significant increase in canalization in the selected lines, there was no significant variation in the fluctuating asymmetry of the same trait. This result seemed to support Waddington's distinction between developmental stability as the ability to withstand developmental noise and canalization as the ability to produce a given phenotype despite genetic or environmental perturbations[29] (see Chapter 5).

Despite the demonstrative character of the results obtained, Waddington was not completely satisfied with his experimental model because it was based on a form of selection that he considered too "artificial".[30] For this reason, several years later, he took up this question by changing the procedure involved. In 1966, this second line of research led to the publication of an article, once again in *Genetics Research*, soberly titled "Selection for Developmental Canalization".[31] It should be noted that this late publication represents one of the very last experimental works of Waddington, who was on the verge of leaving laboratory research for more abstract considerations which were sometimes far removed from science as such. Waddington used the same experimental system and focused on the same character (the number of eye facets of the *Bar* mutant), but this time his experimental design was based on what he presented as a particular form of so-called disruptive selection. The concept of disruptive selection had been clearly identified and codified by Mather in the late 1940s, when he distinguished it from that of directional selection (natural selection in its standard sense) and from that of normalizing selection (in the sense used in the preceding pages, that is, when the selected phenotype corresponds to the average value of the character).[32] Since Mather, disruptive selection (sometimes also termed "diversifying selection") is usually defined as a selection in favor of two or more modal phenotypes and against those intermediate between them.

Waddington set up two selection lines: one ("canalizing line") where the degree of canalization was supposed to increase and the other ("anti-canalizing line") where it was supposed to decrease. To do this he used two populations of *Drosophila*, one raised at 18°C and the other at 25°C. For the "canalizing line", he selected in each generation the individuals raised at 18°C with the smallest eyes (while a low temperature increases the number of facets) and made them reproduce with the individuals raised at 25°C with the largest eyes (while a high temperature decreases the number of facets). Note that it can be argued, against Waddington's presentation, that this was not really a form of disruptive selection as usually defined, since it was the same phenotype (a better developmental buffering to the effects of temperature on eye structure) that was selected in two different forms only because the rearing environments were themselves different.

The results were again quite significant: the measurement of the number of facets showed a clearly reduced gap from the fourth generation onward, which continued to improve at the ninth generation (Table 8.2):

Table 8.2 *Partial results of the canalizing selection experiments performed by Wad-dington: mean number of facets per eye for a different series of flies reared at either 18°C or 25°C (Waddington, 1966b). In the course of the selective process, the sensitivity to temperature variation during development is in the main lowered; that is, the variability of the number of facets per eye is decreasing.*

|  | Males reared at 18°C | Males reared at 25°C | Difference |
|---|---|---|---|
| Foundation stock | 163.9 | 98.5 | 65.4 |
| Anti-canalization lines (F4) | 185.8 | 97.4 | 88.4 |
| Canalization lines (F4) | 175.7 | 116.6 | 59.1 |
| Anti-canalization lines (F9) | 164.1 | 88.7 | 75.4 |
| Canalization lines (F9) | 197.4 | 142.6 | 54.8 |

Three points are of particular interest in the interpretations formulated by Waddington. The first is that here, as in some other texts, Waddington seems to assume that there may be "canalizing genes", that is, genes which, in this case, reduce sensitivity to temperature variations.[33] In this, he contradicted his dominant conception based on a systemic understanding of the functioning of the genome, in which canalization was understood to be a consequence of a form of general epistasis. The second is that Waddington linked his experimental design to ecological considerations related to the possibility of sympatric speciation.[34] We saw in the previous chapter the space the debate on the mechanism of speciation took up in the history of organic selection during the 1940s and 1950s and how Mayr's peripatric model eventually took hold. Here, Waddington wished to emphasize that sympatric speciation could be even more likely when certain environmental parameters, which define an ecological niche, played not only the role of selective agents but also that of inducers of adaptive responses. In doing so, he was eventually clearly in line with the ecological perspective of the organic selection tradition. The last point concerns the extreme complexity of the interpretations that could be drawn from the relatively significant results obtained. In the last lines of the article,[35] Waddington appears a little overwhelmed by this complexity, and perhaps this is one of the aspects that contributed to his scientific withdrawal in the mid-1960s. Perhaps he felt that a satisfactory genetic theory of canalization and genetic assimilation was still out of reach, or at least out of his reach.

## 8.2   Genetic models of canalization

The concepts proposed by Waddington and Schmalhausen had a stimulating effect on the discipline of quantitative genetics. They unquestionably posed a challenge to some practitioners of this science of measurement, even if the experimental mastery of canalization presented formidable problems, as we have just seen. In parallel with these experiments that aimed at testing canalization in its empirical reality, other, more speculative works also came into being which sought to construct genetic models of canalization. These highly theoretical works were nonetheless often accompanied by numerous experiments. While Waddington was ultimately very vague about the genetic nature of developmental canalization, others wished to go much further, even if it meant departing from some of Waddington's own intuitions. I will deal here with the three most significant essays by Mather, Fraser (partly in collaboration with Dun), and finally Rendel whose work was the most extensive and in-depth of the three.

### 8.2.1 *Mather: a polygenic account of canalization*

In Chapter 5, we saw the role played by Mather in the development of studies of fluctuating asymmetry in relation to the quantification of developmental stability. Decidedly very interested in Waddingtonian concepts, he did not stop at this question and also reflected on the genetic nature of canalization. This reflection was rooted in his own conception of "polygenic inheritance", his distinction between "oligogenes" and "polygenes", and his theory of "balanced polygenic combinations". Since these terms have largely become obsolete, I will start with a general reminder of the theoretical framework that Mather developed.

Mather's theoretical activity long predated his positive reception of Waddington's concepts and finds its original impetus in his desire to pursue the synthesis between Mendelian genetics and biometrics initiated by Fisher, as noted in Chapter 5. Early models of population genetics were Mendelian in the standard sense, in that they were concerned with the kinetics of diffusion of a given allele as a function of its fitness value. In this standard framework, alleles had clearly delineated effects, and therefore selection coefficients could be assigned to them. It was these alleles that formal genetics and then (*Drosophila*) chromosomal genetics had highlighted. For Mather, like for Waddington, it was at that time essential to go beyond these first models and their restrictive simplifications. They believed that if such genes with discontinuous phenotypic effects existed, they were rare and did not represent the standard case. Mather therefore postulated an important distinction between what he called "oligogenes", precisely the genes of standard Mendelian genetics, and "polygenes", whose effects were both small and interchangeable. The oligogenes were the object of traditional population genetics, the polygenes of the new quantitative genetics.[36] This distinction, he argued, reinforced the ongoing synthesis between Darwinism and genetics: if heredity was mainly polygenic, then hereditary variation in natural populations had a continuous distribution, which guaranteed the efficiency and creativity of natural selection.[37]

By the early 1940s, Mather had arrived at a relatively complete synthesis of his ideas. He could already build on the results of his extensive and successful experimental program on the artificial selection of the number of chaetae of abdominal segments in *Drosophila*.[38] The high level of response to selection obtained argued for the existence of "stored" genetic variability. To explain how this genetic storage proceeded, he devised a theory of his own in which, like Lerner, he gave the leading role to heterozygosity.[39] His starting point, very Waddingtonian in spirit, was that a given phenotype could be produced on a very wide range of genetic bases:

> The phenotype is produced by the genotype acting as a whole. Since polygenes have effects similar to one another, a given phenotype may correspond to various genotypes some containing one and some another allelomorph of a given polygene.[40]

In his theoretical framework based on the distinction between oligogenes and polygenes, genetic variability was stored due to the polyallelism of the many polygenes involved in the construction of the phenotype. As many different genotypes could produce the same phenotype, this genetic variation was mostly unexpressed ("potential variability").[41] Nevertheless, because of the genetic recombination due to sexual reproduction, this potential variability could sometimes be expressed in the phenotype ("free variability") and gave rise to rapid selective events. Mather took care to summarize his heterodox conceptions in the form of a general model. While mutation was ultimately responsible for the creation of new alleles, natural selection essentially worked on the continuous variation resulting from the partial release of this "potential variability".

If there were a very large number of polygenes involved in the realization of a phenotypic trait, and if their effects were quantitative and interchangeable, then the same phenotype could be evolutionary "stabilized" by multiple combinations of alleles with antagonistic effects. These alleles must be located on the same chromosome so that inter-chromosomal recombination cannot destroy the balanced combinations built by natural selection too quickly.[42] Mather thus saw a chromosome as a highly evolved product, truly designed by natural selection.[43] Stabilizing selection, as he understood it, was responsible for the progressive development of these balanced combinations.[44] These balanced combinations were favored by natural selection because, in addition to allowing the stabilization of the average phenotype (which would have been possible via an increase in homozygosity), it ensured a reserve of variability in the long run, if the environmental conditions were to change. There was a form of compromise between what Mather called the "flexibility" of a population and the stability of the wild phenotype. Not so much that natural selection could predict and anticipate future needs but simply because current populations represented the provisional end of past selective episodes where genetic flexibility already played a causal role in reproductive success:

> The most advantageous arrangement will thus involve a small amount of the former [free variability] and a large store of the latter [potential variability]. Inasmuch as the species of to-day are descended from the successful species of the past, their genetical structure must betray the means by which this balance of free and potential variability is achieved and maintained.[45]

Mather called "balanced combinations" theory this complex set of assumptions. It was thereby in the light of an extremely elaborate theory that Mather read Waddington's first speculations on canalization. To the article published by Waddington in *Nature* in November 1942,[46] Mather responded very quickly with an article of his own, which appeared in January 1943.[47] There is thus no doubt that Waddington's concepts immediately resonated with his own preoccupations. In his *Nature* article entitled "Polygenic Balance in the Canalization of Development", Mather explicitly tried to translate them into his theoretical framework.[48] His distinction between oligogenes and polygenes offered him a powerful tool to give substance to Waddington's less defined ideas. Waddington's embryological conceptions stipulated two main processes during ontogeny: on the one hand, the existence of discrete developmental paths, without intermediaries, and more or less canalized ("chreods"); on the other hand, the existence of switch points between divergent paths. This embryological distinction found an immediate genetic explanation within Mather's model: the switches were under the control of oligogenes (which explained the important phenotypic effects of oligogenic mutations), whereas the canalization of developmental pathways resulted from the coordinated action of balanced combinations of polygenes.

As noted earlier, Mather's balanced combinations theory was not well received at the Edinburgh Institute of Animal Genetics. Nor was Waddington particularly pleased with the interpretation Mather had sought to produce of his ideas within this theory. As early as April 1943, Waddington published another article in *Nature* in which he opposed Mather's interpretation in very strong terms:

> Mather afterwards attempted to develop this idea by identifying the genes which act in a buffering manner with his so-called polygenes and the genes acting by switch mechanisms with "oligogenes" – a new word which he coined to include the genes with comparatively large effects normally studied in genetic laboratories. *I wish to show that this identification cannot be sustained and has only been suggested by extremely confused thinking.*[49]

Unsurprisingly, what Waddington contested was the ontological dimension of the distinction between oligogenes and polygenes. That there are oligogenic phenotypic variations – determined by one or only a few genes – and polygenic phenotypic variations – determined by many genes – was, according to Waddington, a fruitful or at least acceptable view. On the other hand, to reify this at the level of the genes themselves was a further step that he refused to take.[50] It must be stressed that, at least at this point in his career, Mather had indeed gone very far in the molecular understanding of his distinction. He anchored this distinction in the most recent data from cytology[51] and matched the two forms of nuclear chromatins to the two classes of genes he had postulated: oligogenes corresponded to euchromatin, while polygenes corresponded to heterochromatin, which he thought to be less essential.[52]

Waddington never deviated from his initial critique of Mather's central distinction and reiterated it word for word in his book *The Strategy of the Genes*[53] and in his 1961 detailed review on genetic assimilation.[54] For his part, Mather did not give up on the idea that canalization was under polygenic control when he turned to the issue of fluctuating asymmetry[55] (see Chapter 5). However, it is also possible that he attenuated the distinction between oligogenes and polygenes and especially that he gave up, at least in part, on giving it a molecular basis (but this remains to be further documented). Unlike Rendel, as we shall see, he apparently never tried to translate his distinction in the terms of the new molecular biology,[56] even though the Pasteurian concept of the "regulatory gene" offered an obvious solution to what an oligogene could be.[57] He instead remained faithful to his more phenomenalist and quantitative approach, and his final scientific article, published posthumously in 1990, was still about the "Consequences of Stabilising Selection for Polygenic Variation".[58]

### 8.2.2    *Fraser and Dun's model: canalization as the sigmoid relationship between the genotype and the phenotype*

In the late 1950s, research on canalization gravitated away from Edinburgh and toward Sydney. The main reason for this new Australian impetus was the recruitment of James ("Jim") Meadows Rendel (1915–2001) by the CSIRO (the Commonwealth Scientific and Industrial Research Organization) in 1951. Of all the scientists who worked with Waddington at the time of the early phase of the Institute of Animal Genetics in the late 1940s, Rendel was certainly the one who was most interested in the concepts forged by his mentor, foremost among which was that of canalization. He was therefore in some ways the opposite of Robertson, to whom he was nevertheless very close.[59] Rendel's arrival in Australia coincided with the long-lasting establishment of these themes within the CSIRO and in related institutions. Rendel was a recognized leader and was instrumental in the development of many research programs, of which only a few will be mentioned in the present and following section. Moreover, he did not arrive in Australia alone, as he was accompanied by Alex S. Fraser (1923–2002), one of the first PhD students to be trained in Edinburgh under Waddington's "supervision".[60] Rendel and Fraser formed a lasting and fruitful intellectual relationship, and they did much for the development of the Division of Animal Genetics of the CSIRO, of which Rendel became the director in 1959. In the next section, I will detail Rendel's work on *Drosophila* and the model he elaborated about the genetic nature of canalization. This was slightly later than Fraser's and Dun's own work on mice, which resulted in a first version of the same theoretical account of the canalization of a specific metric trait.[61]

In addition to his pioneering work with Dun on canalization in mice, Alex Fraser is best known today for being one of the first to conceive and execute computer simulations of genetic systems as early as the late 1950s and 1960s.[62] This line of research was not independent of his work on canalization, since he inaugurated the first numerical simulations of genetic systems capable of generating developmental canalization in 1960.[63] These results were among the few that caught Waddington's attention, although he remained measured about their biological significance.[64] Fraser's starting point was a critique of the theoretical models developed by Robertson,[65] which did not take epistasis into account.[66] This was an issue that was still difficult to address through laboratory experiments. On the contrary, by using early numerical simulations, it was possible to model the effects of epistasis on the evolution of canalization and this is what Fraser did as early as 1960. What interested Fraser the most, and what would henceforth become central in the genetic models of canalization developed at the CSIRO, was that epistasis allowed, during a selective process, the progressive loss of the linear relationship between the genotype and the phenotype. Thus, there were wide ranges of genotypes corresponding to the same phenotype, which became a canalized phenotype (Figure 8.1).

It was this way of considering canalization as a loss of the linear relationship between the genotype and the phenotype that was at the heart of the models developed at the CSIRO by Fraser, Dun, Rendel, and also Berenice Kindred and others. Given Fraser's central role in the theoretical side of this way of conceiving canalization, it is likely that he, rather than Dun or even Rendel, was the main instigator of this conception.[67] Thus, what is commonly referred to as the "Dun and Fraser model"[68] should in fact be more accurately known as the Fraser model to which Dun contributed his experimental expertise.

While Fraser was inclined toward theory and abstraction, Robert Bruce MacLeay ("Bob") Dun,[69] who first graduated from veterinary science at Sydney University in 1953, was an experimentalist familiar with selection methods and the evaluation of response to selection. It is in this field that he did his PhD at the CSIRO from 1955 to 1959, under the joint supervision of

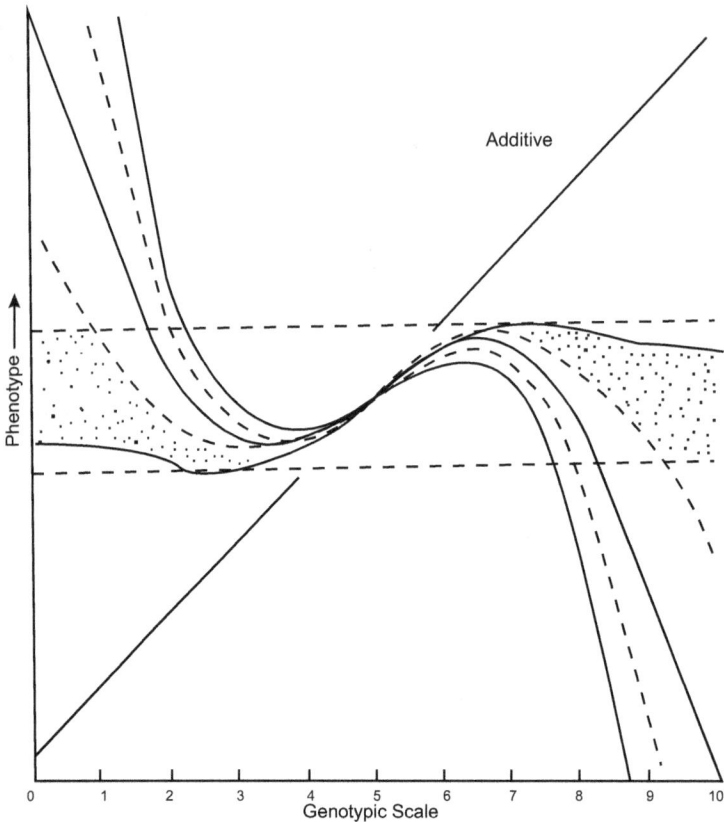

*Figure 8.1 Graphic representation of the numerical results obtained by Fraser that show that epistasis might be responsible for a specific relationship between the genotype and the phenotype that allows some portion of the phenotype to become insensitive to genetic change.* Reproduced from: A.S. Fraser, "Simulation of genetic system by automatic digital computer", *Australian Journal of Biological Sciences*, 1960, 13, pp. 150–162. With permission from CSIRO Publishing.

Fraser and Rendel. His PhD dissertation consisted of work on the artificial selection of the number of vibrissae in mice, a material that would serve as the basis for their model of canalization. The vibrissae are long hairs located at the front of the face and forelimbs which allow for a fine sensory exploration of the environment. Typically, the number of vibrissae is an almost constant character in mice – and therefore probably highly canalized – since it is almost invariably 19 (Figure 8.2).

From the outset, Dun and Fraser placed their work within the continuity of the experiments performed by Waddington in the early 1950s on genetic assimilation.[70] As we have seen, Waddington's work involved two main steps: (1) decanalizing a trait (such as *Drosophila* wing venation) by means of a very powerful environmental treatment (such as heat shock during pupation) and

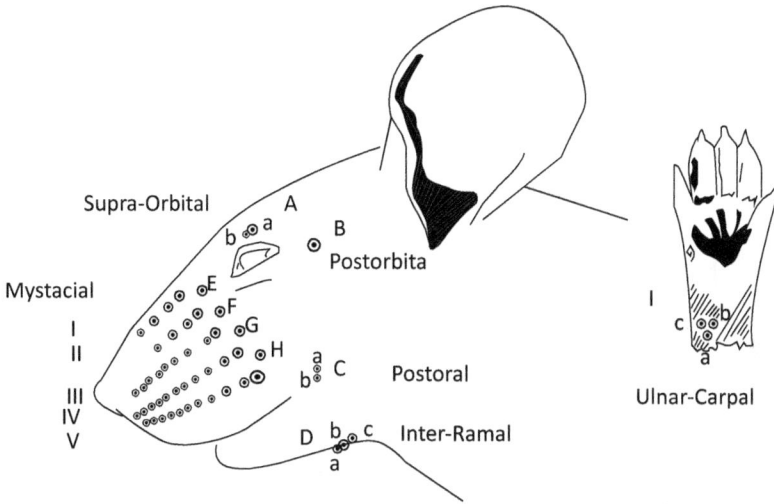

*Figure 8.2 Representation of the head and distal part of the right fore limb of the mouse showing distribution of vibrissae.* Reproduced from: R.B. Dun, A.S. Fraser, "Selection for an invariant character, 'vibrissae number', in the house mouse", *Australian Journal of Biological Sciences*, 1959, 12, pp. 506–523. With permission from CSIRO Publishing.

then (2) carrying out an artificial selection procedure to increase or decrease the penetrance of the trait and/or its metric value. Dun and Fraser adopted this experimental design but substituted a genetic perturbation to Waddington's environmental perturbation. To do this, they used strains of mice with the *tabby* allele (located on the X chromosome), identified by Falconer in 1953,[71] which had a very marked decanalizing effect on the number of vibrissae: carriers of this allele have both a lower (from 8 to 15 on average) and much more variable number of vibrissae. It was then possible to select either for an increase or for a decrease in the vibrissae number. Dun and Fraser began their work in 1957 and, by early 1958, were able to publish their first positive results in *Nature*. In this first phase of their collaboration, they conceived canalization as a genetic property independent from the character on which it acts (i.e., *tabby* was seen as a genuine (de)canalizing gene). Their view was graphically translated in the sigmoid shape of the curve expressing the relationship between the genotype (in the form of a hypothetical "vibrissae substance") and the phenotype ("vibrissae number"; see Figure 8.3).[72]

Fraser and Dun's general interpretation, which Rendel adopted, can be summed up as follows:

(1) Vibrissae formation is dependent on a gene product or morphogenetic substance.

(2) The relation between the amount of morphogenetic substance (the genotype) and vibrissae number (the phenotype) is not linear but is sigmoid (i.e.,

Vibrissa Substance

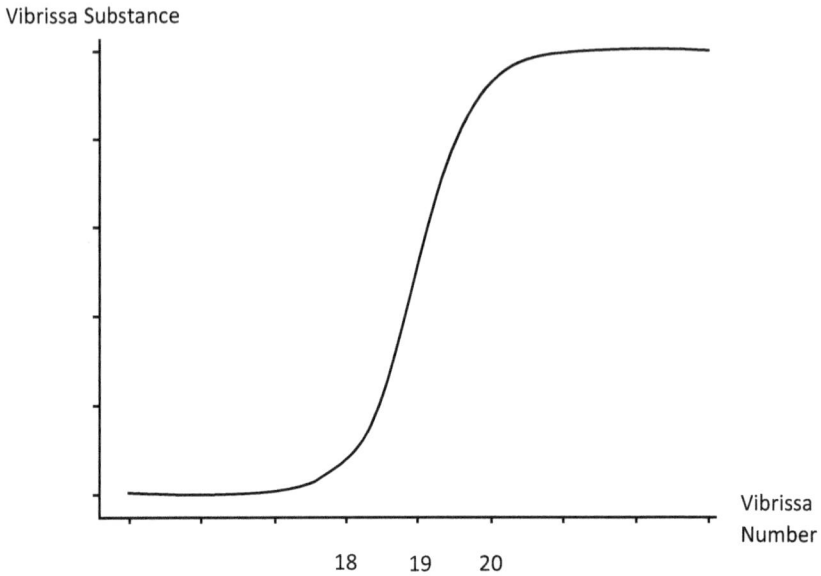

Vibrissa
Number

18      19      20

*Figure 8.3 Graphic representation of Dun and Fraser's account of developmental canalization.* The relationship between the genotype ("vibrissa substance") and the phenotype ("vibrissa number") takes the form of a sigmoid curve centered on value 19. This means that to vary this number, the amount of vibrissa substance must be radically altered. In the presence of the tabby gene, the curve loses its sigmoidal shape and becomes linear, reflecting the loss in canalization.

there is a region of the parameter space in which change in genotype has no phenotypic effect).

(3)  The sigmoid function is itself under the genetic control of canalizing genes, like *tabby*.

It was initially thought that in individuals without the *tabby* allele (++ for females, +. for males), the curve was strongly sigmoid, which meant that only a drastic variation of the amount of "vibrissae substance" would produce a phenotypic effect (i.e., would change the number 19). On the contrary, in individuals carrying a *tabby* allele, and in particular in males, the function became linear, which made the developmental system much more sensitive to the amount of vibrissae substance. At the end of their first article, Dun and Fraser did not hesitate to conflate *tabby* with a "major gene mutation", able "to break down developmental homeostasis and so expose concealed genetic variation to natural selection".[73]

In late 1958, they submitted a much more detailed study which appeared in 1959 in the *Australian Journal of Biological Sciences*. In large experimental populations, they established two lines of selection: an HST (High Selection) line in which the selection pressure was for an increased number of vibrissae and an LST (Low Selection) line where it was the opposite. To avoid fixing the *tabby* allele in the populations, they applied a precise mating scheme that allowed them to keep the + allele[74] while having a selective action (positive or negative) on the alleles determining the quantity of vibrissae substance (at that time they still thought that *tabby* had no direct effect on the amount of vibrissae substance). Again, they obtained positive results in individuals carrying the *tabby* allele, whereas wild-type genotypes showed almost no response to selection (Figure 8.4).

In 1958, they concluded that canalization consisted of the sigmoid relation that linked the phenotype to the genotype and that this function was somehow independent of the specific value of the metric trait itself. However, this time they were already aware of the first results Rendel had obtained with the *scute* allele in *Drosophila* (see the next section). Rendel's results largely converged with theirs, but Rendel's interpretation differed in one significant respect: for Rendel, as we shall see, the *scute* allele (the *tabby* equivalent) was not an allele that acted on canalization but instead, it acted directly on the phenotype under selection. In other words, it did not change the shape of the function linking the phenotype to the genotype (it remained a sigmoid) but shifted the value of the morphogenetic substance along the sigmoid curve. This alternative hypothesis was represented graphically by Dun and Fraser as follows (Figure 8.5).

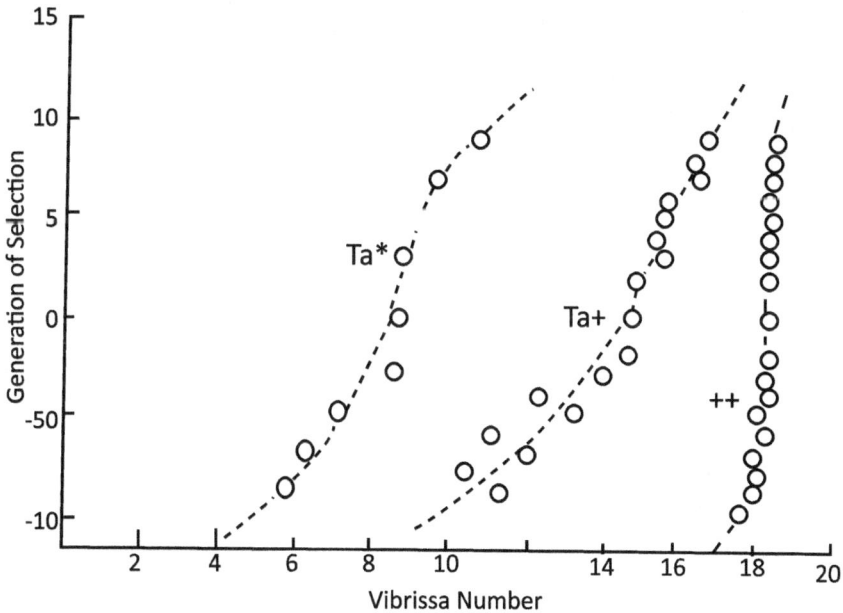

*Figure 8.4 Mean vibrissa number plotted against generations of selection for the ++, Ta+, and Ta genotypes.* Reproduced from: R.B. Dun, A.S. Fraser, "Selection for an invariant character, 'vibrissae number', in the house mouse", *Australian Journal of Biological Sciences*, 1959, 12, pp. 506–523. With permission from CSIRO Publishing.

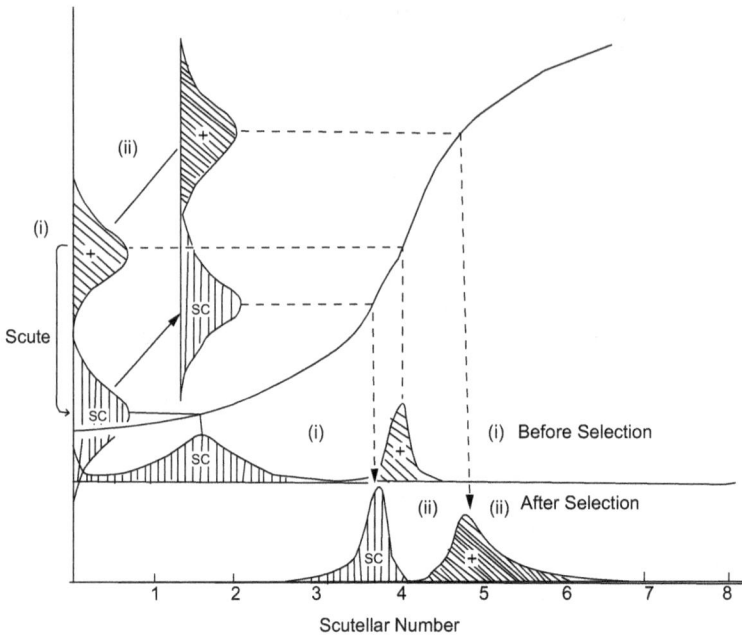

*Figure 8.5 Dun and Fraser's 1959 representation of the results and interpretation of Rendel's selection experiment.* According to them, "Scute is postulated not to affect the sigmoid relationship of genotype/phenotype as has been proposed for the tabby gene but rather to directly reduce the potency of the basic 'scutellar number' genes. The genetic distributions are shown vertically; the phenotypic distribution horizontally". Reproduced from: R.B. Dun, A.S. Fraser, "Selection for an invariant character, 'vibrissae number', in the house mouse", *Australian Journal of Biological Sciences*, 1959, 12, pp. 506–523. With permission from CSIRO Publishing.

In any case, Dun and Fraser admitted that the interpretation of the results obtained was very complex and that, at that time, the experimental data did not allow for a decision to be made between their interpretation, that *tabby* was a canalizing gene, and Rendel's interpretation that it was not.

### 8.2.3  Rendel: toward a molecular and quantitative theory of canalization

It would not be an exaggeration to read the whole of Rendel's scientific work as building upon the concept of canalization. In his conceptual deepening, he both developed experimental designs to demonstrate the evolution of canalization under artificial selection and constructed theoretical models to account for canalization in the most contemporary terms of molecular biology. Note that this distinction reflects the unequal posterity of his work: while his experimental results would remain important and known for decades and still are today, most of his theoretical proposals (and the vocabulary that accompanied them) rapidly fell into oblivion after the 1970s.

Around the time Rendel was appointed Chief of the Division of Animal Genetics of the CSIRO and elected to the Australian Academy of Science, he also produced what would, for years, constitute the most significant results on the reality, nature, and evolvability of canalization. In 1959 he published the first of a series of important articles, entitled "Canalization of the *Scute* Phenotype of *Drosophila*" in *Evolution,* a work which culminated in his 1967 reference book, *Canalisation and Gene Control.* In contrast to Robertson, Falconer, and others, Rendel, like Fraser and Dun, made his link to Waddington's theorization explicit from the start.[75] Since the approaches which applied normalizing selection to experimental populations had proved disappointing, Rendel opted for a different procedure, which was close to the one used by Dun and Fraser. By means of specific alleles which appeared to have a decanalizing effect, it was possible to make a normally constant trait highly variable, and then, by directional selection, to substantially alter its mean value. This is what Rendel did in *Drosophila* by focusing on the number of bristles carried by the scutellum, a part of the fly's thorax. All the experiments of artificial selection performed by Rendel and his collaborators from 1959 onward are very refined and relatively complex. I concentrate here only on their essential principles and results. The results obtained argued in favor of the following points:

(1) The canalization of some phenotypic traits is rather strong: the normal value of traits is buffered at a relatively circumscribed level.
(2) Canalization is not independent of the value of the quantitative trait that is canalized.
(3) Canalization is itself genetically controlled and can therefore be changed.

Normally, the number of bristles carried by the thoracic scutellum is almost invariably fixed at 4, with very rare exceptions. However, the presence of certain alleles of the *scute* locus (like $sc^{sc}$) leads on the one hand to a much lower average number and on the other hand to a much greater variability within the experimental population, exactly like *tabby* does.

Directional selection can then be practiced to increase the number of scutellar bristles, and this is how Rendel started his entire research program. Here again, the selection applied to *Drosophila* was successful: there was a progressive increase in the number of bristles. How was developmental canalization concerned by this type of experimental design? While it was quite easy to increase the number of bristles from 1 (in males) or 2 (in females) to 3, then from 3 to 4, it was much more difficult to exceed this number, which corresponded to the value of the wild type.

To go beyond the wild phenotype "4 bristles" required prolonged selection, but once the number of 5 was reached, there was again an almost linear progression according to the time during which directional selection was applied. Importantly, Rendel was able to design a complex method to quantify the genetic change required to bring about successive modifications in bristle numbers and was able to show that to move from 3 to 4 took about eight times as much "genetic change" as it did to move from

1 to 3.[76] The existence of a plateau of phenotypic invariance around a specific value showed that the "4 bristles" trait was canalized, that is, could resist directional selection. These results formed the basis of Rendel's ideas about canalization, which he would patiently pursue for nearly a decade.

The next step in this research program was to canalize a phenotype that differed from the wild type, which would show that canalization was itself under genetic control and, as such, evolvable. This was achieved a few months later, when Rendel and his collaborators succeeded in canalizing the "2 bristles" phenotype, that is, in considerably reducing the intra-population variance of this trait by artificial selection.[77] To do this, they did not return to a normalizing selection procedure like Falconer and others tried to do in the mid-1950s. Instead, following two experimental lines, the "Low variance line" and the "High variance line", the phenotype was kept constant at around 2 bristles, and parents were chosen in the cultures either with the lowest variance or with the highest one.[78] This form of artificial selection continued for 28 generations. In the Low variance line, they found a decrease in the variance of the phenotype but with a slight difference between males and females (Figure 8.6).

Nevertheless, as we have seen, a decrease in variance was not a sufficient argument for an increase in canalization. Therefore, as the selective process proceeded, from the 24th generation, some eggs were incubated at varying temperatures (15°C, 20°C, and 30°C) to assess the degree of environmental canalization achieved. It was shown that the change in temperature affected both the Low variance and the High variance lines but that individuals from the latter showed three times the variability of those from the former.[79] All in all, the general degree of canalization of this trait was artificially produced,

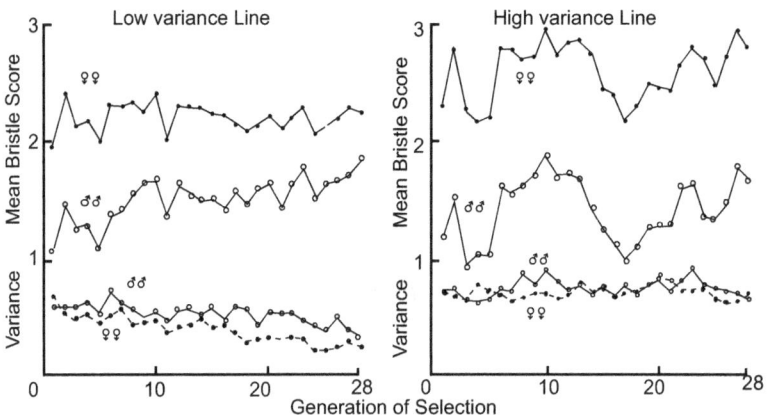

*Figure 8.6 Means and variances of the selection lines plotted against the generation of selection.* Reproduced from: J.M. Rendel, B.L. Sheldon, "Selection for canalization of the scute phenotype in Drosophila melanogaster", *Australian Journal of Biological Sciences*, 1960, 13, pp. 36–47. With permission from CSIRO Publishing.

and the results obtained could not be explained by a simple allelic homogenization of the population under selection:

> Whatever the explanation of the discrepancy between the reduction of variation within lines and their sensitivity to temperature, the selection experiment has shown that variation can be reduced by selection and that this reduction of variation is not accounted for by homozygosis. Further, invariability at the temperature at which selection was carried out has been accompanied by a surprisingly large increase in insensitivity to changes in temperature. The invariability produced by selection, therefore, is general and not specific to the variation brought about by a particular set of genes. Finally, invariability at one temperature has resulted in invariability at all temperatures in which the flies were tested.[80]

Rendel's results were impressive and were acknowledged as such. For example, Prout considered that of all the work done in the past decade or so, only Rendel seemed to show conclusively that canalization could be controlled by artificial selection procedures.[81] Rendel's results also participated in the diffusion of this type of experimental design based on decanalizing effects of specific alleles, which was sometimes used by other colleagues, including, for instance, Maynard Smith.[82] Note that although Waddington did not elaborate on the subject, he accepted the demonstrative nature of the work done at the CSIRO.[83]

Because of Rendel's results, Fraser himself doubted that *tabby* was a canalizing gene and finally had to give up on the idea. In collaboration with Kindred, they show that, exactly like *scute*, the strength of canalization was not controlled by the *tabby* gene but was determined by the phenotypic value of the character "vibrissae number": even in the presence of *tabby*, the action of the directional selection was more and more inoperative when it approached the normal value of 19 vibrissae. They thus concluded (at this time, around 1960, it seems that Dun was already no longer involved):

> Clearly, the original hypothesis, that the *tabby* gene reduces or removes canalization is not valid, since the above data demonstrate that the zone of canalization at 18–19 vibrissae can be detected in *Ta+* mice when their mean vibrissa number is sufficiently high. The correct explanation is that the *tabby* gene reduces vibrissa number to a point where this is not affected by the zone of canalization, i.e. the explanation suggested by Rendel (1959).[84]

It is necessary here to clarify a point that has been a source of confusion in the scientific literature,[85] and still is today.[86] Fraser and Dun's models, on the one hand, and Rendel's, on the other hand, are equivalent from a formal point of view. Both equate canalization with a non-linear relationship that links the genotype to the phenotype. Both postulate that this function is genetically controlled; in other words, there must be canalizing genes. What differed between Fraser's and Dun's initial interpretation and Rendel's was that the former believed that they had shown that *tabby* was one of these canalizing genes (i.e., that *tabby* acted upon the sigmoid form of the function: decanalization is then the consequence of the loss of this sigmoidal portion), whereas Rendel hypothesized from the outset that both *scute* and *tabby* were more likely to be directly involved in the realization of the phenotypic trait under selection (i.e., *tabby* acted on the amount of "vibrissae substance": decanalization is then the consequence of a shift outside the sigmoid zone, while the shape of the function remains intact). In short, while their models were equivalent, the status of the *tabby* gene was not.

Unlike Waddington, Rendel showed extraordinary theoretical and experimental perseverance in exploring the genetic basis of canalization. His 1959 work initiated a vast research program whose results were collected in the only book he published, *Canalisation and Gene Control* (1967). It is not possible here to review all the work he did at the CSIRO during the period 1960–1967 in detail, so I will confine myself to a summary of this book, which, as Rendel's own former colleagues agreed,[87] provides a very reliable view of his definitive conceptions on the genetic nature of canalization.

While Rendel's book provides a reliable summary of his conceptions, it should be pointed out that like Lerner's, it is deceptively simple and contains a rather considerable number of heterodox hypotheses in more or less distant dialogue with a wide range of experimental results. As odd as it may seem, what Rendel proposed was a sort of broad synthesis of the hypotheses put forward successively by Waddington (concerning the centrality of developmental canalization), Mather (concerning the oligogene/polygene distinction), Fraser (concerning the identification of canalization to the form of a non-linear function linking the genotype to the phenotype) and Jacob and Monod (concerning the process of genetic regulation involving two classes of genes, namely "regulatory genes" and "structural genes"[88]). Rendel's theoretical speculations, unlike those of his experimental results, apparently did not have much posterity.

Rendel conceived of each developmental pathway (each chreod in Waddington's terminology) as being primarily under the control of a "major gene"[89] (Mather's oligogene[90]), which was supposed to be a "structural gene" in Jacob's and Monod's sense,[91] itself regulated according to the mechanism promoted by the operon model.[92] To this, he added "a set of minor genes [Mather's polygenes] which are not controlled and have the same effect in kind as the major gene but a smaller one".[93] It should be noted that the status of the genes that operate the regulation is not perfectly clear. They cannot be "minor genes" in the sense of Mather (polygenes), as this would not be consistent with the operon model. However, Rendel never called them "major genes" either. On some occasions, he enigmatically called them "the genotype which is responsible for the control mechanism".[94] Rendel probably thought of these "regulator genes"[95] as major genes, like the structural gene responsible for a developmental path. He made sure to clarify that the "minor genes" he introduced in his theoretical framework had no equivalent in the operon model.[96]

It seems that Rendel identified canalization as a portion of the parameter space in which the control exerted by regulatory genes on the phenotype was particularly effective, which was graphically reflected by the plateau phase.[97] From an experimental point of view, all of Rendel's efforts were directed toward demonstrating the existence of this plateau, bound by a double threshold, when considering the phenotypic variation of a quantitative trait as a function of the quantity of the morphogenetic substance. Unlike Fraser, Rendel chose to plot the amount of morphogenetic substance on the x-axis and the corresponding phenotypic variation on the y-axis. The result was no longer a sigmoidal function but a graphical representation showing the existence of a plateau materializing the canalization of the developmental pathway under study. Rendel called this function "$\phi$". In Figure 8.7, $\phi_1$ shows a linear relationship, in the absence of canalization, while $\phi_2$ shows a double threshold characteristic of the existence of a certain amount of developmental canalization.

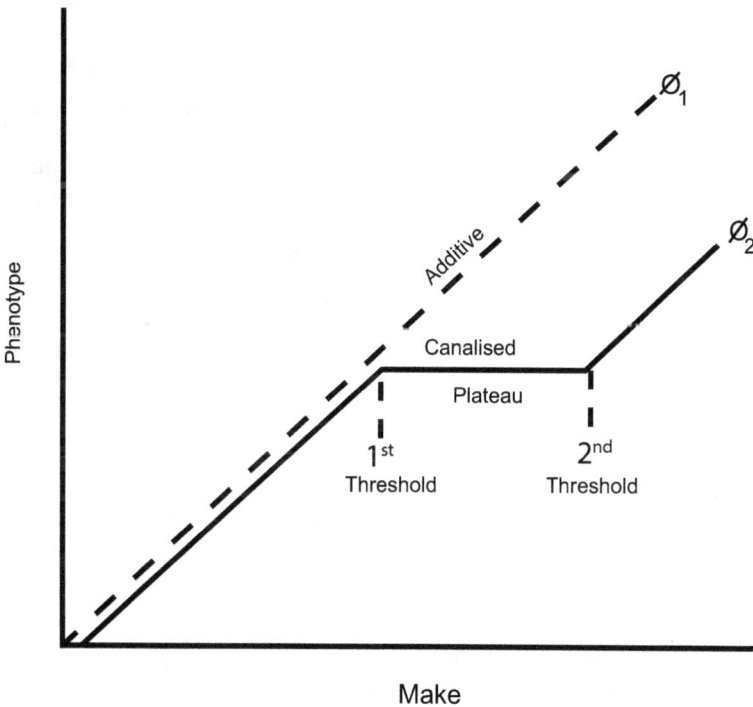

*Figure 8.7 Phenotype plotted against the variable "Make".* $\phi_1$ shows an additive relationship between the amount of "Make" and the value of the character, whereas $\phi_2$ shows a canalized relationship. Reproduced from: J.M. Rendel, *Canalization and Gene Control*, London, Logos Press & Academic Press, 1967. With permission from Elsevier.

In what can be seen as a genuine genetic theory of canalization, "M", that is, "Make", was supposed to represent, in a general way (i.e., without any necessary relation to a particular experimental system such as the *scute* system in *Drosophila*), a certain quantity that could be characterized by its causal action on the phenotype (e.g., the previous "vibrissae substance"). Rendel did not want to get lost in the concrete study of complex causal paths and therefore considered "M" as epistemologically equivalent to the concept of force in physics.[98]

φ was itself supposed to have a genetic basis and could therefore be altered by natural selection, as some results had shown (in his rather short list, Rendel did not include those of Waddington on the sensitivity of *Bar* mutants to temperature,[99] which he judged not to be demonstrative[100]). With the benefit of hindsight, Rendel finally concluded that only his own work and that of his collaborator Berenice Kindred (1928–1985)[101] showed with certainty that φ could have been modified by means of artificial selection protocols specifically designed for this purpose.[102]

Quite unexpectedly, Rendel ended his book with overtly macroevolutionary considerations. On the one hand, Rendel considered that the degree of global canalization of a developmental system was a function of its paleontological ancestry. This was why the fly showed a much more buffered morphology than the human species.[103] On the other hand, the most important consequence of canalization on the macroevolutionary scale was perhaps the alternation it imposed between phases of evolutionary stability and brief moments when, due to major environmental perturbations, "great evolutionary change"[104] was possible. On the basis of a very elaborate theory of developmental canalization, Rendel thus finally came to anticipate the general pattern that Eldredge and Gould famously called "punctuated equilibria" in 1972. It should be noted here, and this may not have been yet fully appreciated, that Eldredge and Gould themselves chose to conclude their long 1972 text by explicitly linking their paleontological model to the fact that species and individuals are "homeostatic systems", that is, "amazingly well-buffered to resist change and maintain stability in the face of disturbing influences".[105] Although neither Waddington nor Rendel was cited on this occasion, Eldredge and Gould explicitly referred to Lerner's book, *Genetic Homeostasis*, and we have seen the structuring role it played regarding these issues in the 1950s and 1960s (see also Section 9.3.2).

## 8.3 Genetic assimilation as threshold selection: on a series of long-lasting conceptual misunderstandings

At the same time that work on developmental canalization was gaining momentum, other researchers, notably in Edinburgh, focused on the other side of Waddingtonian speculations: the microevolutionary process of genetic assimilation, its causality, and its genetic assumptions. This line of research, however, was not comparable to that on canalization since it interested fewer biologists and gave rise to only two main series of experimental works (that of Bateman and that of Milkman), whereas, as we have just seen, several experiments were performed on canalization. However, although less developed, part of this work on genetic assimilation had a remarkable – and problematic – legacy in that it convinced most evolutionary biologists that the mechanism of genetic assimilation did not require the concept of canalization, as Waddington had initially proposed. This judgment, which is regularly found in the subsequent literature, must be strongly revised. On the one hand, a central hypothesis of the present book is that there were two concepts of genetic assimilation in Waddington's work, GA1 and GA2, which coexisted in a more or less complex way. And in my opinion, the threshold selection models developed during the 1950s were a perfectly legitimate interpretation of GA1 only, leaving the issue of the possible genetic conceptualization of GA2 unsolved. On the other hand, as we will see in the present section, even within such threshold selection models, canalization still constitutes an unavoidable dimension of the explanation.

### 8.3.1 *Stern's qualitative reasoning*

American-born German geneticist Curt Stern (1902–1981) was one of the first scientists to propose, independently of K.G. Bateman (see the next section), that genetic assimilation, as a *phenomenon*, could be reinterpreted in terms of a threshold selection model in which canalization apparently played no obvious role. Stern was one of the leading geneticists of the mid-20th century and was influential because of his active participation in the final phase of the emergence of genetics as a discipline.[106] Initially a specialist in *Drosophila* genetics, before turning to human genetics and writing leading textbooks,[107] he was also the editor of *Genetics* from 1947 to 1951. At the time he became interested in the question of the links between canalization and genetic assimilation, Stern was himself active in the emerging discipline of developmental genetics, and, like Waddington, he sought to address the question of the genetic control of development based on work involving *Drosophila*.

As early as 1958, he published a short text in *The American Naturalist* in which he presented what was at the crux of his reasoning. It should be noted that, unlike Bateman, Stern's reasoning was only qualitative (a thought experiment, as it were), and was never backed by experimental research. Stern began by observing that to account for the genetic fixation of plastic adaptations (he used the Waddingtonian term "pseudo-exogenous adaptations") there were two competing mechanisms in the recent literature: the Baldwin effect, which seemed to him to be very unlikely (in the canonical form given to it by Simpson in 1953), and genetic assimilation.[108] In the causal scheme of genetic assimilation, the problem for Stern was the genetic fixation of the accommodation: why would organisms not continue to react indefinitely in a plastic and individual way to the demands of the environment? This seemed to him to be a weak point in Waddington's reasoning, and it was to remedy this that he proposed a model that did not seem to require canalization but was entirely based on the idea of "threshold selection". Stern put it as follows:

> It seems to me that a *single* cause may underly the origin of pseudo-exogenous adaptations. If, in a population, genes are present which produce a certain phenotype in *both* environments A and B, and if selection for these genotypes can be accomplished initially in environment B only, then, by definition, selection in B will accomplish production of the trait in both A and B. The possibility of existence of the postulated genotypes is a matter of common genetic experience. So is the existence of genotypic differences which remain phenotypically indistinguishable in one environment but can be discerned in another. We may speak of them as subthreshold differences in A and supra-threshold differences in B.[109]

Stern's speculative reasoning was in essence GA1: he assumed that the same genes were responsible for both the plastic reaction and the final evolutionary adaptation. It meant that if these genes were revealed and thus selected for in a given environment B, they could be accumulated to a point in which they would allow a constant modification, even in A. In 1959, for the commemoration of the centenary of the publication of *The Origin of Species* in the *Proceedings of the American Philosophical Society*, Stern reiterated his critique and his alternative explanation. For him, while the concept of canalization had played a heuristic role in the way Waddington had set up his experiments (context of discovery), the mechanism of genetic assimilation could be accounted for by a simpler process, that of threshold selection (context of justification).[110]

Because of Stern's academic influence, Waddington was forced to take a stand, and in this lies the main value of Stern's brief foray into these debates. Waddington's response is informative because it already signaled the ambiguity of the threshold selection mechanism – that is, GA1 – with respect to the concept of canalization. Waddington did not oppose Stern's explanation head on, nor did he consider it irrelevant. But he considered that in the form that Stern had given it, the mechanism of genetic assimilation (at least that of GA1) was "oversimplified",[111] mainly because the concept of canalization was missing. How to account for the fact that the same genes have, in environment A, an action below the threshold of expression and, in environment B, an action above this same threshold, if not by using the concept of canalization and the decanalization induced by the passage from A to B? This was essentially Waddington's response to Stern (and, in my opinion, it was a valid one).

### 8.3.2 *Bateman's experimentally based model and its ambiguous legacy*

At about the same time that Stern was speculating on his threshold selection hypothesis, the same idea came to the forefront in a series of experiments in Edinburgh. These experiments were performed by K.G. Bateman, who in 1956 had defended a thesis on genetic assimilation under Waddington's supervision.[112] There are in fact very few resources on Bateman's life and career.[113] Most likely, she remained in Edinburgh as a postdoc after her PhD, and it was in this context that she carried out a series of experimental works that have remained famous and that were published in 1959 as two articles in the *Journal of Genetics*.

Bateman's body of work can be understood as an experimentally rigorous expansion of Waddington's earlier work on the genetic assimilation of venation phenocopies ("crossveinless"). She used the same experimental system (*Drosophila* wing structure, with four identified phenotypes: "pcvl", "acvl", "fpcv", and "smcv"), the same experimental procedures (application of a 40°C heat shock during pupation to reveal genetic standing variation), and the same selective regimes (at first selection with treatment and then selection without treatment). The results she obtained with different strains were spectacular: in only a dozen of generations, the phenotypes under selection became the majority in the experimental populations while their frequency had been almost zero[114] at the beginning.

Bateman had two main starting points. First, she wanted to know whether genetic assimilation was based on standing variation (genetic assimilation in Waddington's sense) or whether environmental treatment produced new mutations that were indispensable for the genetic fixation of the desired phenotype (a hypothesis closer to the Baldwin effect and, for instance, Naumenko's 1941 work (see Section 3.1.3)). Second, she also wanted to clarify the question of the genetic basis of the final adaptation: was it distributed in a large part of the genome, as Waddington thought, or were some specific major genes indispensable?

To answer the first question, highly inbred lines (with a very weak level of standing variation) were subjected to the same treatment and no genetic assimilation was observed (i.e., there was no response to selection).[115] It seemed that without initial genetic polymorphism, the mechanism was inoperative, which went against the Baldwin effect hypothesis. The second point was more complex to address and required thorough genetic analysis of the different populations in which genetic assimilation had been performed. Bateman made several reciprocal crossbreeding in both assimilated and unselected stocks to clarify the issue. Her first results remained ambiguous, because even though one chromosome seemed to play a key role, the others were also involved in the process of assimilation. Despite the ambiguity of the results obtained, she believed she was justified in concluding this first series of works in opposition to Waddington's views. For her, there was little doubt that "the assimilated characters are not controlled merely by an indefinite number of genes of individually small effect, i.e. are not 'polygenic' ".[116]

It was primarily to provide a definitive answer to this question that Bateman embarked on a second[117] series of experiments on the "dumpy" phenocopy. The *dumpy* phenotype consisted of a significantly reduced wing compared to the wild type. Bateman hypothesized that the assimilation of this phenotype required the presence of a specific allele, in this case, $dp^{TP2}$. It was therefore possible to test this hypothesis directly by producing strains in which this allele was absent, and by applying the standard treatment/selection procedures that usually lead to genetic assimilation. The results were very clearly in favor of the idea that there was a need for a "major gene".[118] In addition, it was likely this time that the treatment applied to *Drosophila* pupae resulted in *de novo* production of this mutation.[119] In short, Bateman's results ran counter to Waddington's ideas on these matters.[120] Once again, therefore, the work done in Edinburgh, even when it was the work of Waddington's students, was far from endorsing the director's views.

The question of the genetic basis of the crossveinless phenotype did not end with Bateman's work (Waddington, for his part, did not work on it directly after the publication of his seminal 1953 paper). Clarifications on this aspect came mainly from the efforts of population geneticist Roger D. Milkman (1930–2011). As early as his doctoral work, defended in 1956[121] (like Bateman[122]), he had embarked on the Mendelian analysis of the genetic basis of the crossveinless phenotype. This led to numerous experiments, both to assimilate the phenotype in different populations and, above all, to elucidate its genetic basis. It is not possible in a few lines to do justice to the considerable experimental work that was accomplished by Milkman in over a decade.[123] He succeeded in developing extremely complex and refined Mendelian procedures to precisely identify the loci involved in the process of assimilation.[124] I merely point out here that, overall, and at least for this particular phenotype, his results proved Waddington right: several genetic determinants distributed over all the chromosomes seemed to be necessary for the fixation of the crossveinless phenotype. Moreover, the crossveinless phenotype could be obtained from very different genetic bases depending on the strain used.[125] Strangely enough, while Waddington referred extensively to Bateman's results, he cited very little of Milkman's results[126] which remain, to this day, poorly known.[127]

As far as Bateman was concerned, her main legacy was more theoretical than experimental. In her 1959 paper on the genetic assimilation of the crossveinless phenotype, she developed a threshold explanation that very rapidly became the one favored by most evolutionary geneticists. It is in the "General Discussion" section that she expanded on the mechanism of genetic assimilation and assumed a double threshold of expression of the phenotypic character: a threshold corresponding to a "normal" environment, and a second threshold, easier to reach, when the developmental environment is strongly perturbed. This is how Bateman summarized her explanatory hypothesis:

The relation between the phenocopy, assimilated crossveinless, and "spontaneous" crossveinless, depicted in terms of the changing position of a continuous distribution relative to fixed thresholds for the expression of the character in the normal and in the treated environments, is given in Figure 12. Distribution 1 represents the situation prior to selection. Under normal environmental conditions the threshold for the character is practically outside the range of variation shown by the population so that only a minute proportion of individuals are "spontaneously" crossveinless. The effect of treatment is to lower this threshold so that a good part of the population falls above it. Selection in this environment shifts the population mean (distribution 2) so that an increasingly large area of the distribution falls above the "treatment" threshold, and *pari passu* an increasingly large part of the tail of the distribution falls above the threshold for the normal environment. These are the assimilated genotypes. Selection of assimilated individuals ultimately yields an entirely crossveinless population (distribution 3).[128]

More importantly, she illustrated her explanation with a synthetic diagram showing the effect of the selection allowed by the lowering of the expression threshold of the trait of interest (Figure 8.8). This diagram has had a

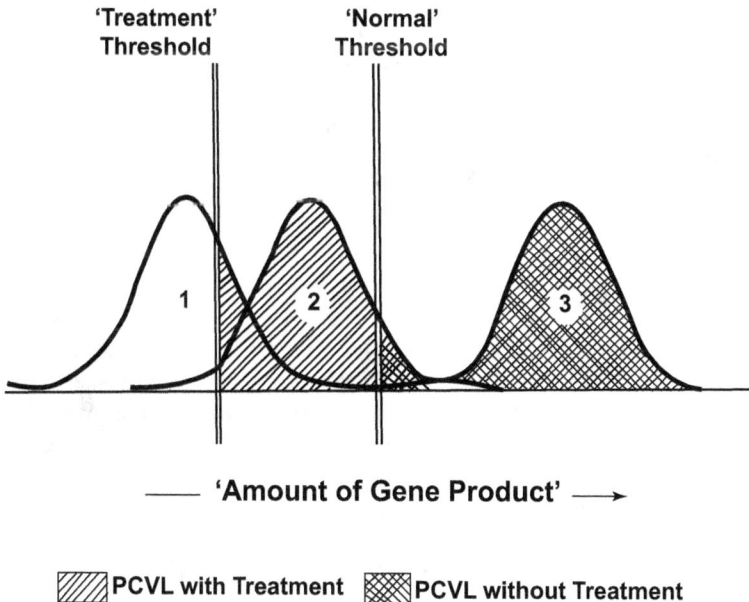

*Figure 8.8 Diagram illustrating the model of threshold selection designed by Bateman to provide a quantitative explanation of genetic assimilation. Reproduced from: K.G. Bateman, "The genetic assimilation of four venation phenocopies", Journal of Genetics, 1959, 56, pp. 443-474. With permission from Springer Nature BV.*

considerable legacy, up to the present day, especially since it was reprinted identically in Falconer's influential textbook,[129] *Quantitative Genetics*, from 1960 onward.[130] From that time on, many geneticists and evolutionary biologists did not think it necessary to include the concept of canalization in the explanation of the mechanism of genetic assimilation. It seemed that threshold selection, as envisioned by Bateman or Stern, offered a simpler and therefore preferable mechanism to account for the same process.

It is surprising that so many important evolutionary geneticists – such as Stern,[131] Robertson,[132] and Scharloo[133] – subscribed to the idea that Bateman's model, or the explanations that could be derived from it, no longer required the concept of canalization. In fact, Bateman's model, which was a slight reformulation of GA1, still does and Bateman herself was fully explicit on the issue. What her diagrammatic representation changed was the scale considered: where Waddington, as a developmental biologist, favored a representation at the scale of the individual organism, Bateman represented the same phenomenon at the scale of a Mendelian population. This time the degree of canalization was no longer represented by the depth of the valleys, like in Waddington's epigenetic landscape but by the spacing between the two thresholds of phenotype expression:

> The process that has brought this about [i.e. genetic assimilation] can be represented in either of two ways. At the level of the population it can be shown as a change in the position of a normal distribution relative to fixed points on some underlying scale; at the level of the individual it can be symbolized, using topographical concepts, as the progressive cutting-away of a ridge separating two valleys representing the normal and an alternative direction of development. The "thresholds" in the two cases can be identified. In the first the value of the threshold is lowered relative to that of the mean genotype of the population; in the second, this lowering of the threshold is represented by the gradually decreasing height of the ridge.[134]

Thus, the diagram proposed by Bateman to illustrate her threshold model did not negate the role of canalization; it simply transformed its visual representation. As the history of the concept of genetic assimilation shows, pictorial representations can have a major impact on the course of conceptual history, and Bateman herself was probably partially misled by her own diagram. Indeed, Bateman chose to represent decanalization by shifting the phenotype expression threshold from right to left (Figure 8.8). This choice is problematic because it led her to leave unchanged the variance of the successive distributions (stages 1, 2, and 3). However, given that the x-axis represents the amount of gene product (i.e., a quantity), a more suitable representation of a decanalization episode would have been to add a second distribution with the same mean but showing a larger variance, which leads to a larger portion of the population to develop the crossveinless phenotype (Figure 8.9,

Fixed threshold of
phenotypic expression

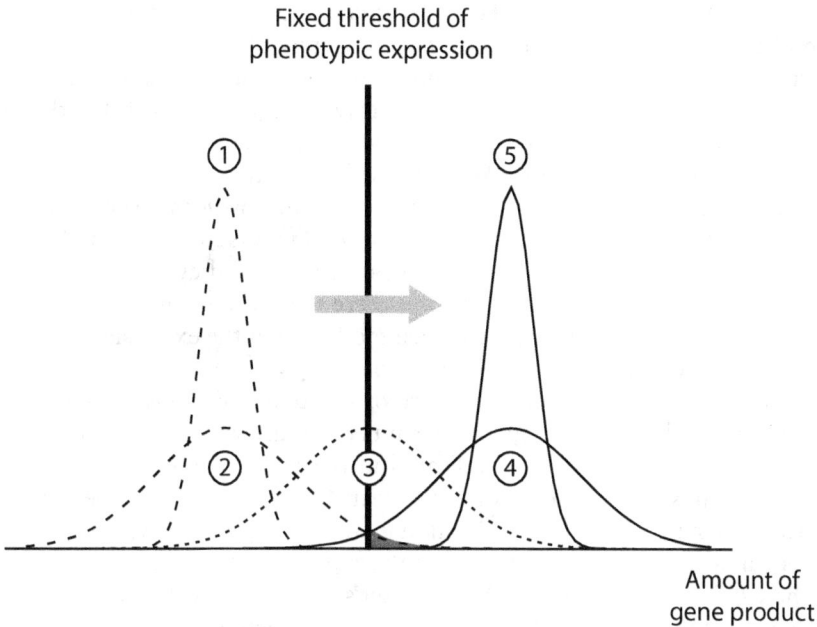

Amount of
gene product

*Figure 8.9 Alternative diagram illustrating the model of genetic assimilation at the population level (GA1). In this representation, the phenotype expression threshold is fixed, which is consistent with the fact that the x-axis represents a quantity (the amount of gene product). The loss of canalization (due to environmental stress) results in an increase in variance within the population (stage 2), which causes a sufficient proportion of individuals (grayed section) to develop the phenotype to give rise to selection (stages 3 and 4). In such a representation, it is conceivable that the continuation of selection could allow the recanalization of the genotype (stage 5) so that the totality of individuals is now beyond the threshold value, even in the absence of an inducing shock. This diagram is a population-level representation of the process pictured by Waddington at the level of a single developmental system.*

striped section). This diagrammatic representation of the phenomenon also has the advantage of making it possible to represent a terminal stage of variance reduction. This recanalization stage is in fact difficult to represent in the diagram proposed by Bateman, and this is precisely another aspect where she explicitly opposed Waddington.[135] For her, there was no need for genetic assimilation to end with what Waddington called the "tuning" of canalization.

## 8.4   Chapter's conclusion

If this chapter is especially extensive,[136] it is because Waddington's (and, to a lesser extent, Schmalhausen's) speculations on stabilizing selection, canalization, and genetic assimilation experienced a substantial reception within English-speaking quantitative genetics, which was in full expansion during the years 1950–1960. This is far from trivial and must be stressed. In both the historiography and contemporary philosophy of biology Waddington is portrayed as an isolated thinker, fighting the genocentrism and reductionism of post-war biology alone.[137] Such a portrayal must be revised. On the one hand, as we saw in Chapter 5, Waddington's conceptions, whatever intellectual postures he liked to take, remained for many of them genocentric. On the other hand, as I have just shown in detail, his ideas were discussed for at least two decades, and gave rise to several works involving mathematical models, multiple artificial selection procedures, and even the first available numerical simulations. Therefore, there is no doubt that Waddington, directly or indirectly, participated (even modestly) in the expansion of a part of quantitative genetics during about two decades.

Were these works faithful to the spirit of Waddingtonian ideas, often elaborated in a strictly verbal or figurative manner but rarely formalized or detailed by Waddington himself? As I indicated in the introduction to this chapter, this question is difficult and requires a clear distinction between the different concepts at stake. As far as canalization is concerned, as we have seen, several genetic models were proposed between the end of the 1950s and the beginning of the 1960s, whether it be Mather's, Fraser's, or eventually Rendel's, the most elaborate of all. Whatever the vocabulary these models used, most of them put forward a very strong, almost ontological distinction between two classes of genes (oligogenes/polygenes, major genes/minor genes, etc.). Clearly, this type of hypothesis, especially when it was rooted in the most recent molecular biology (the operon model of Jacob and Monod), was far removed from the spirit of Waddington's ideas. Waddington, most of the time (yet not systematically), saw canalization as an emergent property, linked to the systemic functioning of the genome where epistatic relationships were the rule. This was not, overall,[138] the direction his contemporaries chose to explore.

As far as genetic assimilation is concerned, things are probably simpler. On the one hand, because GA2 was never considered or discussed: the most heterodox part of Waddington's speculations went completely unnoticed during this period. On the other hand, as I hope to have demonstrated, and despite a persistent misunderstanding, there was indeed a significant conceptual convergence between GA1 and the threshold selection model in the canonical form developed by Bateman in the late 1950s. In this case, at least, the implementation of Waddington's concepts within quantitative genetics was rather faithful to the initial intentions.

There is a third point to be made here. While Waddington had a stimulating effect on quantitative genetics for nearly two decades, it seems to have been interrupted rather abruptly in the early 1970s. For a long time, work on canalization and genetic assimilation was marginalized before Waddington's ideas regained some appeal in the second half of the 1990s. There was thus undeniably something of an eclipse of Waddington's biology, both on the side of "ultimate causes" (evolutionary) biology and on the side of "proximate causes" (molecular) biology.

## Notes

1 Robertson (1977, pp. 612–613).
2 Falconer had very harsh words to describe Waddington's renunciation, which had an impact on the whole scientific activity of the Institute of Animal Genetics in Edinburgh: "Possibly the prime reason was that Waddington's interests moved from genetics and the Institute to futurology. In 1970 he went for two years to the Center for Theoretical Biology at Buffalo, and when he returned he set up a new 'School of the Man-Made Future' in the University. In consequence he was seldom seen in the Institute from then till his death in 1975" (Falconer, 1993, p. 141).
3 This aspect is well documented in Grodwohl (2013, pp. 29–65).
4 Falconer (1993).
5 Robertson (1977); Falconer (1993). Surprisingly, to my knowledge, there is no historiographical work on the golden age of the Institute of Animal Genetics, that is, the 1950s and 1960s. The earlier period, when Francis A.E. Crew (1886–1973) founded and ran the institute for some 20 years, has already been explored: Marie (2004); Button (2018).
6 If no further precision, "Robertson" means "Alan Robertson".
7 Robertson (1977, p. 601).
8 In Chapter 5, we saw how fluctuating asymmetry was seen by some (but not by Waddington) as a reliable proxy for measuring the degree of canalization.
9 Mather (1942); Grodwohl (2013, p. 34).
10 In 1948, Lerner stayed in Edinburgh for several months and used this time to write most of his 1950 book *Population Genetics and Animal Improvement*, the contents of which were thus known and mastered by most young Edinburgh geneticists (Falconer, 1993).
11 For biographical details, see Allard (1996).
12 See, for instance, Lerner (1954, p. 119).
13 Falconer & Robertson (1956).
14 For instance, he discovered the *Tabby* gene (Bowman, 2005, p. 126), which will play a major role in the experiments designed by Dunn and Fraser on the decanalization of vibrissae number in mice (see Section 8.3).
15 Falconer (1957, p. 552).
16 Robertson (1956).
17 Robertson (1956, p. 247).
18 It is most probably because he could not make anything of Waddington's speculations that he turned away from them very quickly.
19 Robertson (1956, pp. 247–248).
20 Falconer (1957, p. 552).
21 Simpson (1944).
22 He also briefly considered Mather's hypothesis of "balanced" linkage relations between genes of opposite actions (see Section 8.3).
23 Falconer (1957, p. 555).
24 Falconer (1957, p. 557).
25 Prout (1962, p. 376).
26 Waddington (1960, p. 141).
27 Waddington (1960, p. 143).
28 Waddington (1960, p. 143).
29 Waddington (1960, pp. 146–147).
30 Waddington & Robertson (1966, p. 303).
31 Waddington (1966b).
32 The seminal paper is "The Genetical Theory of Continuous Variation" (Mather, 1949b).

33 Waddington & Robertson (1966, p. 307).
34 Waddington & Robertson (1966, p. 308).
35 He went as far as concluding that: "There is, therefore, no major generalization to be drawn from the results of the disruptive selections", Waddington & Robertson (1966, p. 311).
36 Grodwohl (2013, p. 35).
37 Mather (1943a, p. 38).
38 Mather (1941).
39 Mather (1943a, p. 45).
40 Mather (1943a, p. 62).
41 Mather (1943a, pp. 44–45).
42 Mather (1941, pp. 190–191).
43 Mather (1943a, p. 62).
44 Mather (1943a, p. 48).
45 Mather (1943a, p. 45).
46 Waddington (1942).
47 Mather (1943b).
48 Mather (1943b, p. 69).
49 Waddington (1943, p. 394). My emphasis.
50 "There is a true distinction between polygenic variation (determined by numerous genes) and oligogenic variation (determined by few genes); but this is certainly not a distinction between the kinds of genes involved, and need not correspond to a distinction between the modes of action of the genes during development" (Waddington, 1943, p. 394).
51 Before turning to population genetics, Mather trained in cytology with Darlington (Lewis, 1992, p. 252). He had thus a first-hand knowledge of this field.
52 Mather (1949a, pp. 19–20).
53 Waddington (1957a, pp. 52–53). Since 1943, Waddington thought that some clear-cut experimental results, like those of Goldschmidt on podoptera, argued against Mather's distinction.
54 Waddington (1961a, p. 283).
55 Mather (1953, p. 328).
56 Lewis (1992, p. 260).
57 Jacob & Monod (1961).
58 Mather (1990). This paper was the second part of a two-part study (Mather, 1987).
59 For biographical details, see Franklin et al. (2004). For an illuminating treatment of Rendel's theoretical framework, see Petino Zappala (2024) (to be published).
60 As one might expect, Waddington's supervision of his students was minimal, similar to his direct interactions with his colleagues in Edinburgh. This often led to delicate situations: "Many of the Ph.D. students came expecting to work under Waddington's supervision. But often he was away on his travels, and then the rest of us had to come to the rescue, hastily think up suitable projects, and find space and facilities for them" (Falconer, 1993, p. 141).
61 Usually Fraser's and Dunn's model and Rendel's are presented in the scientific literature, until today, as two different models. This is a misinterpretation that will be rectified later in this section and the following.
62 Fogel (2002).
63 Fraser (1960).
64 "Fraser (1960) has shown how the conditions for the evolution and maintenance of genotypes determining canalized paths of development can be investigated by Monte Carlo methods on a digital computer. This method seems likely to become a powerful tool for studying the complex theoretical structure of

such situations, but so far no very striking results have appeared" (Waddington, 1961a, pp. 283–284).

65 Fraser (1960, p. 150).

66 It seems that Robertson was never really interested in epistasis throughout his work, for reasons that remain to be known (Hill, 1990, p. 467).

67 Most probably, only an archival work could allow settling the question of the paternity of the class of models that I detail here.

68 See, for instance, Thomas et al. (2016, pp. 443–445).

69 Dun is born in 1930 and might still be alive in 2024.

70 Dun & Fraser (1958, p. 1018).

71 Falconer (1953).

72 Many of their graphs would have been easier to understand if Dun and Fraser had reversed their axes and chosen to represent the amount of "vibrissae substance" on the x-axis instead of the y-axis. They most likely chose to measure the amount of "vibrissae substance" on the y-axis to obtain a sigmoidal representation of the relationship between genotype and phenotype variation.

73 Dun & Fraser (1958, p. 1019).

74 Dun & Fraser (1959, p. 510).

75 Rendel (1959, p. 425).

76 Rendel (1959, pp. 433–436).

77 Rendel & Sheldon (1960).

78 Rendel & Sheldon (1960, p. 37).

79 Rendel & Sheldon (1960, p. 42).

80 Rendel & Sheldon (1960, p. 46).

81 Prout (1962, p. 379).

82 See Maynard Smith & Sondhi (1960) on the number of ocelli, which is normally canalized at three in *Drosophila*.

83 Waddington (1961a, pp. 280–281, 1972a, p. 132).

84 Fraser & Kindred (1960, p. 53).

85 Scharloo (1991, pp. 74–75).

86 Thomas et al. (2016, pp. 443–445).

87 Franklin et al. (2004, p. 274).

88 This distinction was made in 1959. See especially Jacob & Monod (1961).

89 Rendel (1967, pp. 11–12).

90 Rendel (1967, p. 22).

91 Rendel (1967, pp. 148–149).

92 Rendel (1967, p. 14).

93 Rendel (1967, preface).

94 Rendel (1967, p. 14). "Our concept of a developmental process is something carried through as the result of the action of a major gene with its satellite of minor modifying genes controlled by a third genotype, all three kinds of genes being influenced by internal and external environmental factors. The object of this book is to find a descriptive framework which relates the different components of the process in quantitative terms" (pp. 14–15).

95 Rendel (1967, p. 149).

96 Rendel (1967, p. 149).

97 "Between the two thresholds is a plateau representing an area in which *control of development is complete*" (Rendel, 1967, p. 15, my emphasis).

98 Rendel (1967, p. 16).

99 Waddington (1960).

100 Rendel (1967, pp. 131–132).

101 Berenice Kindred began her career at the CSIRO in the late 1950s, where she collaborated on many occasions with Fraser and Rendel. She then continued her

career in various European laboratories where she turned to molecular immunology and abandoned the question of the experimental control of developmental canalization. For a short yet informative biographical note, see Nagy (1985).

102  Rendel (1967, p. 134).
103  Rendel (1967, pp. 152–153).
104  Rendel (1967, p. 153).
105  Eldredge & Gould (1972, p. 114).
106  For biographical details, see Neel (1987).
107  For instance, his *Principles of Human Genetics* which had three successive editions (1949, 1960, 1973).
108  Stern (1958, p. 313).
109  Stern (1958, p. 314).
110  Stern (1959).
111  Waddington (1961a, p. 272).
112  Bateman (1956).
113  Dworkin in his 2005 paper only focuses on some aspects of the content of Bateman's experimental work (Dworkin, 2005).
114  Unlike Waddington, in at least some experimental populations, she successfully identified the phenotype of interest *before* the protocol began. This was an additional argument that genetic assimilation did not require the occurrence of new genetic mutations, as – according to them – in the case of the Baldwin effect model.
115  Bateman (1959b, p. 448).
116  Bateman (1959b, p. 461).
117  In the *Journal of Genetics*, the article on the dumpy phenocopy (referenced 1959a) is published before the one on venation phenocopies (referenced 1959b). This order does not reflect the order of production or even probably the order of submission of the papers to the journal. The article on venation phenocopies was submitted on October 1, 1957, well before its publication. No date of submission is given for the second text, but one can think that it was much later, probably in the second half of 1958.
118  Bateman (1959a, p. 346).
119  Bateman (1959a, p. 348).
120  Waddington, of course, could not ignore the results and interpretations of his former doctoral student. In fact, he referred to them on many occasions, systematically trying to minimize those that did not go in his direction, especially as regards the genetic determinism of the assimilated phenotype. See, for example, Waddington (1961a, pp. 264–266).
121  At Harvard, under the supervision of Paul Levine (1927–2022). See Milkman (1956).
122  It seems that Milkman and Bateman directly met at least once.
123  He published no less than ten successive articles in *Genetics*, from 1960 (Milkman, 1960a) to 1970 (Milkman, 1970).
124  See especially Milkman (1960a, 1960b, 1961).
125  Milkman (1964, p. 625).
126  What might be Waddington's only reference to Milkman's work is a brief mention to his PhD dissertation in his 1961 long review on canalization and genetic assimilation (Waddington, 1961a, p. 261). In the bibliography of the article, in addition to Milkman's PhD dissertation, Waddington mentioned Milkman's first three *Genetics* papers, but there is no reference to this work in the main text.
127  Milkman's articles are cited only two to three dozen times on average in the scientific literature (Google Scholar, January 2023), and only a fraction of these citations are for the 1960–1970 period. Mayr, in his book *Animal Species and*

*Evolution*, was one of the few to acknowledge their importance (Mayr, 1963, p. 190).

128 Bateman (1959b, pp. 465–470).
129 Falconer's textbook ran to four editions in the second half of the 20th century and was with no doubt the most used textbook in quantitative genetics during this period. The first edition (1960) alone sold over 34,000 copies (Bowman, 2005).
130 Falconer (1960, p. 310).
131 Stern (1959, p. 188).
132 Robertson (1977, p. 601).
133 Scharloo (1991, pp. 68–69).
134 Bateman (1959b, p. 470).
135 Bateman (1959b, p. 470).
136 This chapter could have been even further expanded. For example, the work on the links between developmental homeostasis and disruptive selection by John M. Thoday (1916–2008) – who succeeded Fisher in the Cambridge chair of genetics – could have been considered (Thoday, 1958, 1959). Similarly, the normalizing selection experiments of Willem Scharloo (1928–2004) could have been discussed here (Scharloo, 1964).
137 See, for instance, Peterson (2016).
138 Fraser's (1960) simulations are an exception: they made significant room for epistasis in the generation of developmental canalization (Fraser, 1960).

# 9 The complex fate of Waddingtonian concepts in the subsequent history of the Modern Synthesis

In the previous chapter, we saw that within quantitative genetics, itself undergoing a revival in the 1950s, several scientists paid attention to at least some of Waddington's speculations and that this resulted in several theoretical models, experimental trials, and even, at least in the case of Rendel and his colleagues, the setting up of a genuine research program that continued for many years. What happened within the Modern Synthesis? Did the canonical books and articles also attempt to discuss, contest, or appropriate the concepts developed by Schmalhausen and Waddington? This is of course a difficult question, as it requires a minimal agreement on what this corpus was, what a "Waddingtonian concept" is, and how things evolved over the course of two decades, from the mid-1950s to the 1970s.

There have been several attempts to categorize the scientists who participated in the collective enterprise of the synthetic theory over the past decades. Simpson himself, as early as 1949, established the first list of contributors.[1] It is not my intention here to discuss the precise perimeter of such attributions: I simply acknowledge that some scientists participated in the construction of the Modern Synthesis and that there is an obvious consensus to place Dobzhansky, Simpson, and Mayr among them. Within this list, Dobzhansky and Mayr were the ones who, a priori, were most likely to be sympathetic to the ideas of Waddington and Schmalhausen. It was Dobzhansky who was responsible for the American translation of *Factors of Evolution* (1949),[2] and who, in the 1950s, was one of the main supporters of the integrationist and interactionist turn in evolutionary genetics. In a similar perspective, Mayr has remained famous for his critique of beanbag genetics, an orientation that also predisposed him to be interested in the organicist conceptions of Waddington. It is therefore essentially toward these two biologists that my investigation has gravitated, aiming to understand whether their views interacted with the concepts of canalization, genetic assimilation, or stabilizing selection. This investigation, necessarily limited in its scope, says nothing of what might be found in the works of other founders of the Modern Synthesis of the same period.

DOI: 10.4324/9781003422990-17

To Dobzhansky and Mayr's books and articles, it was necessary to add the long review on phenotypic plasticity published by botanist Anthony Bradshaw in 1965.[3] Not because Bradshaw could claim equal importance in the history of evolutionary theory but because this single text, which was surprisingly disconnected from the rest of the author's work, played a major role in the 1980s when the issue of the evolution of plasticity again became a subject of interest in evolutionary ecology.

This chapter, fragmentary as it is, nevertheless offers some interesting results. The first is that Waddington and Schmalhausen were far from being ignored and their concepts were indeed tested and integrated in Dobzhansky and especially Mayr's work. The second is that, overall, the more population-based concepts of Lerner were preferred to the embryological concepts of Waddington and Schmalhausen. This was the case for Dobzhansky, and it was also the case for Mayr, whose model of peripatric speciation based on his hypothesis of genetic revolution is largely grounded on Lerner's theory of genetic homeostasis. The third is that, of Dobzhansky, Mayr, and Bradshaw, it was surprisingly Bradshaw who was most rigorously in line with Waddington. This was unexpected because, to my knowledge, the profoundly Waddingtonian identity of Bradshaw's concept of phenotypic plasticity had never been stated. When, in the 1980s, Schlichting, Scheiner, and West-Eberhard moved on from Bradshaw to put the question of phenotypic plasticity back on the agenda, the link with Waddington had become invisible.

This chapter aims to clarify different levels of use of the concepts of canalization, genetic assimilation, and stabilizing selection, from a real but limited interest (Dobzhansky) to an almost complete appropriation (Bradshaw) and a systematic population reformulation (Mayr). Once these different levels are identified, it becomes possible to evaluate, on a concrete basis, the radicality of Waddington's and Schmalhausen's theses in relation to the theoretical framework of the Modern Synthesis.

## 9.1   Dobzhansky: a minimal consideration

Theodosius Dobzhansky (1900–1975) was one of the central contributors to the Modern Synthesis. Given what we know of his theoretical upbringing, it would have made sense for him to be quite receptive to Waddington's work and especially to his critique of certain simplifications of mathematical population genetics. Trained in entomology and genetics in the very active Russian school, historical studies from the 1990s have helped us better understand what his own conceptions owe to his native intellectual environment.[4] Indeed, in contrast with the reductionist and abstract orientation of Fisherian population genetics, the Russian school, under the impulse of Chetverikov, was more interested in working on natural populations and on interactions between genes.[5] This double inclination is evident in Chetverikov's 1926 seminal paper "On Certain Aspects of the Evolutionary Process from the Standpoint of Modern Genetics",[6] which was regularly cited by Dobzhansky.[7] It was for example in this text that Chetverikov elaborated the notion of "genetic milieu" to explain why the phenotypic effect of a gene could not be treated as a constant[8] and thus why "alleles cannot be assigned fixed 'fitness' values".[9]

Dobzhansky, like Timofeeff-Ressovsky, Dubinin, and others, shared this double inclination with Chetverikov: he too wanted to test the theory of population genetics using empirical data obtained from the careful study of natural populations; he too was particularly receptive to the idea that genes should not be isolated but should, on the contrary, always be considered as part of a larger whole, including the other genes of a Mendelian population. However, despite this non-reductionist take, he only gave a very limited space to Waddingtonian ideas in his writings, and this point requires further explanation.

### 9.1.1   From *Genetics and the Origin of Species* (1937) to *Genetics of the Evolutionary Process* (1970)

Regarding Dobzhansky's limited interest in developmental plasticity and other Waddingtonian themes, the comparison of the contents of the successive reissues of his magnum opus is illuminating. In 1937, based on his Jesup Lectures at Columbia University in the fall of 1936,[10] he published the first edition of *Genetics and the Origin of Species*, considered one of the founding works of the Synthesis. A second edition was published in 1941, and a third in 1951. In 1970, a completely revised edition appeared with the new title *Genetics of the Evolutionary Process*. Here I will give a brief overview of Dobzhansky's treatment of the questions of plasticity, canalization, and stabilizing selection in the 1937, 1951, and 1970 editions.

Unsurprisingly, these issues were virtually absent from the first edition. This can be explained by the context of this first moment of the Synthesis: in the mid-1930s, it was necessary to *constrict* the field of acceptable hypotheses, which required, among other things, to make no concessions to Lamarckism. Dobzhansky's preface was marked by this required antagonism.[11] While the concept of reaction norms was not absent,[12] it was at that time still too early for plasticity to be endowed with a causal role within a Darwinian framework. Like others, Dobzhansky expressed a clear-cut alternative between Lamarckism and Darwinism, that is, between the postulation of a direct role through the inheritance of acquired characters or the complete absence of effective causality:

> To suppose that geographical races were originally modifications that have subsequently been fixed by heredity is contrary to the whole sum of our knowledge, just as it would be absurd to assume that mutations are phenocopies that have become hereditary.[13]

The 1951 edition, which came after Gause's (1947) and Schmalhausen's (1949) important texts, and after Waddington's first articles (but before his experimental program in Edinburgh), significantly differs from the 1937 one. This time, the question of plasticity, in the broadest sense of the term, is the object of serious, though limited, consideration. In the preface, Dobzhansky now had enough distance to give a first historical overview of the Modern Synthesis. Besides the essential roles played by Mayr, Stebbins, Huxley, and Simpson, he mentioned White and Schmalhausen. We have already seen that Dobzhansky held Schmalhausen's work in high esteem, and it was his name that was primarily mentioned in the few paragraphs of the 1951 edition in which Dobzhansky discussed the concepts of canalization,[14] stabilizing selection[15] and, on one occasion only, organic selection.[16] While Gause's work was also mentioned, Waddington did not appear anywhere in the book. As we have seen in Chapter 5, Waddington was distressed by what he saw as an inequality of treatment[17] and addressed this directly with Dobzhansky in a letter dated July 25, 1959.[18] Dobzhansky justified his choice using considerations that stepped outside the realm of science: as a victim of Lysenkoism, Schmalhausen should be massively supported.[19]

In addition to this reason, at least two complementary reasons probably led Dobzhansky to place greater emphasis on Schmalhausen's work. The first is that Dobzhansky was always careful to stress the role played by Russian and Ukrainian scientists in the history of evolutionary theory in the 20th century. Thus, in numerous publications, he was careful to emphasize the merits of Chetverikov in the development of population genetics, alongside (but not behind) the famous Fisher-Haldane-Wright triad.[20] Most likely, the same was true of Schmalhausen. The second reason has to do with the content of the concepts. Even though the notions developed by Schmalhausen and Waddington were indeed remarkably convergent, there remained slight differences between them. One of them was not a clear-cut conceptual difference but more something like a divergence of perspective. In a nutshell, although *Factors of Evolution* is a very obscure and difficult book to read, Schmalhausen presents a more populational perspective in the way he links the issues of embryonic development, the evolutionary genesis of reaction norms, and the action of stabilizing selection.[21] My hypothesis is that this more openly populational treatment was closer to Dobzhansky's own theoretical intentions.

Almost 20 years after the final edition of *Genetics and the Origin of Species*, when *Genetics of the Evolutionary Process* was published, biology as a whole had undergone major transformations, most importantly the advent of molecular biology and the identification of genes with DNA segments. However, these upheavals did not alter the general economy of the book and Dobzhansky's own themes still dominated: the mechanisms of reproductive isolation and their evolution, balanced polymorphism and balancing selection, the issue of the genetic load, and so on. This time, Waddington was no longer absent, though Dobzhansky still referenced Schmalhausen more. When it came to the various forms of stabilizing selection or to the experimental work conducted during the 1950s, particularly in Edinburgh, Dobzhansky adopted a relatively distant and informative stance: he mentioned these works but did not really take sides and especially did not integrate them into his own conception of evolutionary dynamics.[22]

Dobzhansky was particularly insistent when it came to underlining the genetic cohesion of either gene pools or Mendelian populations and regarding the fact that species were above all "genetic systems".[23] Here, neither Waddington nor even Schmalhausen was mentioned, but references to Lerner abounded.

### 9.1.2   Dobzhansky's holistic view of population genetics

As we briefly saw in Chapter 8, population genetics itself underwent significant transformations in the late 1940s and throughout the 1950s. Under the leadership of the new generation but also with the contribution of Sewall Wright, the focus shifted from genes as atoms whose individual kinetics could be traced by means of fitness estimation, to genomes and gene pools whose primary characteristics were the dynamic and complex integration of their constituent elements. Mather and Lerner were at the forefront of this shift in perspective, in which Dobzhansky also played an important role.[24]

From the late 1940s onward, his constant use of the concept of Mendelian population (forged by Wright), his promotion of the concept of gene pool, and his emphasis on the genetic cohesion of the gene pool[25] led Dobzhansky to develop an increasingly holistic view of the understanding of genetic change in populations which was expressed in his model of balancing selection and balanced polymorphism. This population thinking significantly differed from, for instance, Goldschmidt's developmental take,[26] and explained why Waddington was not a necessary resource for Dobzhansky. While evolution still ultimately consisted of the change of allele frequencies within populations, it was no longer possible to only consider one-locus cases. In 1955, four years before Mayr launched his full-scale attack on beanbag genetics (see Section 9.3.2), Dobzhansky, in a more nuanced way, considered that this new form of population genetics constituted an undeniable progress:

> In recent years it is becoming increasingly appreciated that Mendelian populations, though, of course, composed of individuals, have an internal genetic cohesion which is a property of a population, not that of any individual. Here we face the old problem of the relationships between the whole and its parts, with which the philosophers have been struggling at least since Plato. In the biological materials this problem is particularly clear-cut. An individual is not identical with the whole to which it belongs; but a population as a whole is what it is only because

of the individuals of which it is composed. The properties of a population transcend those of the individuals. In the words of Lerner (1954), "This change of outlook has caused the integrative properties of populations to assume paramount importance in evolutionary thought, not only for problems dealing with differentiation between populations, but with even greater force for those relating to genetic phenomena within populations". This new outlook or change of emphasis is perhaps a sign of coming of age of population genetics. A Mendelian population represents a level of organic integration which obeys its own laws and contains its own regularities. These laws and regularities population genetics is called upon to study and to describe.[27]

This "new outlook" that put the emphasis on the Mendelian population[28] instead of individual genes or genotypes probably explains why Dobzhansky was not interested in organic selection (either before or after Simpson), or even in Waddington's embryological conceptions. Indeed, in organic selection and genetic assimilation (or any other related concept), the causal role falls primarily on the plasticity of individual organisms, even though they are considered elements of a population and even though canalization was endowed with population consequences (a high level of standing genetic variation). In Dobzhansky's increasingly overtly population-based and holistic perspective, the plasticity *of a population* was a much more powerful causal factor than the developmental plasticity of individual organisms:

> It is hardly a rash assumption that no one genotype, either homozygous or heterozygous, is likely to be a paragon of adaptability, superior to all other genotypes in all environments which the species is confronted with. The adaptedness of a Mendelian population may, then, be advanced if it contains a variety of genotypes suited to different adaptive niches and facies of the environment which the population inhabits.[29]

It was also for this reason that Dobzhansky attributed great value to Lerner's theory. Indeed, Lerner's genetic homeostasis provides a simpler link (compared with Waddington's or Schmalhausen's hypotheses) between the buffering abilities of an individual and the evolutionary plasticity of a population: in both cases, it is the degree of heterozygosity that plays the role of leading factor. The question here is not to argue that Dobzhansky was "influenced" by Lerner. I am instead interested in noting that when Dobzhansky himself took part in this holistic turn in experimental population genetics, he found in the theory of genetic homeostasis an intellectual resource that was particularly in tune with his own concerns. He saw Lerner's 1954 book as a "brilliant"[30] work, and stressed that Lerner had to "be credited more than anyone else with the authorship of the balance hypothesis of population structure".[31]

## 9.2   Interlude: a closer look at Lerner's theory of genetic homeostasis

Given the central role that some of Lerner's ideas played in the second phase of the history of the Modern Synthesis, and thus in the way Schmalhausen's and Waddington's works were themselves read and understood by Dobzhansky, and even more so by Mayr (see Section 9.3), it is essential to elaborate on the content of Lerner's conception. The theory of genetic homeostasis, briefly addressed in the previous chapter, is indeed deceptively simple and, as emphasized by Hall,[32] articulated no less than seven "main theses".[33] Allen has shown how, at the turn of the 1950s, Lerner gradually formulated his ideas in constant interaction with the hypotheses developed at the same time by Waddington and Schmalhausen but also Mather, Dobzhansky, and even Fisher.[34] His first meeting with Dobzhansky, which took place as early as 1931 at the University of British Columbia,[35] where Lerner specialized in poultry science,[36] was probably decisive in Lerner's career, and certainly contributed to drawing his attention to the heterosis issue and to the high rate of heterozygosity in natural populations. In the following paragraphs, rather than reconstructing the intellectual path that led Lerner to write, in the space of a few months in 1953,[37] his most important book, I will limit myself to summarizing its content, to better understand why, for approximately 15 years,[38] it interested so many of his contemporaries, among whom Dobzhansky and Mayr.

This small book (120 pages of text) consists of a disparate set of more or less speculative and heterodox hypotheses that Lerner brought together under the heading of "genetic homeostasis". The book consists of two parts of equal size. The first one gathers a large number of facts about various experiments on artificial selection, on inbreeding degeneration, and on the selective advantage of heterozygotes. The second part, entitled "Interpretation", presents several hypotheses that shed light on these complex facts. Lerner's main hypothesis was that Mendelian populations were capable of some form of "self-regulation", that is "the property of the population to equilibrate its genetic composition and to resist sudden changes".[39] To account for this self-correcting capacity of Mendelian populations, which explained why certain selective practices were inefficient,[40] Lerner stipulated nothing other than the mechanisms classically acknowledged in genetics. Although his attention was focused on populations, he never adopted the idea, which was then becoming fashionable, of a possible selection acting at this level.[41] Among these mechanisms ensuring the genetic homeostasis of populations, Lerner gave the major role to heterozygosity, which was for him the foundation of developmental homeostasis (first-order consequence, at the individual level) and therefore of genetic homeostasis (second-order consequence, at the population level).

Lerner thus took from Waddington and Schmalhausen (both explicitly cited) his starting point: the existence of a form of developmental canalization allowing the attainment of the wild-type phenotype despite the vagaries of the environments and in a wide range of genetic backgrounds. But he gave a very restrictive and narrow explanation of it, based only on heterozygosity. Heterozygosity provided developing organisms with more chemical possibilities and then more alternative pathways to the same wild-type final stage. More importantly, heterozygosity served as a *means* for Lerner to consider what interested him most: genetic homeostasis on a population level. Here is how he summarized the core of his hypothesis:

> The most likely mechanism for both types of homeostasis [developmental, i.e. canalization, and genetic, i.e. populational] lies in the superiority with respect to fitness of the heterozygous over the homozygous genotypes:
>
> (a) ontogenetic self-regulation (*developmental homeostasis*) is based on the greater ability of the heterozygote to stay within the norms of canalized development;
> (b) self-regulation of populations (*genetic homeostasis*) is based on natural selection favouring intermediate rather than extreme phenotypes.[42]

The central point of Lerner's theory, its starting point, was therefore that heterozygous individuals were better canalized than homozygous ones. As a result, they should show much less phenotypic variance. To make this aspect of his hypothesis clear, Lerner contrasted the predictable consequences of his hypothesis with those of the opposite hypothesis, that is, "the viewpoint that phenotypic variance is smaller in homozygotes than in heterozygotes":[43] "the basic test of the hypothesis advanced [the highest buffering capacities of heterozygotes] should be whether the environmental variance of heterozygotes is greater or smaller than that of homozygotes, particularly with respect to the traits of adaptive significance".[44] Unfortunately, most of the available experimental results were ambiguous and conclusions were therefore uncertain.

Nevertheless, there were also some demonstrative results, and here Lerner gave prominence to the joint work by Dobzhansky and his former student who later became one of his favored collaborators, Bruce Wallace (1920–2015). In a 1953 paper entitled "The Genetics of Homeostasis in *Drosophila*", they made a first assessment of a large comparative work involving four species of *Drosophila*: *Drosophila pseudoobscura, persimilis, prosaltans,* and *melanogaster*. Their main result was that homozygous individuals showed very high variability in their survival rates in non-uniform laboratory culture conditions (density, amount of food, etc.) while, on the contrary, cultures in which genetic heterozygosity was high showed a much higher homogeneity. Although they did not consider their results to be fully demonstrative,[45] they inferred that "homeostatic adjustments are superior in heterozygotes to those in homozygotes"[46] and concluded in a marked holistic-populational tone:

> The gene complexes carried in the homologous chromosomes in sexual and cross-fertilizing populations are coadapted by natural selection to give high fitness in heterozygotes with most other chromosomes of the same population. The genotype of a Mendelian population is an integrated system, the parts of which are fitted together in the process of evolution.[47]

For Lerner, Dobzhansky and Wallace's results were no less than "the strongest link in the chain of evidence we have on the role of heterozygosity in developmental homeostasis".[48] This example shows how the geneticists involved in this holistic turn in experimental population genetics were in constant interaction at that time.[49] It also explains why Dobzhansky was particularly interested in Lerner's theory, since it was based in part on his own experimental research and gave a maximal extension to his interest in heterozygosity.

How was heterozygosity supposed to produce developmental canalization? This was an important question and there was little data to constrain the models that could be developed. Lerner distinguished between two, not mutually exclusive, possibilities: "Is heterozygosity obligate for specific loci, or is it its general level (i.e. percentage of loci in that state) that is significant in determining whether an individual is well buffered or not?"[50] In general, Lerner found the second alternative sounder, and he tended to favor it.

Lerner's book quickly became a landmark among population geneticists.[51] Waddington was delighted to see that population geneticists themselves were now interested in the relationship between development and evolution and praised Lerner for his highly stimulating synthesis.[52] However, he also worried that the use of the term "homeostasis" might be confusing outside the field of physiology,[53] and, more importantly, he saw the explanation of developmental canalization based on the degree of global heterozygosity as unnecessarily limiting and weakly founded on an empirical level.[54] As we saw in the previous chapter, Robertson and Falconer were also highly interested in Lerner's proposal, which had a major impact (more than Waddington's own conceptions) on their theoretical and experimental work on the consequences of normalizing selection.

In the years that followed, Lerner gained considerable fame and maintained his main hypothesis, that of a direct causal link between the mechanism responsible for developmental canalization and the genetic homeostasis of populations.[55] In his "Concluding Survey" of the 1955 Cold Spring Harbor symposium devoted to population genetics,[56] he noted, like Mayr and Dobzhansky, that a revolution was underway in this new scientific discipline which was becoming more holistic and more integrative and, in his opinion, more "sophisticated".[57] Within this transformation of population genetics, Lerner's theory of genetic homeostasis undoubtedly played, for a time at least, one of the leading roles.

## 9.3 Lerner rather than Waddington? Mayr and his model of "genetic revolution"

Like Dobzhansky, Ernst Mayr was one of the founders of the Modern Synthesis most likely to interact with some of Waddington's theoretical conceptions. Mayr and Waddington seemed to converge in their criticism of the theory of population genetics and of certain simplifications that they considered unfounded. For example, both disputed the validity of assigning fixed selective coefficients to specific alleles. Both were inclined to oppose atomistic views of the genotype and to favor a more integrated understanding of genes. Thereby two related questions arise. Did their positions actually overlap, and if so to what extent? Was this possible convergence a sign of Waddington's "influence" on Mayr (or the other way around)? These questions are especially difficult because Mayr's attitude toward Waddington was ambiguous. Indeed, while in some respects Mayr progressively made more room for Waddington's ideas (and mainly for his concept of canalization) than Dobzhansky had done, in other respects he was also extremely critical of the mechanism of genetic assimilation, seeing it as mere verbiage with no new conceptual content. Mayr also seemed at times to be annoyed, as were many others (Alan Robertson primarily), by Waddington's theoretical nonchalance and the vagueness with which he developed his concepts and experimental designs.[58] The purpose of the following pages is thus to clarify, as much as possible, this ambiguous personal and theoretical relationship.

### 9.3.1 *Mayr against beanbag genetics: a reappraisal*

Mayr's critique of beanbag genetics from the late 1950s onward is well known and has already given rise to several historical, philosophical, and theoretical treatments.[59] What Mayr criticized was the atomism he perceived in the (theoretical) population genetics of the 1930s, according to which genes should be conceived as independent entities, allowing them to be unequivocally assigned a fixed selective value. In his famous inaugural address ("Where Are We?") to the 1959 Cold Spring Harbor symposium devoted to "Genetics and Twentieth Century Darwinism", he launched the attack, before some of the major evolutionary geneticists of the moment:[60]

> Yet, this period [Fisher's, Haldane's & Wright's] was one of gross oversimplification. Evolutionary change was essentially presented as an input or output of genes, as the adding of certain beans to a beanbag and the withdrawing of others. This period of "beanbag genetics" was

a necessary step in the development of our thinking, yet its shortcomings became obvious as a result of the work of the experimental population geneticists, the animal and plant breeders, and the population systematists, which ushered in a third era of evolutionary genetics.[61]

Mayr criticized these early models for not taking into account the interactions between genes (epistasis) nor the fact that the same gene could participate in the determination of many phenotypic traits (pleiotropy). In a detailed 1997 examination of the beanbag controversy, De Winter produced some much-needed theoretical clarification, showing that the concept of atomism which Mayr had used was ambiguous. It is necessary to distinguish between two meanings of the term "atomism" in the context of the theoretical basis of population genetics. On the one hand, atomism can be understood in connection with the *transmission* of hereditary material (*transmission-atomism*). In this respect, the theoretical population genetics of Fisher, Haldane, and Wright is indeed atomistic: what is transmitted through generations, according to a Mendelian pattern, is a sufficiently stable unit of heredity (at least at the timescale of a few generations), the gene. Nevertheless, as Dawkins has repeatedly claimed, this atomism does not imply that, as far as gene *expression* is concerned, each gene can only have an additive action in the realization of the phenotype (*expression-atomism*). In other words, the fact that genes are the atoms of heredity does not prohibit either epistasis or pleiotropy. This was already true for Fisher: if his genetic theory of selection "assumes a particulate mode of inheritance, i.e. . . . assumes that genes are *inherited* independently, . . . it does *not* assume, however, that genes are expressed independently during epigenesis, or that they *function* independently".[62] It is therefore certain that Mayr's vehement criticism, which he himself understood to be provocative, went too far and produced a caricature of the population genetics of the 1930s.[63]

Should we leave it at that? Did Mayr simply get it (largely) wrong, taking a whole community with him in his mistake? The matter seems more complex, and it is perhaps appropriate to distinguish here between what the theory of population genetics assumed and the way in which this theory was (and still largely is) embodied in concrete models. From a theoretical point of view, with a few exceptions,[64] epistasis was not ignored. But overall, it was considered, notably by Fisher, to be of negligible importance on the scale of long-term evolution because of the impact of recombination: it was thought that only additive genetic variance could give rise to the cumulative action of natural selection. Wright was not fully convinced by this Fisherian conception and was more interested in the issue of epistasis. Hence his vehement reaction to Mayr's criticism, which he considered unfair (at least as far as he was concerned) and simply wrong.

Nevertheless, even if the theory developed by Fisher and Wright had taken into account the question of epistasis (with largely different aims), it remains that *in practice*, one-locus models were the norm[65] (and still are) and that it was the experimental population geneticists themselves, such as Lerner, Dobzhansky, or Wallace, who, *before* Mayr, wanted to take into account the interactions between genes in their models. This debate, which started out internal to population genetics, appeared to reach a kind of resolution with the theoretical model elaborated by Richard C. Lewontin and Ken-ichi Kojima at the end of the 1950s. This model showed that, except when extremely strong, epistasis could indeed be neglected.[66] Yet, at the same time, Michael J. D. White's results from his careful chromosomal study of different populations of *Moraba scurra* grasshoppers showed precisely the opposite: epistasis did appear to be involved in determining fitness.[67] Like Dobzhansky and Mayr, White deduced "that genotypes were organized systems presenting a high level of integration".[68] These results were of direct interest to Lewontin, so much so that he began a collaboration with White, which did not however lead to clear conclusions as to the role that could be attributed to epistasis. A few years later, notably because of the massive use of simulations, the question reappeared,[69] and certain results seemed to indicate that epistasis could have an evolutionary impact, to the great astonishment of certain population geneticists (such as Alan Robertson), and to the great satisfaction of Mayr.[70] The question has in fact never been satisfactorily resolved and remains open today. So Mayr's insights, despite being qualitative and verbal in nature, cannot be ignored simply because he was an outsider. It is in the light of the internal debates in evolutionary genetics in the 1950s and 1960s that his own position must be evaluated. It is this interpretative line that is favored here, because it allows us to better situate Mayr in relation to both Waddington and Lerner.

At first, it could seem that Mayr's positioning was equivalent to or at least close to Waddington's critical stance. In Mayr's 1959 text, some passages are extremely Waddingtonian, for instance when he posits that "what is exposed to natural selection is not the individual gene nor the genotype but rather the phenotype, the product of the interactions of all genes with each other and with the environment".[71] In other places, Mayr directly refers to *The Strategy of the Genes*.[72] This is why, in a recent work, Peterson sees evidence of an "influence" of Waddington on Mayr.[73] Nonetheless, this reading seems problematic mainly because it tends to erase the specificities of Mayr's position compared to Waddington's.

The term sometimes used to describe Mayr's theoretical positioning is "organicism",[74] a term that Mayr himself never used in his scientific publications and that, in my view, does not do justice to his ideas. In contrast to Waddington and other organicist biologists, Mayr's opposition to genetic *expression-atomism* was not primarily centered on the biological organism and its irreducible individuality but rather on the individuality/cohesiveness of Mendelian populations. This is why, following De Winter,[75] I prefer to use "gene interactionism" to label Mayr's own perspective. This form of holism must first be understood as a major component of his own intellectual trajectory, without reducing it to the importation of an exogenous thesis, like Waddington's. Already in 1942, in his major book *Systematics and the Origin of Species*, Mayr criticized "the tendency of workers to consider the actions of genes entirely as those of *separate* units"[76] and, crucially, emphasized the need to understand the genetic "integration" of species and subspecies.[77] Well before he read *The Strategy of the Genes* (1957), Mayr already positively considered genes interactions and systems thinking but at the level of the population, not at the level of the organism.

Thereby, if my reading is correct, the points of contact between his critique and Waddington's are in fact quite incidental and, at least to some extent, superficial, meaning that they are not underpinned by the same theoretical perspective. In a nutshell, Waddington speaks from the perspective of developmental biology, whereas Mayr speaks from the perspective of systematics. Both were interested in the interactions of genes but not in the same way. As a developmental biologist, Waddington focused primarily on individual genomes, that is, the interactions between the gene products at work during embryogenesis that allow the formation of the phenotype. As we have seen, it was in this developmental sense that he conceived the genome as an integrated system. His epigenetic landscape diagram refers to this very conception. Mayr did not deny this dimension, but he brought another form of genetic integration to the forefront, the one operating at the scale of a Mendelian population. This was already the case in *Systematics and the Origin of Species* and was again crystal clear in his 1959 address, in which he heavily relied on Dobzhansky's results about the internal balance of the gene pool[78] and on Lerner's theory of genetic homeostasis.[79]

Mayr's views were therefore fully consistent with his populational credo and his fieldwork as a systematist and they were also largely congruent with the Russian school of evolutionary genetics, as Adams pointed out in 1968,[80] hence his proximity to Lerner and Dobzhansky and the conceptual gap that distinguishes his populational take from Waddington's more developmental one.

### 9.3.2   *The model of genetic revolution and its implications*

Mayr's opposition to beanbag genetics, his appropriation of the gene interactionism of the 1950s, and his strong interest in Lerner's theory are all evident in his peripatric model of speciation which was based on what he called a "genetic revolution". He always believed that this model was his most significant contribution to evolutionary theory.[81]

At the time he wrote *Systematics and the Origin of Species*, Mayr's expertise in evolutionary genetics was rather sketchy, and his references to Wright and genetic drift were only second hand (it was only when he wrote *Animal Species and Evolution*, published in 1963, that Mayr read Wright directly[82]). However, as we have seen, as early as 1942, Mayr was interested in the interactions between genes on a population scale, that is, insofar as individual genes were parts of a whole that was not yet called a gene pool. Provine has shown how, especially through regular exchanges with Bruce Wallace in the spring of 1950, Mayr became aware of the "new" population genetics, and in particular of the work of Dobzhansky, Lerner, and Mather.[83]

By 1951, Mayr made use of this set of new insights to develop a genetic model that could account for speciation in geographically isolated, peripheral natural populations. The result was his concept of genetic revolution, first published in a 1954 volume honoring Julian Huxley, nearly three years after Mayr wrote a first draft of the text.[84] In this paper, entitled "Change of Genetic Environment and Evolution", Mayr highlighted a primordial evolutionary factor in speciation, according to him largely ignored: the evolutionary consequences of a drastic change in what he called the "genetic environment". Relegating drift to the rank of a secondary factor, Mayr was first interested in the "conservative" and "stabilizing" effect of gene flow between populations, which prevents the emergence of a new "gene-complex" fully adapted to local biotic and abiotic constraints.[85] Already in 1951, Mayr opposed what he called "beanbag thinking", that is, the idea that a selective coefficient could be assigned to a given allele. His criticism therefore seems to be fully in line with Waddington's. However, he quickly deviated from this criticism, insisting on the population dimension of his conception:

> Still more important than are the multiple alleles is the fact that during development all genes are members of a team. Not only has every gene that has been thoroughly studied been found to have pleiotropic effects, but it has also been found that every character is produced by the joint action of many genes. It is immaterial in this connection what particular genetic theory one adheres to: major genes and modifiers, genes and polygenes, switch genes and gene complexes, position effects, and non-localized genes. They all agree in the essential point which is that the action of a given gene is strongly influenced by its genetic background, its genetic "co-actors". And what is true for the function of a gene is true also for its selective value. A gene which is of high selective advantage on one genetic background may be selected against on another genetic background. *The selective value or viability of a gene is thus not an intrinsic property but is the sum-total of the viabilities on all the genetic backgrounds that occur in a population.*[86]

This holistic and interactionist conception of the gene pool constituted the basis of his model of genetic revolution: if the coefficients of selection were not fixed parameters, if they depended on the frequencies of the other genes within the gene pool (genetic environment), then, thanks to stochastic events, frequent in peripheral populations with small numbers of individuals, and provided that the stabilizing effect of the gene flow was interrupted enough, there could be a radical change in the selective value of the alleles that were still present, and thus with the rebuilding of the gene pool as a whole. Such an event was extremely random, most of the time it would lead to the extinction of the population because of its very low number. But on some occasions, the gene pool of the daughter population could be completely rebuilt and the alleles could acquire new selective values in this radically altered "genetic environment". After the critical phase of a strong decrease in its number, the population would be able to develop again, which would stabilize from an evolutionary point of view, its new "gene-complex".

In the early 1950s, now fully aware of the holistic orientation of population genetics, Mayr was able to take advantage of these interactionist and population-based conceptions (especially that of Lerner) to postulate, in a speculative[87] and strictly qualitative manner, the genetic stages of a peripatric speciation process. In doing so, Mayr, as a systematist, was considering the question of evolutionary diversification, not adaptation. As Thorpe had done in the 1940s about sympatric speciation (building on Mather's "balanced combinations" model, as we have seen), Mayr used the concepts of genetic homeostasis, integration of the genotype,[88] and gene complex to rethink the dynamics of geographic speciation.

The reception of Mayr's model is complex[89] and exceeds the scope of this section. It is known[90] that Eldredge and Gould's model of punctuated equilibria derives very directly from this model, as the conclusion of Mayr's 1954 text clearly shows.[91] Mayr was very upset about this reappropriation of his own ideas because Gould and Eldredge minimized their debt to him.[92] Beyond these attribution disputes, what I am most interested in here is that, as we have already seen (Chapter 8), Eldredge and Gould also massively relied on Lerner's theory of genetic homeostasis in the conclusion of their seminal 1972 text.[93] What was then at stake for Eldredge and Gould, as for Mayr, was the genetic individuality and cohesiveness of the species. This required holistic models of population genetics precisely like those provided by Lerner.

### 9.3.3   *Mayr's final synthesis:* Animal Species and Evolution *(1963)*

Unlike Dobzhansky, Mayr never rewrote or republished his seminal work, *Systematics and the Origin of Species*. Instead, in the late 1940s,[94] he began working on an even more ambitious book, which was supposed to show the significance of speciation for evolutionary theory, and which was finally[95] published in 1963 under the title *Animal Species and Evolution*.[96] This work is probably the most important book of this second phase of the history of the Modern Synthesis,[97] characterized in particular by a more adaptationist take, the so-called hardening emphasized by Gould.[98] It should be noted that as early as 1963, decades before the project of an Extended Synthesis,

Mayr saw the synthetic theory of evolution as an ongoing process of which he already distinguished at least two major episodes since the end of the 1930s.[99] The purpose of this section is not to give a detailed account of all the views expressed by Mayr in this particularly rich and dense synthesis but rather to stress that Mayr was particularly receptive to Waddington's conceptions, including those considered the most radical. In many respects, *Animal Species and Evolution* can be read as the only attempt ever made to integrate (most of) Waddington's concepts into the framework of the Modern Synthesis.

To begin with, it is necessary to point out that Mayr always opposed the value of Waddington's concept of genetic assimilation. Following Stern, Bateman, and Falconer, he thought that the "simplest interpretation",[100] that is, a threshold model, should be preferred and did not see what the terminology invented by Waddington could add to an explanation. Mayr was particularly firm on this issue, especially in *Animal Species and Evolution*:

> It seems to me, furthermore, that it beclouds the issue to introduce a separate term, "genetic assimilation" (Waddington), for the accumulation of such polygenic threshold genes by selection. The unwary might think, as Baldwin did for his organic selection, that this is an alternative to natural selection or at least a special and rare case of selection. Actually, what Waddington designates as genetic assimilation is one of the normal aspect of natural selection. Since (a) virtually all characters are highly polygenic, and (b) all genotypes tend to vary phenotypically according to the existing environmental conditions, natural selection will act most strongly on the extreme phenotypes (of the character in question), whatever genetic or environmental constellation produced them. It will result in an accumulation and integration of all the genes that will produce the favored phenotype to an optimal extent in the greatest number of encountered environments. This, I believe, is the normal process of selection of a polygenic character and requires no special terminology.[101]

Unquestionably, then, it was not on the question of genetic assimilation that Mayr joined Waddington. However, simultaneously, other aspects of Waddington's speculations resonated with Mayr's intellectual trajectory. Archives show that Mayr's correspondence with Waddington started in 1951,[102] shortly after he began work on his book. From then until Waddington's death in 1975, they were in regular contact and exchanged about on the then crucial issue of gene interactionism and the role of phenotypic plasticity in evolution. Mayr was thus perfectly aware of Waddington's embryological conceptions and had read *The Strategy of the Genes*. In *Animal Species and Evolution*, he directly built upon Waddington's most important articles on developmental canalization and genetic assimilation, including Waddington's 1942 seminal paper and the results of the experiments performed in Edinburgh. Mayr used several Waddingtonian viewpoints and concepts throughout the book, which gives his *Animal Species* an indisputable Waddingtonian flavor, especially in comparison to *Systematics and the Origin of Species*. In what follows, I detail the ideas Mayr used that were most probably of Waddingtonian origin, or at least corresponded to ideas which Waddington supported.

## (1) Natural selection acts on phenotypes

This aspect is certainly the least Waddingtonian of the six conceptions listed here, not because Waddington did not support it repeatedly but because many others had also made it a necessary postulate for a more realistic under-standing of the way natural selection worked. Indeed, in his final letter to Waddington, dated July 15, 1975, Mayr emphasized this aspect and pointed toward a convergence rather than an influence:

> I was particularly pleased with your stress on the phenotype. Ever since 1955, I have emphasized that the target of natural selection is the whole individual and not single genes. This has made me quite unpopular with H.J. Muller, Sewall Wright, and others.[103]

Unsurprisingly, we find this idea regularly throughout the pages of *Animal Species*.[104]

## (2) The phenotype is the developmental consequence of the integrated genome

It was, as we have seen, in this developmental sense that Waddington con-ceived his gene interactionism. This point was not, for Mayr, as central to his own conception of the evolutionary process, but he did fully endorse it, which had not yet been the case in most of his texts of the 1950s. Moreover, from the very first pages of the book, he presented it as one of the major advances in biological theory since the advent of the Modern Synthesis, and credited Waddington very explicitly:

> A comparison of current evolutionary publications with those of only 20 or 25 years ago shows what a great conceptual progress has been made in this short period. Since much of this volume is devoted to reporting this progress, I will barely mention some of these advances in this introductory discussion. Our ideas on the relation between gene and character have been thoroughly revised and the phenotype is more and more regarded not as a mosaic of individual gene-controlled char-acters but as the joint product of a complex interacting system, the total epigenotype (Waddington, 1957).[105]

### (3) Canalization masks the phenotypic effect of genetic mutations, allowing genetic assimilation to be conceived in terms of threshold selection

We previously saw that Waddington conceived of canalization in two distinctly different but often intermingled senses. According to the first meaning, the one still accepted today, canalization acts as a mask which attenuates the phenotypic effects of mutations. Mayr used this meaning in his book as it allowed him to interpret the concept of genetic assimilation in terms of threshold selection, which was a perfectly legitimate reading of some of Waddington's writings. Mayr thus contested the need for a new terminology to describe the standard functioning of natural selection.[106]

### (4) The same phenotype can be realized based on many different allelic combinations

Mayr used Waddington's central idea of multi-realizability of the phenotype in a different sense. Waddington's idea was based almost exclusively on developmental canalization. Mayr's was closer to Lerner's hypothesis of developmental homeostasis[107] and, perhaps even more so to Mather's elaboration on quantitative heredity, a work which Mayr considered "revolutionary".[108] Within a polygenic conception of heredity, it was easy to explain the non-injectivity of the function linking genotype to phenotype without having recourse to the concept of canalization:

> Polygenic inheritance confers two great advantages. First, it provides a storage system that not only has an enormous capacity but can readily respond to the slightest shift in selection pressure. Second, in view of the similar action of the factors in a polygenic set, the same phenotype can be produced by many different combinations, for instance *Abcd*, *aBcd*, *abCd*, and *abcD*.[109]

### (5) Developmental canalization results from interactions between genes

We have seen that the issue of the nature of the genetic determinism of canalization was a difficult one for Waddington: although he seemed to favor an emergentist conception, certain ambiguous passages may also suggest that he envisaged the existence of canalizing genes. Mayr did not go into much detail about this question, but when he did address it, it was clearly from a systemic and emergentist perspective, one in line with the general spirit of Waddington's theses. Mayr understood canalization as the result of numerous interactions between the products of the genes of a developmental system:

> Canalization is a developmental phenomenon, but does not differ in its genetics from other cases of polygenic inheritance. To assume that each dominant and overdominant gene, each isoallele, and developmental

feedback has its own *ad hoc* modifiers and polygenes would lead to the postulation of a number of genes far in excess of what can be accommodated on the chromosomes. The other alternative, the one in which I firmly believe, is that each gene serves as a modifier for many, if not most, other gene loci. Any interference with such a system will lead to chain reaction. *The multiplicity of interactions, including multiple pathways, gives such a system great plasticity as well as stability.*[110]

## (6) Canalization directs the phenotypic effect of mutations and thus may be at the origin of evolutionary trends

As we have seen, the hypothesis according to which developmental canalization could take charge of the phenotypic effect of mutations, and thus at least partially decouple phenotypic evolution from its genetic record, was probably the most radical and original point of Waddington's theories (Chapter 6). This aspect was largely ignored by his contemporaries in the following decades. It is therefore quite remarkable that Mayr not only understood it well but also made it his own. Indeed, in the last part of the book, when he addressed the question of macroevolution and the "emergence of evolutionary novelties", Mayr made a very Waddingtonian use of the concept of canalization. He no longer saw canalization as an exclusively negative buffer restricting the expression of genetic variation, but he also saw it as a positive constrain *directing* the phenotypic change recorded by fossils:

> Only part of these differences can be explained by the differences in selection pressures to which the organisms are exposed; the remainder are due to the developmental and evolutionary limitation set by the organism's genotype and its epigenetic system. Earlier authors (for example, Haecker, 1925) had a far greater interest in the phenotypic potential bestowed on an evolutionary line by its epigenetic heritage than recent evolutionists. This must be kept in mind when we speak of the randomness of mutations. *Mutations are random with respect to the environmental constellation. However, the epigenotype sets severe limits to the phenotypic expression of such mutations; it restricts the phenotypic potential.*[111]

This very original use of the concept of canalization allowed Mayr to explain certain parallel evolutions, showing the same pattern, in the forefront of which was the evolution of the "grade" of mammals (Figure 9.1):

> A well-knit system of canalization tends to narrow down evolutionary potential quite severely. It accounts for parallel evolution which, for instance, induced several separate lines of mammallike reptiles to cross the borderline to the mammals independently. Which gene will mutate and at what time, is "random," but the subsequent fate of such a mutation is strongly controlled by the gene complex in which it occurs. The

Known Occurrence

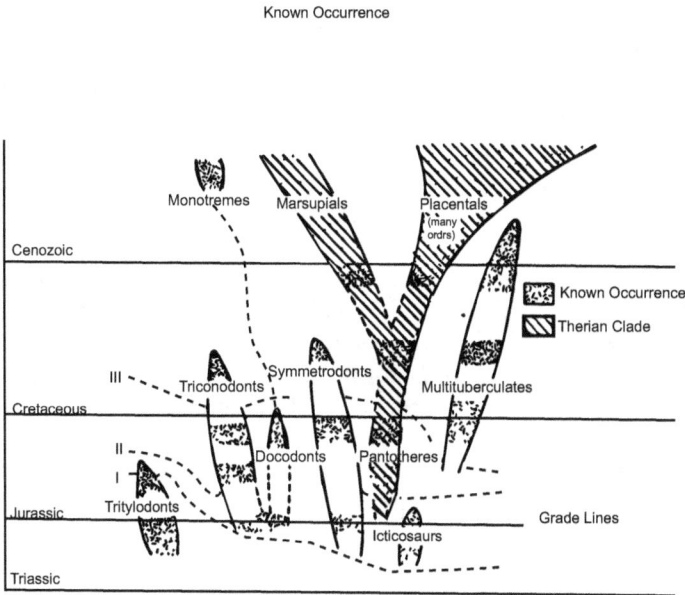

*Figure 9.1 Mayr's diagrammatic representation of the "Repeated and independent acquisition of mammalian characters (grades I, II, and III) by various lines of mammallike reptiles (therapsids)". This macroevolutionary pattern was explained by the long-term consequences of canalization. Reproduced from: E. Mayr, Animal Species and Evolution, Cambridge, The Belknap Press of Harvard University Press, © 1963 by the President and Fellows of Harvard College. Used by permission. All rights reserved.*

direction of evolution is therefore, not random, and yet it is not predictable either.[112]

Thus, of all the founders of the Modern Synthesis, Mayr was probably by far the one who did the most to integrate Waddington's concepts into the neo-Darwinian theoretical framework. Should we see this as an importation of developmental logic into a theory that we know made little room for embryological phenomena? The answer to this question is overall negative. If Mayr was interested in Waddington's developmental take, as shown earlier (e.g., in point *(1)*), he did not wish to only import Waddington's concepts (canalization, genetic assimilation, stabilizing selection) into the Modern Synthesis. Instead, he *transformed* them quite radically to his own ends, in this case, his population-based perspective.

In the previous section, we saw that, overall, Mayr preferred Lerner to Waddington precisely for this reason. In *Animal Species and Evolution*, Mayr went further in this direction: he "populationalized" the Waddingtonian categories, giving them a meaning that did not exist in Waddington's usual accounts. Thus, the concept of canalization, which was at the core of Waddington's entire theoretical enterprise, was in Mayr's case reused, not on the scale of the developmental system but of the gene pool. In a nutshell, in Waddington's account, canalization was mostly a property of an individual genome; in Mayr's, it became a property – and perhaps even the most important property – of a particular gene pool. Mayr came to define the biological species by the canalization system that characterized it, exactly as though he were attributing a specific epigenetic landscape to the *species*:[113]

> What, then, makes a species different from an intraspecific variant? [. . .] We might say that it is the total system of developmental inter-actions, the totality of feedbacks and canalizations, which makes a species. Two individuals of *Drosophila melanogaster* that differ in five conspicuous mutations affecting eye color, pigmentation, wing shape, bristle structure, and haltere formation may look strikingly differ-ent from each other, yet they still share their "modifiers," their total developmental system, and are thus still *Drosophila melanogaster*. Two wild-type individuals of *D. melanogaster* and *D. simulans*, which are hardly distinct visibly, nevertheless differ from each other by hundreds if not thousands, of genes and are the possessors of totally different developmental systems. *The important point is that different species are different systems of gene interaction or different epigenetic systems.*[114]

In such a conception, speciation required overcoming this "well-buffered sys-tem of developmental canalizations [plural]" of the "epigenotype".[115] Mayr came to understand speciation as a break in the "well-balanced epigenotype" of the gene pool, through a genetic revolution:

> Morphological uniformity and evolutionary stagnation are often ascribed to a "depletion" of genetic variability. It is far more probable that a too well-balanced epigenotype is responsible for such phenotypic and genotypic stability. Indeed, it appears to me that to store genetic variability is perhaps not nearly as great an evolutionary problem as to escape the strait jacket of a too well-balanced genotype. This is the problem that Goldschmidt sensed quite properly, even though his solu-tion ("systemic mutations") was unrealistic. Since the new characters are not produced by mutations (as the typologists thought) but by a reorganization of the genotype, it may require a "genetic revolution" (Chapter 17) to break up the perfectly buffered genotype.[116]

Thus, even though Mayr made a significant, or at least regular, use of Wadding-tonian concepts, he subverted them for his own integrally population-based and anti-typological perspective of evolutionary phenomena. Mayr was explicit about the fact that the level of the individual, that is, that of the developmental system, was far too ephemeral to have any real evolutionary impact. For him, evolution played out at the population level, the unit of evolution (but not of selection) par excellence:

> [T]he individual is only a temporary vessel, holding a small portion of the gene pool for a short time. It may, through mutation, contribute one or two new genes. It may, if it has a particularly viable and productive combination of genes, somewhat increase the frequency of certain genes in the gene pool, yet in sum its contribution will be very small indeed compared to the total contents of the gene pool. It is the entire effective population that is the temporary incarnation and visible manifestation of the gene pool. It is in the population that the genes interact in numerous combinations, genotypes. Here is the proving ground of new genes and of novel gene combinations. The continued interaction of the genes in a gene pool provides a degree of integration that permits the population to act as a major unit of evolution.[117]

Hence Mayr's lack of interest in the Baldwin effect, which in his understanding referred to an outdated typological conception of the mutation-selection couple,[118] hence also his reluctance to consider plasticity as an evolutionary factor as such, except, remarkably when it came to behavioral plasticity.[119] For the remainder, plasticity was understood as a product of adaptive evolution, that is, as a proximate mechanism controlled by natural selection, and not as an ultimate cause of evolutionary change per se.[120]

## 9.4 Bradshaw on phenotypic plasticity, a Waddingtonian perspective tailormade for the Modern Synthesis

In 1965, British botanist Anthony ("Tony") D. Bradshaw (1926–2008) published a long and dense review on phenotypic plasticity, entitled "Evolutionary significance of phenotypic plasticity in plants".[121] This text, published in *Advances in Genetics*, considered a landmark and widely quoted since,[122] is thought to have introduced the concept and the term "phenotypic plasticity" in the framework of the Modern Synthesis.[123] It was with reference to this article that Carl Schlichting, Samuel M. Scheiner, and Mary-Jane West-Eberhard, among others, positioned themselves and developed their own ideas about plasticity and evolution. In 2014, Scheiner deplored the fact that Bradshaw did not even refer to Baldwin and blamed Simpson for this lack of interest.[124] Such a judgment is problematic: while there is no doubt that Simpson's paper on the Baldwin effect was not intended to convince evolutionary biologists they should consider this mechanism as important, the conceptual content of Bradshaw's review was in fact far removed from the logic of organic selection. On the other hand, and this has gone mostly unnoticed, it was first by making use of Waddington's categories that Bradshaw brought plasticity into the theoretical framework of the Synthesis. But Bradshaw reflected upon plasticity as a botanist, and this transformed Waddington's hypotheses to a significant extent. The objective of this section is therefore twofold. First, I will try to understand how Bradshaw, starting from botany, used Waddington's concepts. This will allow me, in the second place, to better understand why this review became a reference and had such an impact on the late 20th-century evolutionary theory.

In a recent article,[125] Eric Peirson reconstructed in an extremely precise and documented manner the experimental and institutional context that allowed Bradshaw to write his famous review in the summer of 1964.[126] From this very rigorous work, I retain the following essential points:

(a)  As early as the late 1950s, Bradshaw had embarked on a series of experiments on plasticity in plants, by taking advantage of the phenotypic flexibility of the genus *Agrostis*.[127] While most of this work was not published,[128] this shows that the 1965 review was not just an abstract assessment but was linked to a solid experimental project that was ultimately abandoned.

(b)  From 1968 onward, only three years after his seminal paper, Bradshaw turned away from the question of plasticity.[129] His appointment as Professor of Botany at the University of Liverpool appears to have marked the end of his involvement in this field.[130]

(c)  Peirson is to some extent aware that Waddington is unavoidable in the context of this history, but he struggles to characterize Waddington's significance and his account is therefore, in my reading, substantially incomplete. It is a simplification to reduce Waddington's holistic conception of canalization to a mere obstacle (i.e., a negative factor) that had to be overcome to think, as Bradshaw did, of the differential plasticity of independent traits.[131]

Bradshaw in fact relied on Waddington (and to a lesser extent Schmalhausen) in such a massive and obvious way in his conception of phenotypic plasticity that it is surprising that neither biologists nor historians of science appear to have noticed this. In fact, Bradshaw almost entirely adopted Waddington's theoretical framework with two differences, to which we will return: he substituted the concept of plasticity for that of canalization, and he atomized the biological organism by considering its phenotypic traits as independent entities from the point of view of development.

Like Waddington, Bradshaw began with a critique of the vocabulary of homeostasis when used outside the realm of physiology.[132] For him, developmental homeostasis did not offer a sound description of the studied phenomena, and the term "phenotypic plasticity" was to be preferred. His entire effort was devoted to adaptive plasticity, in other words, the plasticity that allows a better adjustment of phenotypic traits to the requirements of the environment. From this point of view, plants constituted a privileged material of study, since, lacking mobility, they could not adapt through behavioral plasticity, their adaptability therefore consisted essentially of their morphological plasticity.[133] The thesis that underpinned Bradshaw's whole argument, was that adaptive plasticity was under genetic control, exactly in the same way that, as Waddington had been able to show, canalization, in animals, was also under genetic control. Bradshaw was indeed full of praise for the experimental work carried out in Edinburgh:

> *The influence that selection may have on the stability or plasticity of a character is perhaps most elegantly demonstrated in the investigations on genetic assimilation by Waddington and others (Waddington, 1961).* In *Drosophila* the ease of production of venation phenocopies by high temperature shock was radically altered by about a dozen generations of selection for ease of production (Waddington, 1953; Bateman, 1959). Similarly the ease of production of bithorax by ether treatment was radically altered by selection. In a character not involving a threshold effect Waddington (1959) was able to improve, by selection, the capacity of a strain of *Drosophila* to react to increased salt content in the medium by the development of large papillae. This is a particularly significant experiment since in this case the changes appear to be of adaptive significance. The altered strains show increased survival in high salt media. In all these experiments the environmentally induced effect eventually became assimilated and determined genetically. *Nevertheless, the significance of the results to the present consideration is that the plasticity of the genotype in one specific respect and direction could be radically altered by selection in a few generations.*[134]

Like Waddington and using the same vocabulary ("exogenous adaptation", "pseudo-exogenous adaptation", "endogenous adaptation"[135]), Bradshaw envisaged the progressive evolution from a plastic accommodation to a genetically fixed adaptation. For him, genetic assimilation also consisted in the progressive internalization of the *cause* at the origin of the phenotypic trait: from a true environmental cause, when the intensity of the plastic response is proportional to the variation of the parameter, to a simple exogenous signal or cue, and then to an endogenous signal when the phenotype is automatically set up during ontogeny. There was thus, for Bradshaw also, developmental homology between the plastic response and its genetic equivalent.[136] While Waddington and Schmalhausen used such reasoning to account for the evolution of calluses in different animal species, as we have seen, Bradshaw used it in the case of the timing of seed germination in plants.[137]

The theoretical framework depicted in the epigenetic landscape diagram allowed Bradshaw to account for this progressive genetic assimilation. Bradshaw then adapted Waddington's metaphor to the case of plants. Like Waddington and Schmalhausen, he insisted on the fact that the reactivity of a

developmental system, its capacity to show a certain form of adaptive plasticity, fundamentally depended on the "pattern of the epigenetic landscape".[138]
Like them, he considered that this pattern could be altered by natural selection, which explained the phenomenon of genetic assimilation:

> At the one extreme is the character which shows a continuous range of
> modification dependent on the intensity of the environmental stimulus.
> This has been termed "dependent morphogenesis" by Schmalhausen
> (1949). The pathway of development can be considered to be a broad
> flat delta, so that the individual or character can follow any number of
> different paths to the final adult state. Such characters as height and
> seed number follow this sort of pathway. In these characters adjust-
> ment can be continuous. This is shown by the remarkable manner in
> which many species can adjust seed output per plant to density so that
> the seed output per unit area is constant over a wide range of densities
> (Harper, 1961).
>
> At the other extreme is the character which shows a series of discrete
> modifications, often only two, with no intermediates. In this case the
> pathway can be envisaged as two divergent steep-sided valleys with
> the environmental stimulus merely acting as switch. This type of plas-
> ticity has been termed "autoregulatory dependent morphogenesis" by
> Schmalhausen (1949). He justifiably argues that it is due to the occur-
> rence of stabilizing selection in each phase of the regularly varying envi-
> ronment. Such a character as leaf form in *Ranunculus peltatus* follows
> this pattern of development. The canalization of the two types of leaf
> shape is considerable; there are no intermediates.[139]

These passages give an idea of the fundamentally Waddingtonian nature of
Bradshaw's reflections on phenotypic plasticity, so much so that the concept
of plasticity in Bradshaw's account occupied the same theoretical space as
that of canalization in Waddington's theory of genetic assimilation. It is, for
example, because of developmental plasticity that the same phenotype could,
according to Bradshaw, have multiple different genetic bases.[140]

There are, however, at least two significant differences between the way in
which these concepts were put to work in Waddington's and Schmalhausen's
synthesis and the way in which Bradshaw reused them. On the one hand,
because he was dealing with plants and not *Drosophila*, Bradshaw was more
sensitive to the relative independence of phenotypic traits, which may each
have their own degree of plasticity (canalization) as a function of a specific
environmental parameter, both within an organism but also when comparing
the same phenotypic trait in closely related species. Bradshaw *atomized* the
organism into a sum of traits, each of which had its own degree of plasticity
controlled by natural selection.[141]

The second difference, which converges with the first, is that Bradshaw
considered plasticity almost exclusively as a *product* of evolution and was
barely interested in the possibility that plasticity might play a *causal role*

in adaptive evolution. We have already seen that Waddington and Schmalhausen focused on the developmental consequences of the process of genetic assimilation, to the detriment of the evolutionary reasons that might drive such a process. In their framework, and this is evident in Waddington's experimental designs, the increase in fitness was taken for granted. Bradshaw radicalized this perspective and viewed plasticity as a selected adaptation whose degree of intensity was optimized over the course of evolutionary history. He restricted plasticity to the domain of proximate mechanisms, explicitly adhering to the conception developed by Mayr in *Animal Species and Evolution*.[142]

These two specificities of Bradshaw's use of Waddingtonian concepts largely explain why his review was so well received. At a time when the Modern Synthesis was turning toward adaptationism, such a conception of phenotypic plasticity could immediately be integrated into the standard theoretical corpus and this is exactly what happened in the 1980s when, in Bradshaw's wake, the question of the evolution of plasticity was taken up again. This understanding of the links between plasticity and evolution was still the one put forward, a few years ago, by most evolutionary geneticists when this question resurfaced during the debate about the possibility and the merits of an Extended Synthesis. Most of these scientists are convinced that Mayr's proximate-ultimate distinction remains central in biology[143] and that plasticity is best conceived as a proximate mechanism designed by natural selection to optimize the calibration of phenotypes.

## 9.5   Chapter's conclusion

The reception of Waddington's and Schmalhausen's ideas within the framework of the Modern Synthesis was never going to be simple, and this is confirmed in my historical survey. But despite the complexity of the conceptual shifts that took place in the period 1950–1970 between the works of Dobzhansky, Mayr, Waddington, and Schmalhausen, it is possible to summarize this reception as follows:

(1)   Neither Dobzhansky nor Mayr was interested in the mechanism of genetic assimilation as formulated by Waddington. For reasons similar to those developed by Stern, Bateman, or even Williams in 1966, they saw in this mechanism nothing more than the normal play of natural selection, and therefore it seemed counterproductive to adopt Waddington's many neologisms.

(2)   What interested them more, especially Mayr, was the concept of canalization, which put the interactions between genes at the forefront. Dobzhansky made moderate use of it, while Mayr went remarkably far in his appropriation, making it one of the explanatory keys to certain macroevolutionary patterns, thereby adopting the most subversive meaning of the concept (canalization as an emergent property capable of orienting the phenotypic effect of mutations).

(3)  Throughout this history, both Dobzhansky and Mayr continued to prefer Lerner's theory of genetic homeostasis to Waddington and Schmalhausen's developmental perspective. The former offered a population understanding of the notion of developmental stability, and the plasticity of a population seemed to Dobzhansky and Mayr a more powerful factor than the plasticity of individual organisms alone. Mayr, again, went very far in this direction, since in *Animal Species and Evolution* he radically transformed the concept of canalization, to the point of making it an attribute of a population and even the main criterion for defining a species. The biological species then became a set of genomes sharing the same epigenetic landscape. In the process Waddington was, in a way, "Lernerized", in that his ideas were made operational within the theoretical framework that Mayr had been building since the end of the 1930s, and within which the question of speciation (and not that of adaptation) was played the leading role.

Bradshaw's foray into this history was far from showing the same complexity, and in this case it probably makes sense to speak of an extremely significant *influence* of Waddington's conceptions on his way of thinking about phenotypic plasticity. Bradshaw borrowed from Waddington the idea that plasticity was strongly genetically determined, and thus controlled by natural selection, that it could give rise to a process of genetic assimilation (though this aspect did not interest him much), and that all these phenomena could be explained with the metaphor of the epigenetic landscape. He differed from Waddington in his more analytical conception of the organism, which he broke down into phenotypic traits, the degree of plasticity of each trait being thought to be independent of the degree of plasticity of the other traits. Bradshaw's adaptationist conception, which focused essentially on plasticity as a selected adaptation, could directly be integrated into the Modern Synthesis and was very much in line with the view advocated by Mayr at the same time, as Bradshaw himself acknowledged. It is therefore not surprising that when this question was resurrected in the 1980s, it was in the form presented in Bradshaw's 1965 review. That this integration was so easy also indicates that Schmalhausen's and Waddington's conceptions were probably not as controversial as the latter liked to repeat throughout his career.

## Notes

1  In *The Meaning of Evolution*, Simpson's first list included 18 names, to which he later added Schmalhausen's (Simpson, 1949b, p. 278). This retrospective mapping exercise was at the origin of the series of colloquiums that Mayr organized and that led in 1980 to the publication of the landmark collective work *The Evolutionary Synthesis, Perspective on the Unification of Biology* (Mayr & Provine, 1980).

2  Adams (1988). In the third edition of *Genetics and the Origin of Species* (1951), Dobzhansky included Schmalhausen among the founders of the Modern Synthesis. According to him, together with Rensch they "generalized the facts of

comparative morphology and comparative and experimental embryology, and integrated them with genetics" (Dobzhansky, 1951, Preface to the Third Edition, p. ix).

3 "Evolutionary Significance of Phenotypic Plasticity in Plants" (Bradshaw, 1965).

4 Notably because of the 1994 collective volume edited by Mark B. Adams, *The Evolution of Theodosius Dobzhansky* (Adams, 1994). On the importance of Dobzhansky's Russian background, see especially the chapters by Alexandrov (Alexandrov, 1994), Konashev (Konashev, 1994) and Krementsov (Krementsov, 1994).

5 Adams (1968).

6 Because of the foundational dimension of this text in the history of the genetic theory of evolution, it was translated into English in 1961 in the *Proceedings of the American Philosophical Society* at the instigation of Lerner (and most probably with Dobzhansky's full support), who edited the text and accompanied it with a biographical note on Chetverikov.

7 For example, Dobzhansky (1950, p. 403, 1959, p. 252).

8 Chetverikov (1961, pp. 189–191).

9 Adams (1968, p. 38).

10 On the context of his invitation to give this set of lectures, see Provine (1994).

11 "Considerations of space have forced us to refrain from a detailed discussion of some of the objections that have been advanced against the genetic treatment of evolutionary problems. Thus, Lamarckian doctrines find but a brief mention. The treatment had to be made assertive rather than polemic, dogmatic rather than apologetic" (Dobzhansky, 1937, p. xi).

12 Dobzhansky (1937, pp. 169–170).

13 Dobzhansky (1937, p. 170).

14 Dobzhansky (1951, p. 24).

15 Dobzhansky (1951, pp. 106–107).

16 Dobzhansky (1951, pp. 154–155).

17 Gilbert (1994).

18 The letters between Waddington and Dobzhansky on this subject were reproduced by Waddington in *The Evolution of an Evolutionist* (Waddington, 1975a, pp. 96–98).

19 "Concerning Schmalhausen, let us also remember that he is one of the victims of Lysenko, and any support which we may give him, especially while he is still alive, is a good deed". Dobzhansky, August 15, 1959, reproduced in Waddington, 1975a, p. 98.

20 For example, Dobzhansky (1950, p. 403). In 1959, he wrote: "The work of Chetverikov was published in Russian and remained little known outside Russia. However, this is not sufficient reason to ignore his contribution, as some authorities are doing" (Dobzhansky, 1959, p. 252).

21 This can be seen in Schmalhausen's choice of titles. For example, the title of chapter 2 is "Dynamics of the Historic Variability of Populations". See also Gilbert (1994, p. 153).

22 Dobzhansky (1970, pp. 95–96, 211).

23 Dobzhansky (1970, pp. 365–367).

24 Grodwohl (2017a, pp. 581–582).

25 See, for instance, his 1950 "Mendelian Populations and Their Evolution" (Dobzhansky, 1950).

26 Lamm (2015).

27 Dobzhansky (1955, p. 14).

28 This emphasis on the population level, however, never led Dobzhansky to seriously consider that there could be a form of natural selection *between* Mendelian

populations (Dobzhansky, 1955, pp. 12–13). To put it briefly, population was a unit of evolution but not a unit of selection.

29  Dobzhansky (1955, p. 10).

30  Dobzhansky (1970, p. 208).

31  Dobzhansky (1959, p. 259).

32  Hall (2005, p. 189).

33  Lerner (1954, p. 6).

34  Allen (1983, pp. 95–96).

35  Sophia Dobzhansky Coe, Dobzhansky's daughter, recalled the rather dramatic circumstances that led her parents to travel to Vancouver in 1931 to try to obtain papers to avoid being deported to the USSR. It was in this difficult context that they met Lerner there, who was to remain a close friend of the Dobzhanskys (Dobzhansky Coe, 1994, p. 20).

36  Hall (2005, pp. 187–188).

37  Lerner was then on a research stay in Italy, at the Institute of Genetics at the University of Pavia (Hall, 2005, p. 187).

38  The issue of the neglect of Lerner's ideas from the 1980s onward is briefly addressed by Hall (Hall, 2005, pp. 193–195).

39  Lerner (1954, p. 2). This idea was quite close to the idea of "genetic inertia" proposed shortly before by Darlington and Mather (1949). However, Lerner found the term "inertia" to be a poor choice because it does not refer to anything like auto-regulation (Lerner, 1954, p. 2).

40  Lerner (1954, p. 83).

41  Lerner (1954, pp. 4–5).

42  Lerner (1954, p. 6). Emphasis in the original.

43  Lerner (1954, p. 44).

44  Lerner (1954, p. 46).

45  Dobzhansky & Wallace (1953, p. 169).

46  Dobzhansky & Wallace (1953, p. 170).

47  Dobzhansky & Wallace (1953, p. 170).

48  Lerner (1954, p. 49).

49  At the end of their 1953 article, Dobzhansky and Wallace thank Lerner for the stimulating discussions they were able to have with him about genetic homeostasis (Dobzhansky & Wallace, 1953, p. 171). Lerner and Wallace exchanged on various occasions from 1952 onwards about the results of ongoing experiments on the four species of *Drosophila*. In a letter dated November 26, 1952, Lerner told Wallace how necessary it was for his theory for such results of genetic homeostasis to be more than just speculation: "The data you have published, particularly in the PNAS, and those mentioned in your letters to Carl, Dempster and me, are all very vital to bolster up the highly speculative writing I intend to undertake in Italy. In fact they are basic to some aspects of my theorizing, and I shall feel very inhibited in not being able to refer to them. I wonder if it will be possible for you to send me advance copies of any manuscripts you will prepare in the next six months, which you feel you can permit me to cite? (. . .) I hope that something of this sort will be possible, because my speculations would be rather empty without good evidence from actual experiments" (Lerner to Wallace, November 26, 1952, Lerner's archives).

50  Lerner (1954, p. 66).

51  Allard (1996, p. 171).

52  Waddington (1955).

53  Waddington (1957a, pp. 42–47; 1961, p. 278).

54  Waddington (1957a, p. 54).

55 Lerner (1959, p. 178).
56 The precise title of the symposium was "Population Genetics: The Nature and Causes of Genetic Variability in Populations". The fact that Lerner was invited to give the closing lecture at this prestigious event shows how much his work, and, in particular, his 1954 book, had earned him the esteem of his contemporaries.
57 Lerner (1955, p. 335).
58 Historian of biology Jonathan Hodge, whose doctoral dissertation Mayr co-supervised, told me that he had heard Mayr "denouncing Waddington's practice of writing his papers on the train between London and Edinburgh" (personal communication, email, 23 July, 2022).
59 De Winter (1997); Crow (2009); Rao & Nanjundiah (2011); Peterson (2016, pp. 189–199).
60 Like Wright, Dobzhansky, Wallace, and Lerner.
61 Mayr (1959, p. 2).
62 De Winter (1997, p. 158). Emphasis in the original.
63 At this time, as he admitted decades later, Mayr did not have a first-hand knowledge of Fisher and Wright (Provine, 2004, p. 1043).
64 Haldane to some extent performed true beanbag genetics, as he himself acknowledged (De Winter, 1997, pp. 154–155).
65 De Winter made this point very clear: "In fairness to Mayr, it should be noted that for reasons of mathematical convenience but probably also because of a primary interest in problems of adaptation rather than diversification, most implementations of the genic approach were restricted to conditions of additive fitness and avoided complications caused by epistatic interactions" (De Winter, 1997, p. 161).
66 In a remarkable synthesis, Grodwohl reconstructed the many theoretical debates internal to population genetics in the period 1935–1980. On the challenge for Lewontin in the 1950s to take epistasis into account, see Grodwohl (2017a, pp. 581–584) and also Grodwohl (2017b).
67 On the importance of this empirical work in the history of population genetics, see Grodwohl (2017b).
68 Grodwohl (2017b, p. 763).
69 Grodwohl (2017a, pp. 594–595).
70 Mayr (1975).
71 Mayr (1959, pp. 7–8).
72 Mayr (1959, p. 2).
73 "Waddington's epigenetics did not sweep the field after the publication of *The Strategy of the Genes*. It did make an impression on some significant individuals, including one of the foremost neo-Darwinians, Ernst Mayr (1904–2005), who had just put his hand to the plow writing the history and philosophy of evolutionary biology. *Influenced* by Waddington's work, Mayr delivered one of the most influential warnings against the gene-centric mechanistic thinking that he saw spreading into evolutionary theory" (Peterson, 2016, p. 189, my emphasis).
74 Milam (2010).
75 De Winter (1997, p. 150).
76 Mayr (1942, p. 68). My emphasis.
77 Mayr (1942, p. 69). On these occasions, Mayr referred positively to Goldschmidt's own ideas.
78 Mayr (1959, p. 8).
79 Mayr (1959, p. 5, p. 9).
80 Adams (1968, p. 38).
81 Provine (2004, p. 1041).

82  Provine (2004, p. 1045).

83  Provine (2004, pp. 1043–1044).

84  Mayr (1954).

85  Mayr (1954, pp. 161–163).

86  Mayr (1954, p. 165). My emphasis.

87  At the end of his 1954 text, he writes (Mayr, 1954, p. 178): "Much of the above discussion is frankly speculative. It was prepared in order to call attention to previously neglected aspects of the evolutionary process. May it stimulate further research in this field".

88  In 1955, despite not being himself a population geneticist, he participated in the Cold Spring Harbor Symposium devoted to the theme "Population Genetics: The Nature and Causes of Genetic Variability in Populations". The title of his talk was "Integration of genotypes: Synthesis" (Mayr, 1955).

89  See especially Plutynski (2019).

90  Futuyma (1998, p. 690).

91  Mayr considers very directly the macroevolutionary consequences of his model, in terms extremely close to those of Eldredge and Gould (Mayr, 1954, pp. 175–177).

92  As late as 2002 (June 17), in a letter to his friend molecular biologist François Jacob, he wrote: "However, suppression of priority happens also within the United States! Whole sentences of Gould's original version of his punctuated equilibria theory are almost identical with my 1954 theory of rapid speciation in peripheral isolates", Jacob archives (JAC.I.10), Pasteur Institute, Paris.

93  Eldredge & Gould (1972, p. 114). In *The Structure of Evolutionary Theory*, Gould fully acknowledged the importance of Lerner's and Mayr's ideas in the setting up of their model: "In our original paper on punctuated equilibrium, Eldredge and I (1972), basing our arguments partly on Mayr's (1954, 1963) concept of genetic revolutions in speciation of peripherally isolated populations but *more* on Lerner's notions (1954) of ontogenetic or developmental but *especially of genetic, "homeostasis"*, proposed such constraint as the primary reasons for stasis" (Gould, 2002, p. 879, my emphasis).

94  In the preface, Mayr notes that the nearly 800-page book was written between 1949 and 1961 (Mayr, 1963, p. vii).

95  As he told Waddington, Mayr made a considerable effort to synthesize the scientific literature, which explains the unusually long time spent writing the book: "My evolution book is nearing completion. It was an immense labor because I have been attempting (alas, not too successfully) to synthetize the modern literature. The mere checking of all the references has been a major task, as you will appreciate when you see the completed job" (Letter to Waddington, January 29, 1962, Mayr archives, HUGFP 74.7 Box 9, Folder 798, Harvard University Archives, Pusey Library).

96  An abridged version appeared in 1970: *Populations, Species, and Evolution* (Mayr, 1970).

97  Gould stated that this book "served as the closest analog to a "bible" for graduate students of his generation" (Gould, 2002, p. 797).

98  Gould (1983).

99  He distinguished an "early Modern Synthesis" from a "recent Modern Synthesis" on the basis of a number of criteria, including constraints on genetic mutations (Mayr, 1963, p. 2).

100  Mayr (1963, p. 190).

101  Mayr (1963, p. 612). See also p. 190.

102 The first letter, dated March 17, 1951, is an invitation from Waddington to come and give a lecture in Edinburgh (Mayr archives, HUGFP 14.7 Box 9, Folder 397, Harvard University Archives, Pusey Library).

103 Mayr, letter to Waddington, July 15, 1975 (Mayr archives, HUGFP 74.7 Box 24, Folder 1243, Harvard University Archives, Pusey Library).

104 Mayr (1963, for example, p. 8, p. 170, p. 178, pp. 184–185, p. 282).

105 Mayr (1963, p. 6).

106 See especially Mayr (1963, pp. 189–190).

107 Mayr (1963, for example, p. 58).

108 Mayr (1963, p. 266).

109 Mayr (1963, p. 268).

110 Mayr (1963, p. 221). My emphasis.

111 Mayr (1963, p. 608). My emphasis.

112 Mayr (1963, pp. 281–282).

113 See also Mayr (1963, pp. 526–527).

114 Mayr (1963, p. 544). My emphasis.

115 Mayr (1963, p. 614).

116 Mayr (1963, p. 615).

117 Mayr (1963, p. 137).

118 Mayr (1963, p. 611).

119 These are perhaps the only cases where Mayr accepted that the Baldwin effect could have played a role: Mayr (1963, p. 604).

120 Mayr (1963, p. 139).

121 Bradshaw (1965).

122 In 2010, Fitter reported 1500 citations (Fitter, 2010, p. 31). In 2022, according to Google Scholar, the number of 3900 citations was exceeded.

123 Nicoglou (2015).

124 Scheiner (2014, p. iii).

125 Peirson (2015b). This article stems from a doctoral work also defended in 2015 (*Evolution under our Feet. Anthony David Bradshaw (1926–2008) and the Rise of Ecological Genetics*, PhD, May 2015, Arizona State University).

126 On a ship, on his way back from the USA (Fitter, 2010, p. 31).

127 Peirson (2015b, pp. 55–57).

128 Peirson does not really explain why Bradshaw did not publish his results. On the basis of the existing archives, he shows how Bradshaw carried out his experiments and gives some of the results he obtained (unfortunately, many results seem to be missing from the preserved archives).

129 In January 2006, Bradshaw was invited to give the introductive address at the 14th New Phytologist Symposium on "Plant ecological development", which was published as a *Letter*, his last article. This text, entitled "Unravelling Phenotypic Plasticity – Why Should We Bother?", shows no difference or novelty with respect to the content of his 1965 review. Bradshaw does not even seem to have kept up with recent work in this area, and essentially still holds to the same idea: plasticity is an evolved property, an adaptation, whose genetic basis is under the control of natural selection (Bradshaw, 2006, p. 645).

130 Peirson (2015b, p. 63).

131 Peirson (2015b, p. 62).

132 Bradshaw (1965, p. 117).

133 Bradshaw (1965, p. 125).

134 Bradshaw (1965, pp. 123–124). My emphasis.

135 Bradshaw (1965, pp. 139–140).

136 Bradshaw (1965, pp. 139–140).

137  Bradshaw (1965, p. 144).
138  Bradshaw (1965, p. 140).
139  Bradshaw (1965, p. 140).
140  Bradshaw (1965, p. 137).
141  Bradshaw (1965, pp. 145–146).
142  Bradshaw (1965, p. 148).
143  See, for example, Dickins & Barton (2013).

# Conclusion of Part III

This part focused on the reception of organic selection, genetic assimilation, and canalization within the Modern Synthesis during the period 1940–1970. "Reception" is to some extent a term of convenience, which does not adequately account for the fact that there were reciprocal and sometimes significant conceptual exchanges and displacements between the different theoretical forms involved. For example, it was from Mather and quantitative genetics that Thorpe borrowed the genetic mechanism that allowed him to reflect on how phenotypic plasticity could initiate sympatric speciation. Mayr, against all odds, gave increasing prominence to the possibility that behavioral plasticity could generate new selective pressures and thus drive trans-specific evolution. Therefore, there was no such thing as an absolute distinction between, on the one hand, proponents of a view in which plasticity could play a causal role in evolution and, on the other, the orthodoxy of the Modern Synthesis. However, there was no co-construction either, because it is possible to retrospectively identify positions that remain different without being disjointed.

There appears to be a glaring asymmetry in the way evolutionary theory seized upon organic selection and the concepts built by Waddington and Schmalhausen. As far as organic selection is concerned, Simpson's conceptual treatment of it in 1953 had considerable consequences. The Baldwin effect, as he defined it, set aside the question of the polarization of selection pressures, speciation, and gene flow and even obscured to some extent the extinction issue. The conceptual impoverishment was thus considerable, and, as an inevitable consequence, the question of the causal impact of phenotypic plasticity on natural selection (i.e., plasticity as an evolutionary factor) was set aside for several decades. It was not that organic selection could not be integrated into the theoretical framework of the Modern Synthesis. But organic selection was ignored because of the redefinition Simpson produced. It is possible to consider the question of extinction (OS1) by working on populations whose size is understood as a finite parameter, even if very few models have so far been developed in this direction. From a theoretical and formal point of view, it is also feasible to explore within population genetics

DOI: 10.4324/9781003422990-18

the idea that plasticity plays an active role in the orientation of selection pressures (OS2). This is what has been happening since the end of the 1980s with the rise of the Niche Construction Theory. Thereby, if organic selection was not further integrated into the Modern Synthesis, it was not because of a theoretical incompatibility but because of a lack of opportunity. Simpson's 1953 paper, in a way, constituted the missed opportunity for this meeting.

As far as genetic assimilation is concerned, the history is very different. On the one hand, GA1 was not only integrated into the MS but literally dissolved in it. As Stern proposed from the outset, GA1 could be conceived in terms of threshold selection. Most of Waddington's interpretations were along these lines, and it is not insignificant that it was in his own laboratory, because of the work produced by Bateman, that a model involving threshold selection was fully developed. This model was taken up by Falconer in his influential textbook on quantitative genetics, and it was soon to become the preferred explanation of the phenomenon of genetic assimilation by most of the biologists interested in these questions. Mayr, in particular, did not hesitate to support it and criticized the Waddingtonian statements for making unreasonable use of neologisms that contributed nothing to the understanding of the causality involved. From one end to the other, was this not the standard mode of action of natural selection operating on a set of intra-population variants? However, it must be emphasized, as Bateman did, that GA1, even in the form of threshold selection, could not dispense with the concept of canalization. At a minimum, canalization was what allowed the storage of cryptic variation whose release, due to a major environmental change, initiated the selection process itself. Thus, even when genetic assimilation is understood as a form of threshold selection, its authentically Waddingtonian dimension cannot be escaped.

Overall, the quantitative genetics of the 1950s and 1960s were receptive to the concepts developed by Waddington and Schmalhausen, as we have seen. It was the concepts of canalization and stabilizing selection (in the sense of canalizing selection, i.e., supposed to produce developmental canalization) in particular, rather than that of genetic assimilation as such, that were the object of theoretical and experimental treatments, which were sometimes punctual (Falconer and Robertson), sometimes more developed (Rendel). For nearly two decades, the possible scope of these concepts was tested in different configurations. The results were, on the whole, mixed. The specificity of Waddington's hypotheses was finally replaced by more directly operative orientations: normalizing rather than stabilizing selection, developmental homeostasis rather than canalization. During this period, one had to constantly arbitrate between the stimulating dimension of Waddington's speculations but also their vagueness and the difficulty that there was (and still is) to be able to derive rigorous and testable hypotheses from them, within the framework of evolutionary genetics. After that, there was no rejection or integration but rather a form of lassitude, and during the 1970s evolutionary geneticists lost interest in these issues. Even Rendel, the most faithful of all the Waddingtonians (and there were very few of them, including in Edinburgh), failed to secure the posterity of his genetic models of canalization.

As for what was probably the most radical Waddingtonian concept, GA2, the history is quite simple: since Waddington himself had not well highlighted it within his own theoretical framework, this understanding of canalization and genetic assimilation was almost totally ignored, both by population geneticists and by the founders of the Modern Synthesis. Under these conditions, there could be neither integration nor rejection. Remarkably, as far as I can tell, only Mayr understood what it was all about. In *Animal Species and Evolution*, he used it (strictly verbally and qualitatively) in a few places to account for certain identifiable evolutionary trends at the macroevolutionary scale. The degree of radicality of this concept thus remains an open question. To what extent is canalization, understood as the property of a developmental system to *direct* the phenotypic effect of mutations, a realistic hypothesis, on the one hand, and on the other, one that is likely to challenge the random mutation/selection couple? This question takes us far beyond the scope of the present work.

# General conclusion

The history of theories that give a causal role to adaptive plasticity in evolution, within a Darwinian framework, is surprisingly rich. On the fringes of dominant conceptions, there have been biologists, and often leading biologists, who have been interested in the way adaptive plasticity could impact evolutionary dynamics. These attempts took many forms, and I have tried to show that four main concepts organize this history, two variants of organic selection (OS1 and OS2) and two meanings of genetic assimilation (GA1 and GA2). These concepts draw a double perspective, still relevant for today's science. If my reconstruction is valid, then organic selection and genetic assimilation are neither two radically different concepts nor two different formulations of the same mechanism but rather *two explanations that do not focus on the same aspect of natural reality*. What organic selection seeks to explain is the selective causality that accounts for the genetic fixation of plastic variation ("Why?"). Whereas genetic assimilation focuses on the developmental causality that accounts for the transformation of the epigenetic landscape that leads to the progressive canalization of plastic variation ("How?"). The problem of the evolutionary significance of plasticity can, indeed, be posed both from the point of view of ecology and from the point of view of embryology, and these two paths have been trodden many times in the history of biology.

The ecological point of view focuses primarily on the relations between plastic variation and natural selection. It is this question that was at the basis of the different forms of organic selection. Plastic variation played a causal role in two ways: either by allowing the (temporary) survival of a population subjected to a new environment (OS1) or by promoting a new selective regime (OS2). In the first case, plasticity plays a reactive role of accommodation in response to extrinsic variations. In the second case, plasticity plays more of a proactive role by producing a new selective environment. In both cases, a decisive question was to understand why the phenotypic response was ultimately constitutively stabilized. In other words, why was the plasticity of the considered trait finally lost? As we have seen, Baldwin asked this question from the start. The answer progressively elaborated during the first

DOI: 10.4324/9781003422990-19

half of the 20th century consisted of thinking that consecutive genetic mutations allowed the attainment of an adaptive peak beyond the reach of plastic
variation. It was therefore the increase in fitness that explained why organic
selection could lead to the genesis of a genetically fixed adaptation. It must
be emphasized here that this problematic issue was considerably simplified
when the concept of the "cost of plasticity" began to be taken into consideration, most probably in the late 1970s.

Thus, organic selection covers a family of related concepts that focus on the
causal relationships between plastic variation and natural selection. Organic
selection states that plasticity alters competitive interactions and that this
transitional phase ends when a fitness peak is reached through the accumulation of genetic variation. In this respect, the perspective promoted by organic
selection was blind to the developmental aspects of phenotype production.
What is more, the difference in fitness between what plastic accommodation
allows and the final stabilized genetic state seemed to require no structural
homology between the developmental paths that lead to one and the other.
This point had already been clearly identified within the Russian school in
the 1940s and by Gause in particular. In a nutshell, history teaches us that
organic selection focuses on natural selection (ecology) and puts aside the
developmental dimension (embryology) of the process.

The exact opposite perspective was at the core of Waddington's and
Schmalhausen's speculations on genetic assimilation. From the embryological point of view, they tried to understand how certain (difficult to define)
selection pressures could reorganize the epigenetic landscape to make constitutive a phenotype initially conditioned by certain environmental variations. They somehow evaded the question of fitness and took for granted
that there was a form of selection, stabilizing selection (Schmalhausen) or
canalizing selection (Waddington), capable of effecting such a reshaping of
the developmental phase-space. This being the case, the concept of canalization (and no longer that of fitness) became the theoretical key to account
for this reconfiguration. Most likely due to Lysenkoism, Schmalhausen was
not able to pursue this developmental perspective, whereas Waddington had
the opportunity, within the very stimulating context of the Edinburgh genetics department, to develop several series of experiments on these questions.
As the 1950s progressed, based on this series of experiments involving the
response to selection of populations of *Drosophila*, he developed two very
intertwined conceptions of genetic assimilation associated with two meanings of the concept of canalization.

In GA1, the role of canalization was to *mask* the phenotypic effect of
the mutations, which nevertheless retained a relatively clearly defined effect.
The environmental stress then only served to *reveal* this effect, which initiated the selection process. Since, in this model, Waddington believed that the
same genes were responsible for both plastic response and genetically fixed
adaptation, he thought that their accumulation should allow the latter to be

achieved. It was thus a form of threshold selection which still required the concept of canalization for the explanation to be complete. In a sense GA1 can be seen as a form of experimental design that allows the recruitment of certain genes whose effect is simply too weak under normal conditions to give rise to selection. This aspect was already sometimes understood and emphasized in the early 1960s, notably by Milkman.[1] If this is the case, then it is difficult to see how genetic assimilation could be the starting point of a renewed version of the Modern Synthesis.

From this point of view, GA2 is much more heterodox and subversive but also not as well defined by Waddington and extremely speculative. This second mechanism was based on a different understanding of canalization, which was no longer a mere filter which could mask the phenotypic effect of mutations but a positive constraint which could give genetic variation its phenotypic expression. In this perspective, phenotypic traits were above all the consequence of the dynamics of the developmental system which was no longer reducible to the alleles contained in the genome. This version of the mechanism of genetic assimilation was only sketched by Waddington, in the space of a few paragraphs spread over a few dozen publications. He did not test it in the work involving *Drosophila* populations. It was not discussed by the quantitative geneticists of the time. But this was the view Waddington emphasized when he wanted to oppose the logic of the Modern Synthesis in a confrontational manner. Whether this conception can find a scientific breath in present-day biology is of course an open question, which goes far beyond what this historical investigation can reasonably determine. But let us imagine that phenotypic traits are not fundamentally determined by specific alleles but by the topography of the epigenetic landscape which, because of canalization, is decoupled (to some extent) from the alleles carried by a genome. The limitation of such a conception, as is often the case, comes from the difficulty of imagining how developmental canalization itself (i.e., the topography of the epigenetic landscape) could be inherited independently from the genes that underlie it. One would have to uncover another "channel of inheritance", which is still an extremely speculative hypothesis.

### Back to the Extended Evolutionary Synthesis

In the past 20 years or so, the question of the possible causal role of phenotypic plasticity in the evolution of species has once again become central.[2] Against the supposed geno-centrism of the Modern Synthesis, scientists and philosophers are reflecting on how to properly consider the issue of the agency of organisms.[3] Phenotypic plasticity (as an evolutionary factor) and niche construction are presented as original contributions that traditional theory has never seriously taken into account. The present book shows that such a judgment needs to be revised. Twice in the history of evolutionary theory, in the period 1894–1905 and again in the period 1935–1967, phenotypic plasticity and organisms' agency were seriously considered as possible evolutionary factors. History shows that, in both these cases, the hypotheses under consideration ultimately failed to significantly integrate into the corpus of Darwinian theory.

However, the reasons for this non-integration seem to have been varied. Around 1900, organic selection, in the form elaborated by Baldwin, Lloyd Morgan, and Gulick, was a strictly verbal speculation, with no empirical basis, expressed in an often extremely obscure pre-Mendelian language. At this point in history, it was thereby for incidental and accidental reasons that this type of hypothesis was quickly eclipsed. In other words, things could have been quite different, and we could imagine an alternative history in which Gulick's stimulating considerations gave rise, directly, and probably much earlier, to something close to Niche Construction Theory.

The situation is quite different for the second moment of this discontinuous history, which corresponds to the consolidation phase of the Modern Synthesis. At this point in the history of evolutionary theory, the proposals of Gause, Thorpe, Schmalhausen, and Waddington were taken seriously and discussed within the framework of quantitative genetics and then by some of the main founders of the Modern Synthesis, such as Huxley, Simpson, Dobzhansky, and Mayr. What emerged sometimes amounted to a mutilation of certain concepts, most notably when Simpson set aside almost all of the ecological content of organic selection when he developed his version of the Baldwin effect. Here again, the historical thread was broken. But this second debate also brought about, for theoretical reasons, a gradual sidelining of developmental plasticity as an evolutionary factor. To most of the evolutionary biologists of the time, the adaptability of a population, that is, of a gene pool, seemed to be a far more significant dimension with far greater causal power. In this regard, the Modern Synthesis was perfectly in line with Fishers's fundamental theorem of natural selection, which shows that the ability of a population to respond (adaptively) to natural selection depends on its additive genetic variance in fitness at that moment.[4] One of the most fundamental principles of neo-Darwinism therefore quantified the adaptability of a population in genetic terms and gave this adaptability the leading role. This explains, in my view, why Lerner's populational theory of genetic homeostasis was preferred to Waddington's more developmental theory of genetic assimilation.

If evolution is still measured by the frequency of allelic variants in a population, then not much has changed since. The Darwinian scheme remains the regulating principle that organizes biological knowledge. If this is the case, then the term "extended" in the label "Extended Synthesis" might be ill-chosen. In the past 30 years or so, we have not so much witnessed an extension of the theory of evolution but, rather, with the major framework still in place, a deepening of its biological content: How do genes reproduce? How do organisms interact with their surrounding conditions? How does the environment condition developmental cycles? And so on. These questions enrich biological knowledge and especially evolutionary ecology. On the other hand, it is not certain that these research programs, however fruitful they may be, have the capacity to alter the main lines of the Darwinian framework that was established in the middle of the 20th century. If genes remain the main atoms of biological heredity, developmental plasticity will be a central issue for ecological theory, and for the design of more realistic models but probably not for evolutionary theory. Like other phenotypic characteristics, plasticity is a dimension of fitness, as, somewhat ironically, Baldwin himself envisioned as early as 1902:

> It may be said, indeed, quite truly, that this value of accommodation is implicit in the theory of natural selection; for, according to that theory, there is continued selection of certain fit individuals, and their *fitness may consist in their being plastic* or "accommodating".[5]

## Notes

1 "We can now see a mechanism of *genetic recruitment*, a *means of speeding evolution*, once it has begun. Once a developmental process loses some of its buffering, genes with previously negligible effects on this process join the team and speed the change. In this way natural selection capitalizes on existing genetic variation, providing grist for its own mill" (Milkman, 1961, p. 36, my emphasis).
2 Odling-Smee et al. (1996); West-Eberhard (2003); Pigliucci & Müller (2010).
3 For example, Walsh (2015).
4 Fisher (1930). For a more recent treatment of the theorem, see Edwards (1994).
5 Baldwin (1902, p. 46). My emphasis.

# References

## Unpublished sources

*François Jacob (1920–2013) archives*, Pasteur Institute, Archives department (CeRIS) Paris, France
  Letter, Mayr to Jacob, June 17, 2002 (JAC.I.10).
*I. Michael Lerner (1910–1977) archives*, Mss.B.L563, American Philosophical Society, Philadelphia, US
  Letter, Lerner to Bruce Wallace, November 26, 1952
  Letter, Mather to Lerner, October 6, 1955
*Ernst Mayr (1904–2005) archives*, Harvard University Archives, Pusey Library – Harvard Yard, US
  Letter, Waddington to Mayr, March 17, 1951(HUGFP 14.7 Box 9, Folder 397).
  Letter, Mayr to Waddington, January 29, 1962 (HUGFP 74.7 Box 9, Folder 798).
  Letter, Mayr to Waddington, July 15, 1975 (HUGFP 74.7 Box 24, Folder 1243).
*Conrad Hal Waddington (1905–1975)* archives, Centre for research collection, Edinburgh University Library, Scotland
  C.H. Waddington – Publications 1927–1978, 19 pages (EUA IN1/ACU/A/5/4)


## Published sources

Adams M.B., 1968, "The founding of population genetics: Contributions of the Chetverikov school 1924–1934", *Journal of the History of Biology*, 1, pp. 23–39.
Adams M.B., 1970, "Towards a synthesis: Population concepts in Russian evolutionary thought, 1925–1935", *Journal of the History of Biology*, 3, pp. 107–129.
Adams M.B., 1980a, "Severtsov and Schmalhausen: Russian morphology and the evolutionary synthesis", in E. Mayr, W.B. Provine (eds.), *The Evolutionary Synthesis, Perspectives on the Unification of Biology*, Cambridge and London, Harvard University Press, pp. 193–225.
Adams M.B., 1980b, "Sergei Chetverikov, the Kol'tsov Institute, and the evolutionary synthesis", in E. Mayr, W.B. Provine (eds.), *The Evolutionary Synthesis, Perspectives on the Unification of Biology*, Cambridge and London, Harvard University Press, pp. 242–278.
Adams M.B., 1988, "A missing link in the evolutionary synthesis", *Isis*, 79, pp. 281–284.
Adams M.B. (ed.), 1994, *The Evolution of Theodosius Dobzhansky*, Princeton, Princeton University Press.
Alexandrov D.A., 1994, "Filipchenko and Dobzhansky: Issues in evolutionary genetics in the 1920s", in M.B. Adams (ed.), *The Evolution of Theodosius Dobzhansky*, Princeton, Princeton University Press, pp. 49–62.

Allard R.W., 1996, "Israel Michael Lerner, 1910–1977", in *National Academy of Sciences, Biographical Memoir*, Washington, National Academic Press, pp. 167–175.

Allen G.E., 1974, "Opposition to the Mendelian-chromosome theory: The physiological and developmental genetics of Richard Goldschmidt", *Journal of the History of Biology*, 7, pp. 49–92.

Allen G.E., 1983, "The several faces of Darwin: Materialism in nineteenth and twentieth century evolutionary theory", in D.S. Bendall (ed.), *Evolution from Molecules to Men*, Cambridge, Cambridge University Press, pp. 81–102.

Ancel L., 1999, "A quantitative model of the Simpson-Baldwin effect", *Journal of Theoretical Biology*, 196, pp. 197–209.

Ancel L., 2000, "Undermining the Baldwin expediting effect: Does phenotypic plasticity accelerate evolution?", *Theoretical Population Biology*, 58, pp. 307–319.

Avital E., Jablonka E., 2000, *Animal Traditions: Behavioural Inheritance in Evolution*, Cambridge, Cambridge University Press.

Bachelard G., 1938, *La Formation de l'esprit scientifique, Contribution à une psychanalyse de la connaissance objective*, Paris, Vrin.

Baedke J., 2013, "The epigenetic landscape in the course of time: Conrad Hal Waddington's methodological impact on the life sciences", *Studies in History and Philosophy of Biological and Biomedical Sciences*, 44, pp. 756–773.

Baldwin J.M., 1896a, "Heredity and instinct", *Science*, 3, pp. 438–441, 558–561.

Baldwin J.M., 1896b, "A new factor in evolution", *The American Naturalist*, 30, pp. 441–451, 536–553.

Baldwin J.M., 1896c, "On criticisms of organic selection", *Science*, 4, pp. 724–727.

Baldwin J.M., 1897a, "Organic selection", *Science*, 5, pp. 634–636.

Baldwin J.M., 1897b, "Determinate variation and organic selection", *Science*, 6, pp. 770–773.

Baldwin J.M., 1897c, *Le développement mental chez l'enfant et dans la race*, Paris, Alcan.

Baldwin J.M., 1902, *Development and Evolution*, New York, The Macmillan Company.

Bard J.B.L., 2008, "Waddington's legacy to developmental and theoretical biology", *Biological Theory*, 3, pp. 188–197.

Baron A.B., 2001, "The life and death of Hopkins' host selection principle", *Journal of Insect Behavior*, 14, pp. 725–737.

Bateman K.G., 1956, *Studies on genetic assimilation*, PhD Dissertation, University of Edinburgh.

Bateman K.G., 1959a, "The genetic assimilation of the dumpy phenocopy", *Journal of Genetics*, 56, pp. 341–352.

Bateman K.G., 1959b, "The genetic assimilation of four venation phenocopies", *Journal of Genetics*, 56, pp. 443–474.

Bateson G., 1963, "The role of somatic change in evolution", *Evolution*, 17, pp. 529–539.

Bateson P., 2004, "The active role of behaviour in evolution", *Biology and Philosophy*, 19, pp. 283–298.

Beatty J., 2008, "Chance variation and evolutionary contingency: Darwin, Simpson, *The Simpsons*, and Gould", in M. Ruse (ed.), *Oxford Handbook of the Philosophy of Biology*, Oxford, Oxford University Press, pp. 189–210.

Beatty J., 2016, "The creativity of natural selection? Part I: Darwin, Darwinism, and the Mutationists", *Journal of the History of Biology*, 49, pp. 659–684.

Beatty J., 2019, "The creativity of natural selection? Part II: The synthesis and since", *Journal of the History of Biology*, 52, pp. 705–731.

Begg M., 1952, "Selection of the genetic basis for an acquired character", *Nature*, 169, p. 625.

Bennour M., Vonèche J., 2009, "The historical context of Piaget's ideas", in U. Müller, J.I.M. Carpendale, L. Smith (eds.), *The Cambridge Companion to Piaget*, Cambridge, Cambridge University Press, pp. 45–63.

Bergson H., 1907, *L'Evolution créatrice*, Paris, Presses Universitaires de France.

Boesiger E., 1980, "Evolutionary biology in France at the time of the evolutionary synthesis", in E. Mayr, W.B. Provine (eds.), *The Evolutionary Synthesis, Perspectives on the Unification of Biology*, Cambridge and London, Harvard University Press, pp. 309–321.

Bowler P.J., 1983, *The Eclipse of Darwinism*, Baltimore and London, The Johns Hopkins University Press.

Bowman J.C., 2005, "Douglas Scott Falconer", *Biographical Memoirs of the Fellows of the Royal Society*, 51, pp. 119–133.

Bradshaw A., 1965, "Evolutionary significance of phenotypic plasticity in plants", *Advances in Genetics*, 13, pp. 115–155.

Bradshaw A., 2006, "Unravelling phenotypic plasticity – why should we bother?", *New Phytologist*, 170, pp. 644–648.

Brazhnikova M.G., 1987, "Obituary", *The Journal of Antibiotics*, 40(7), pp. 1079–1080.

Burian R., Gayon J., 1999, "The French School of genetics: From physiological and population genetics to regulatory molecular genetics", *Annual Review of Genetics*, 33, pp. 313–349.

Burian R., Gayon J., Zallen D., 1988, "The singular fate of genetics in the history of French biology, 1900–1940", *Journal of the History of Biology*, 21, pp. 357–402.

Button C., 2018, "James Cossar Ewart and the origins of the animal breeding research Department in Edinburgh, 1895–1920", *Journal of the History of Biology*, 51, pp. 445–477.

Cain J., 2009, "Rethinking the synthesis period in evolutionary studies", *Journal of the History of Biology*, 42, pp. 621–648.

Ceccarelli D., 2019, "Between social and biological heredity: Cope and Baldwin on evolution, inheritance, and mind", *Journal of the History of Biology*, 52, pp. 161–194.

Chang H., 2012, *Is Water H$_2$O? Evidence, Realism and Pluralism, Boston Studies in the Philosophy of Science*, Dordrecht, Springer.

Chang H., 2021, "Dead or 'undead'? The curious and untidy history of Volta's concept of 'contact potential'", *Science in Context*, 34, pp. 227–247.

Chetverikov S.S., 1961, "On certain aspects of the evolutionary process from the standpoint of modern genetics", *Proceedings of the American Philosophical Society*, 105, pp. 167–195.

Crispo E., 2007, "The Baldwin effect and genetic assimilation: Revisiting two mechanisms of evolutionary change mediated by phenotypic plasticity", *Evolution*, 61, pp. 2469–2479.

Crow J.F., 2009, "Mayr, mathematics and the study of evolution", *Journal of Biology*, 8, p. 13, doi: 10.1186/jbiol117.

Cuénot L., 1911, *La genèse des espèces animales*, Paris, Alcan.

Cuénot L., 1925, *L'Adaptation*, Paris, Gaston Doin.

Cuénot L., 1932, *La genèse des espèces animales* (3rd edition), Paris, Alcan.

Cuénot L., 1941, *Invention et finalité en biologie*, Paris, Flammarion.

Cushing J.E., 1941a, "Non-genetic mating preference as a factor in evolution", *The Condor*, 43, pp. 233–236.

Cushing J.E., 1941b, "An experiment on olfactory conditioning in *Drosophila guttifera*", *Proceedings of the National Academy of Science*, 27, pp. 496–499.

Cushing J.E., 1944, "The relation of non-heritable food habits to evolution", *The Condor*, 46, pp. 265–271.

Darwin C., 1868, *The Variation of Animals and Plants Under Domestication*, Vol. II, London, John Murray.

Darwin C., 1872, *The Origin of Species* (6th edition), London, Murray.

Dawkins R., 2004, "Extended Phenotype – But Not *Too* Extended. A Reply to La-land, Turner and Jablonka", *Biology and Philosophy*, 19, pp. 377–396.

De Puytorac P., 1965, "Le Professeur Raymond Hovasse", in *Livre jubilaire pour le 70ᵉ anniversaire du Professeur Raymond Hovasse, Annales de la Faculté des sciences de l'Université de Clermont*, Clermont-Ferrand, G. de Bussac, vol. 26, pp. 5–7.

De Puytorac P., 1990, "In memoriam: Raymond Hovasse (1895–1989)", *European Journal of Protistology*, 25, pp. 391–395.

De Winter W., 1997, "The beanbag genetics controversy: Towards a synthesis of opposing views of natural selection", *Biology and Philosophy*, 12, pp. 149–184.

Deacon T., 1997, *The Symbolic Species: The Co-Evolution of Language and the Brain*, New York, Norton.

Debat V., David P., 2001, "Mapping phenotypes: Canalization, plasticity and developmental stability", *Trends in Ecology and Evolution*, 16, pp. 555–561.

Defrance L., 1902, "Facteurs de la formation des espèces", *L'année biologique*, 7, pp. 414–417

Delage Y., Goldsmith M., 1909, *Les Théorie de l'Evolution*, Paris, Flammarion.

Delage Y., Goldsmith M., 1912, *The Theories of Evolution*, London, Frank Palmer.

Delage Y., Poirault G., 1897, "L'origine des espèces", *L'année biologique*, 3, pp. 511–513.

Dennett D., 2003, "The Baldwin effect: A crane, not a skyhook", in B.H. Weber, D.J. Depew (eds.), *Evolution and Learning, The Baldwin Effect Reconsidered*, Cambridge, The MIT Press, pp. 69–79.

Depew D.J., 2003, "Baldwin and his many effects", in B.H. Weber, D.J. Depew (eds.), *Evolution and Learning, the Baldwin Effect Reconsidered*, Cambridge, The MIT Press, pp. 3–31.

Dickins T.E., Barton R.A., 2013, "Reciprocal causation and the proximate-ultimate distinction", *Biology and Philosophy*, 28, pp. 747–756.

Dietrich M.R., 1995, "Richard Goldschmidt's heresies' and the evolutionary synthesis", *Journal of the History of Biology*, 28, pp. 431–461.

Dobzhansky T., 1937, *Genetics and the Origin of Species*, New York, Columbia University Press.

Dobzhansky T., 1949 [1986], "Foreword", in I.I. Schmalhausen (ed.), *Factors of Evolution, The Theory of Stabilizing Selection*, Chicago and London, The University of Chicago Press, pp. xv–xvii.

Dobzhansky T., 1950, "Mendelian populations and their evolution", *The American Naturalist*, 84, pp. 401–418.

Dobzhansky T., 1951, *Genetics and the Origin of Species* (3rd edition), New York, Columbia University Press.

Dobzhansky T., 1955, "A review of some fundamental concepts and problems of population genetics", in *Cold Spring Harbor Symposia on Quantitative Biology, Volume XX, Population Genetics: The Nature and Causes of Genetic Variability in Populations*, New York, The Biological Laboratory, Cold Spring Harbor, pp. 1–15.

Dobzhansky T., 1959, "Variation and evolution", *Proceedings of the American Philosophical Society*, 103, pp. 252–263.

Dobzhansky T., 1970, *Genetics of the Evolutionary Process*, New York and London, Columbia University Press.

Dobzhansky T., 1980, "The birth of the genetic theory of evolution in the Soviet Union in the 1920s", in E. Mayr, W.B. Provine (eds.), *The Evolutionary Synthesis, Perspectives on the Unification of Biology*, Cambridge and London, Harvard University Press, pp. 229–242.

Dobzhansky T., Wallace B., 1953, "The genetics of homeostasis in *Drosophila*", *Proceedings of the National Academy of Science*, 39, pp. 162–171.

Dobzhansky Coe S., 1994, "Theodosius Dobzhansky: A family story", in M.B. Adams (ed.), *The Evolution of Theodosius Dobzhansky, Essays on His Life and Thought in Russia and America*, Princeton, Princeton University Press, pp. 13–28.

Dun R.B., Fraser A.S., 1958, "Selection for an invariant character – 'vibrissa number' in the house mouse", *Nature*, 4614, pp. 1018–1019.

Dun R.B., Fraser A.S., 1959, "Selection for an invariant character, 'vibrissae number', in the house mouse", *Australian Journal of Biological Sciences*, 12, pp. 506–523.

Dworkin I., 2005, "Towards a genetic architecture of cryptic genetic variation and genetic assimilation: The contribution of K.G. Bateman", *Journal of Genetics*, 84, pp. 223–226.

Edwards A.W.F., 1994, "The fundamental theorem of natural selection", *Biological Reviews*, 69(4), pp. 443–474.

Eldredge N., Gould S.-J., 1972, "Punctuated equilibria: An alternative to phyletic gradualism", in T.J.M. Schopf (ed.), *Models in Paleobiology*, San Francisco, Freeman, Cooper & Co, pp. 82–115.

Esposito M., 2016, *Romantic Biology, 1890–1945*, London, Routledge.

Ewer R.F., 1960, "Natural selection and neoteny", *Acta Biotheoretica*, 13, pp. 161–184.

Fabris F., 2018, "Waddington's processual epigenetics and the debate over cryptic variability", in D.J. Nicholson, J. Dupré (eds.), *Everything Flows, Toward a Processual Philosophy of Biology*, Oxford, Oxford University Press, pp. 246–263.

Falconer D.S., 1953, "Total sex linkage in the house mouse", *Zeitschrift für inbdukt. Abstammungs und Vererbungslehre*, 85, pp. 210–219.

Falconer D.S., 1957, "Selection for phenotypic intermediates in *Drosophila*", *Journal of Genetics*, 55, pp. 551–561.

Falconer D.S., 1960, *Introduction to Quantitative Genetics*, Edinburgh and London, Oliver and Boyd.

Falconer D.S., 1993, "Quantitative genetics in Edinburgh: 1947–1980", *Genetics*, 133, pp. 137–142.

Falconer D.S., Robertson A., 1956, "Selection for environmental variability of body size in mice", *Zeitschrift für inbduktive Abstammungs und Vererbungslehre*, 87, pp. 385–391.

Fanti L., Piacentini L., Cappucci U., Casale A.M., Pimpinelli S., 2017, "Canalization by selection of de Novo induced mutations", *Genetics*, 206, pp. 1995–2006.

Fisher R.A., 1918, "The correlation between relatives on the supposition of Mendelian inheritance", *Transactions of the Royal Society of Edinburgh*, 52, pp. 399–433.

Fisher R.A., 1930, *The Genetical Theory of Natural Selection*, Oxford, Clarendon Press.

Fitter A.H., 2010, "Anthony David Bradshaw, 1926–2008", *Biographical Memoirs of Fellows of the Royal Society*, 56, pp. 25–39.

Fogel D.B., 2002, "In Memorian, Alex S. Fraser (1923–2002)", *IEEE Transactions on Evolutionary Computation*, 6, pp. 429–430.

Franklin I., Grigg G., Mayo O., 2004, "James Meadows Rendel 1915–2001", *Historical Records of Australian Science*, 15, pp. 269–284.

Fraser A.S., 1960, "Simulation of genetic system by automatic digital computer", *Australian Journal of Biological Sciences*, 13, pp. 150–162.

Fraser A.S., Kindred B.M., 1960, "Selection for an invariant character, vibrissa number, in the house mouse. II. Limites to variability, *Australian Journal of Biological Sciences*, 13, pp. 48–58.

Futuyma D.J., 1998, *Evolutionary Biology*, Sunderland, Sinauer Associates, Inc.

Gans C., 1968, "Preface to the English edition", in I.I. Schmalhausen (ed.), *The Origin of Terrestrial Vertebrates*, New York and London, Academic Press, pp. vii–ix.

Gause G.F., 1932, "Experimental studies on the struggle for existence", *The Journal of Experimental Biology*, 9, pp. 389–402.

Gause G.F., 1934, *The Struggle for Existence*, Baltimore, The Williams & Wilkins Company.

Gause G.F., 1941, "The effect of natural selection in the acclimatization of Euplotes to different salinities of the medium", *Journal of Experimental Zoology*, 87, pp. 85–100.

Gause G.F., 1942, "The relation of adaptability to adaptation", *The Quarterly Review of Biology* 17, pp. 99–114.

Gause G.F., 1947, "Problems of evolution", *Transactions of the Connecticut Academy of Sciences*, 37, pp. 17–68.

Gause G.F., Alpatov W.W., 1945, "On the inverse relation between inherent and acquired properties of organisms", *The American Naturalist*, 79, pp. 478–480.

Gause G.F., Smaragdova N.P., Alpatov W.W., 1942, "Geographic variation in *Paramecium* and the role of stabilizing selection in the origin of geographic differences", *The American Naturalist*, 76, pp. 63–74.

Gayon J., 1989, "Génétique, psychologie génétique et épistémologie génétique dans l'œuvre de Jean Piaget (1896–1980): une ambiguïté remarquable", in C. Bénichou (ed.), *L'ordre des caractères*, Paris, Vrin, pp. 147–173.

Gayon J., 1992, *Darwin et l'après-Darwin, Une histoire de l'hypothèse de sélection naturelle*, Paris, Kimé.

Gayon J., 1995, "La préadaptation selon Lucien Cuénot (1866–1951)", *Bulletin de la Société zoologique de France*, 120, pp. 335–346.

Gayon J., 1998, *Darwinism's Struggle for Survival: Heredity and the Hypothesis of Natural Selection*, Cambridge, Cambridge University Press.

Gayon J., Huneman P., 2019, "The modern synthesis: Theoretical or institutional event?", *Journal of the History of Biology*, 52, pp. 519–535.

Gayon J., Mengal P., 1992, "Théorie de l'évolution et psychologie génétique chez Jean Piaget", in D. Andler, P. Jacob, J. Proust, F. Récanati, D. Sperber (eds.), *Epistémologie et cognition*, Mardaga, Colloque de Cerisy, pp. 41–58.

Gayon J., Veuille M., 2001, "The genetics of experimental populations: L'Héritier and Teissier's populations cages", in R. Singh, C. Krimbas, D. Paul, J. Beatty (eds.), *Thinking about Evolution: Historical, Philosophical, and Political Perspectives*, Cambridge, Cambridge University Press, pp. 77–102.

Gibson G., Hognes D.S., 1996, "Effect of polymorphism in the *Drosophila* regulatory gene *Ultrabithorax* on homeotic stability", *Science*, 271, pp. 200–203.

Gilbert S.F., 1994, "Dobzhansky, Waddington, and Schmalhausen: Embryology and the modern synthesis", in M.B. Adams (ed.), *The Evolution of Theodosius Dobzhansky*, Princeton, Princeton University Press, pp. 143–154.

Gilbert S.F., Sarkar S., 2000, "Embracing complexity: Organicism for the 21st century", *Developmental Dynamics*, 219, pp. 1–9.

Gillespie N.C., 1990, "The interface of natural theology and science in the ethology of W.H. Thorpe", *Journal of the History of Biology*, 23, pp. 1–38.

Gliboff S., 2011, "The golden age of Lamarckism, 1866–1926", in S. Gissis, E. Jablonka (eds.), *Transformations of Lamarckism, From Subtle Fluids to Molecular Biology*, Cambridge and London, The MIT Press, Vienna Series in Theoretical Biology, pp. 45–55.

Godfrey-Smith P., 2003, "Between Baldwin skepticism and Baldwin boosterism", in B.H. Weber, D.J. Depew (eds.), *Evolution and Learning, The Baldwin Effect Reconsidered*, Cambridge, The MIT Press, pp. 53–67.

Goldschmidt R., 1935, "Gen und Außeneigenschaft (Untersuchungen an Drosophila)", *Zeitschrift für inductive Abstammungs und Vererbungslehre*, 69, pp. 38–69.

Goldschmidt R., 1938, *Physiological Genetics*, New York and London, McGraw-Hill Book Company, Inc.

Gould S.J., 1983, "The hardening of the modern synthesis", in M. Grene (ed.), *Dimensions of Darwinism*, Cambridge, Cambridge University Press, pp. 71–93.

Gould S.J., 2002, *The Structure of Evolutionary Theory*, Cambridge and London, The Belknap Press of Harvard University Press.

Gray P.H., 1967, "Spalding and his influence on research in developmental behavior", *Journal of the History of Behavioral Sciences*, 3, pp. 168–179.

Griffiths P.E., 2003, "Beyond the Baldwin effect: James Mark Baldwin's "social heredity," epigenetic inheritance, and Niche construction", in B.H. Weber, D.J. Depew (eds.), *Evolution and Learning, The Baldwin Effect Reconsidered*, Cambridge, The MIT Press, pp. 193–215.

Griffiths P.E., 2006, "The Baldwin effect and genetic assimilation. Contrasting explanatory Foci and gene concepts in two approaches to an evolutionary process", in T. Simpson, S. Stich, P. Carruthers, S. Laurence (eds.), *The Innate Mind, Volume 2: Culture and Cognition*, Oxford, Oxford University Press, pp. 91–101.

Grodwohl J.-B., 2013, *Les nouveaux domains de la sélection naturelle. Hamilton, Maynard Smith, Williams*, PhD Dissertation, Université Paris 7 Denis-Diderot.

Grodwohl J.-B., 2017a, " 'The theory was beautiful indeed'. Rise, fall and circulation of maximizing methods in population genetics (1930–1980)", *Journal of the History of Biology*, 50, pp. 571–608.

Grodwohl J.-B., 2017b, "Natural selection, adaptive topographies and the problem of statistical inference: The *Moraba scurra* controversy under the microscope", *Journal of the History of Biology*, 50, pp. 753–796.

Grodwohl J.-B., 2019, "Animal behavior, population biology and the modern synthesis", *Journal of the History of Biology*, 52, pp. 597–633.

Gulick A., 1924, "John T. Gulick, a contributor to evolutionary thought", *The Scientific Monthly*, 18, pp. 83–91.

Gulick J.T., 1872, "On the variation of species as related to their geographical distribution, illustrated by the *Achatinellinae*", *Nature*, 6, pp. 222–224.

Gulick J.T., 1873, "On diversity of evolution under one set of external conditions", *Journal of the Linnean Society of London*, 11, pp. 496–505.

Gulick J.T., 1888, "Divergent evolution through cumulative segregation", *Zoological Journal of the Linnean Society*, 20, pp. 189–274.

Gulick J.T., 1891, "Intensive segregation, or divergence through independent transformation", *The Journal of the Linnean Society (London), Zoology*, 23, pp. 312–380.

Gulick J.T., 1905, *Evolution, Racial and Habitudinal*, Washington, Carnegie Institution of Washington.

Haldane J.B.S., 1954, "Introducing spalding", *Journal of Animal Behaviour*, 2, pp. 1–2.

Hall B.K., 1992, "Waddington's legacy in development and evolution", *American Zoologist*, 32, pp. 113–122.

Hall B.K., 2001, "Organic selection: Proximate environmental effects on the evolution of morphology and behaviour", *Biology and Philosophy*, 16, pp. 215–237.

Hall B.K., 2003, "Baldwin and beyond: Organic selection and genetic assimilation", in B.H. Weber, D.J. Depew (eds.), *Evolution and Learning, The Baldwin Effect Reconsidered*, Cambridge, The MIT Press, pp. 141–167.

Hall B.K., 2005, "Fifty years later: I. Michael Lerner's *Genetic Homeostasis* (1954) – a valiant attempt to integrate genes, organisms and environment", *Journal of Experimental Zoology*, 304B, pp. 187–197.

Hall B.K., 2006a, " 'Evolutionist and missionary,' the reverend John Thomas Gulick (1832–1923). Part I: Cumulative segregation – geographical isolation", *Journal of Experimental Zoology*, 306B, pp. 407–418.

Hall B.K., 2006b, "'Evolutionist and missionary,' the reverend John Thomas Gulick (1832–1923). Part II: Coincident or ontogenetic selection – the Baldwin effect", *Journal of Experimental Zoology*, 306B, pp. 489–495.

Hardy A., 1965, *The Living Stream*, London, Collins.

Headley F.W., 1900, *Problems of Evolution*, London, Duckworth.

Herbert S., 1919, *The First Principles of Evolution* (2nd edition), London, Black.

Hill W.G., 1990, "Alan Robertson, 21 February 1920–25 April 1989", *Biographical Memoirs of Fellows of the Royal Society*, 36, pp. 463–488.

Hinde R.A., 1987, "William Homan Thorpe", *Biographical Memoirs of Fellows of the Royal Society*, 33, pp. 620–639.

Hinton S.J., Nolan G.E., 1987, "How learning can guide evolution", *Complex Systems*, 1, pp. 495–502.

Ho M.W., Tucker C., Keeley D., Saunders P.T., 1983, "Effects of successive generations of ether treatment on penetrance and expression of the bithorax phenocopy in *Drosophila melanogaster*", *Journal of Experimental Zoology*, 225(3), pp. 357–368.

Hovasse R., 1941, "Adaptation et changement de milieu: préadaptation ou postadaptation?", *Bulletin biologique de la France et de la Belgique*, 75, pp. 410–420.

Hovasse R., 1943, *De l'adaptation à l'évolution par la sélection*, Paris, Hermann.

Hovasse R., 1950, *Adaptation et évolution*, Paris, Hermann.

Huxley J., 1942, *Evolution, The Modern Synthesis*, London, Allen & Unwin.

Huxley J., 1951, "Discussion", *Proceedings of the X^th International Ornithological Congress*, pp. 124–125.

Huxley J., 1963, *Evolution, The Modern Synthesis* (3rd edition), London, Allen & Unwin.

Jablonka E., Lamb M.J., 1999, *Epigenetic Inheritance and Evolution. The Lamarckian Dimension* (2nd edition), Oxford, Oxford University Press.

Jacob F., Monod J., 1961, "Regulatory mechanisms in the synthesis of proteins", *Journal of Molecular Biology*, 3, pp. 318–356.

Kirpichnikov S., 1947, "The problem of non-hereditary adaptive modifications (coincident or organic selection)", *Journal of Genetics*, 48, pp. 164–175.

Konashev M.B., 1994, "From the archives: Dobzhansky in Kiev and Leningrad", in M.B. Adams (ed.), *The Evolution of Theodosius Dobzhansky*, Princeton, Princeton University Press, pp. 63–83.

Krementsov N.L., 1994, "Dobzhansky and Russian entomology: The origin of his ideas on species and speciation", in M.B. Adams (ed.), *The Evolution of Theodosius Dobzhansky*, Princeton, Princeton University Press, pp. 31–48.

Laland K., Matthews B., Feldman M.W., 2016, "An introduction to niche construction theory", *Evolutionary Ecology*, 30, pp. 191–202.

Lamarck, 1802, *Recherches sur l'organisation des corps vivans*, Paris, Maillard.

Lamarck, 1809, *Philosophie zoologique*, Paris, Dentu.

Lamm E., 2015, "Systems thinking versus population thinking: Genotype integration and chromosomal organization 1930s–1950s", *Journal of the History of Biology*, 48, pp. 641–677.

Laplane L., 2016, *Cancer Stem Cells, Philosophy and Therapies*, Cambridge and London, Harvard University Press.

Laplane L., Solary E., 2019, "Towards a classification of stem cells", *eLIFE*, doi: 10.7554/eLife.46563.

Leamy L., Klingenberg C.P., 2005, "The genetics and evolution of fluctuating asymmetry", *Annual Review of Ecology, Evolution, and Systematics*, 36, pp. 1–21.

Lerner I.M., 1954, *Genetic Homeostasis*, Edinburgh and London, Olivier and Boyd.

Lerner I.M., 1955, "Concluding survey", in *Cold Spring Harbor Symposia on Quantitative Biology, Volume XX, Population Genetics: The Nature and Causes of*

*Genetic Variability in Populations*, New York, The Biological Laboratory, Cold Spring Harbor, pp. 334–340.

Lerner I.M., 1959, "The concept of natural selection: A centennial view", *Proceedings of the American Philosophical Society*, 103, pp. 173–182.

Lesch J.E., 1975, "The role of isolation in evolution: George J. Romanes and John T. Gulick", *Isis*, 66, pp. 483–503.

Levit G.S., Hossfeld U., Olsson L., 2006, "From the 'modern synthesis' to cybernetics: Ivan Ivanovich Schmalhausen (1884–1963) and his research program for a synthesis of evolutionary and developmental biology", *Journal of Experimental Zoology*, 306B, pp. 89–106.

Lewin P.D., 1998, *Embryology and the Evolutionary Synthesis: Waddington, Development and Genetics*, PhD Dissertation in Philosophy, University of Leeds.

Lewis D., 1983, "Extrinsic properties", *Philosophical Studies*, 44, pp. 197–200.

Lewis D., 1992, "Kenneth Mather", *Biographical Memoirs of Fellows of the Royal Society*, pp. 249–266.

Limoges C., 1976, "Natural selection, phagocytosis, and preadaptation: Lucien Cuénot, 1886–1901", *Journal of the History of Medicine and Allied Sciences*, 31, pp. 176–214.

Lindegren C.C., 1966, *The Cold War in Biology*, Ann Arbor, Planarian Press.

Loison L., 2010, *Qu'est-ce que le néolamarckisme? Les biologistes français et la théorie de l'évolution des espèces, 1870–1940*, Paris, Vuibert.

Loison L., 2011, "French roots of French neo-Lamarckisms, 1879–1985", *Journal of the History of Biology*, 44, 713–744.

Loison L., 2012, "Le projet du néolamarckisme français (1880–1910)", *Revue d'histoire des sciences*, 65(1), pp. 61–79.

Loison L., 2019, "Canalization and genetic assimilation: Reassessing the radicality of the Waddingtonian concept of inheritance of acquired characters", *Seminars in Cell & Developmental Biology*, 88, pp. 4–13.

Loison L., 2020a, "Disentangling genetic assimilation from the Baldwin effect, a philosophical perspective", *Paradigmi*, 38, pp. 441–462.

Loison L., 2020b, "De la nécessité au hasard et à la finalité. Les transformations du concept de gratuité dans l'itinéraire intellectuel de Jacques Monod", *Revue d'histoire des sciences*, 73, pp. 205–236.

Loison L., 2022, "The environment: An ambiguous concept in Waddington's biology", *Studies in History and Philosophy of Science*, 91, pp. 181–190.

Lukin E., 1936, "On the substitution of non-hereditary variations by hereditary ones from the point of view of the selection theory" [in Ukrainian], *Proceedings of the Kharkiv State University*, 6–7, pp. 199–209.

Lukin E., 1940, *Darwinism and Geographic Regularities in Variation of Organisms* [in Russian], Moscow, Academy of Sciences of USSR.

Lull R.S., 1917, *Organic Evolution*, New York, Macmillan.

Mach E., 1893, *The Science of Mechanics, A Critical and Historical Account of Its Development*, Chicago and London, The Open Court Publishing Company.

Marie J., 2004, *The importance of place: A history of genetics in 1930s Britain*, PhD Dissertation, University College London.

Marillier L., 1897, "Préface", in J.M. Baldwin (ed.), *Le développement mental chez l'enfant et dans la race*, Paris, Alcan, pp. V–XIV.

Masel J., 2004, "Genetic assimilation can occur in the absence of selection for the assimilating phenotype, suggesting a role for the canalization heuristic", *Journal of Evolutionary Biology*, 17, pp. 1106–1110.

Mather K., 1941, "Variation and selection of polygenic characters", *Journal of Genetics*, 41, pp. 159–193.

Mather K., 1942, "Genetics and the Russian controversy", *Nature*, 149, pp. 427–430.

Mather K., 1943a, "Polygenic inheritance and natural selection", *Biological Reviews*, 18, pp. 32–64.

Mather K., 1943b, "Polygenic balance in the canalization of development", *Nature*, 151, pp. 68–71.

Mather K., 1949a, *Biometrical Genetics, the Study of Continuous Variation*, London, Methuen & Co.

Mather K., 1949b, "The genetical theory of continuous variation", *Hereditas*, 35, pp. 376–401.

Mather K., 1953, "Genetical control of stability in development", *Heredity*, 7, pp. 297–336.

Mather K., 1987, "Consequences of stabilising selection for polygenic variation", *Heredity*, 58, pp. 267–277.

Mather K., 1990, "Consequences of stabilising selection for polygenic variation. II. Any number of loci", *Heredity*, 65, pp. 127–133.

Mayley G., 1997, "Guiding or hiding: Explorations into the effects of learning on the rate of evolution", in P. Husbands, I. Harvey (eds.), *Proceedings of the Fourth European Conference on Artificial Life*, Cambridge, MIT Press, pp. 156–173.

Maynard Smith J., Sondhi K.C., 1960, "The genetics of a pattern", *Genetics*, 45, pp. 1039–1050.

Mayr E., 1942, *Systematics and the Origin of Species*, New York, Columbia University Press.

Mayr E., 1947, "Ecological factors in speciation", *Evolution*, 1, pp. 263–288.

Mayr E., 1951, "Speciation in birds", *Proceedings of the X$^{th}$ International Ornithological Congress*, pp. 91–131.

Mayr E., 1954, "Change of genetic environment and evolution", in J. Huxley, A.C. Hardy, E.B. Ford (eds.), *Evolution as a Process*, London, Allen and Unwin, pp. 157–180.

Mayr E., 1955, "Integration of genotypes: Synthesis", in *Cold Spring Harbor Symposia on Quantitative Biology, Volume XX, Population Genetics: The Nature and Causes of Genetic Variability in Populations*, New York, The Biological Laboratory, Cold Spring Harbor, pp. 327–333.

Mayr E., 1958, "Behavior and systematics", in A. Roe, G.G. Simpson (eds.), *Behavior and Evolution*, New Haven, Yale University Press, pp. 341–362.

Mayr E., 1959, "Where are we?", in *Cold Spring Harbor Symposia on Quantitative Biology, Volume XXIV, Genetics and Twentieth Century Darwinism*, New York, The Biological Laboratory, Cold Spring Harbor, pp. 1–14.

Mayr E., 1961, "Cause and effect in biology", *Science*, 134, pp. 1501–1506.

Mayr E., 1963, *Animal Species and Evolution*, Cambridge, The Belknap Press of Harvard University Press.

Mayr E., 1970, *Populations, Species and Evolution, an Abridgement of Animal Species and Evolution*, Cambridge and London, Harvard University Press.

Mayr E., 1975, "The unity of the genotype", *Biologisches Zentralblatt*, 94, pp. 377–388.

Mayr E., 1982, *The Growth of Biological Thought, Diversity, Evolution and Inheritance*, Cambridge and London, The Belknap Press of Harvard University Press.

Mayr E., Provine W.B. (eds.), 1980, *The Evolutionary Synthesis, Perspectives on the Unification of Biology*, Cambridge and London, Harvard University Press.

Merlin F., 2013, *Mutations et aléas, Le hasard dans la théorie de l'évolution*, Paris, Hermann.

Mery F., Kawecki T.J., 2002, "Experimental evolution of learning ability in fruit flies", *Proceedings of the National Academy of Sciences*, 99(22), pp. 14274–14279.

Messerly J.G., 2009, "Piaget's biology", in U. Müller, J.I.M. Carpendale, L. Smith (eds.), *The Cambridge Companion to Piaget*, Cambridge, Cambridge University Press, pp. 94–109.

Milam E.L., 2010, "The equally wonderful field: Ernst Mayr and organismic biology", *Historical Studies in the Natural Sciences*, 40, pp. 279–317.

Milkman R.D., 1956, *The Crossveinless complex, a genetic system in natural populations of Drosophila melanogaster*, PhD Thesis, Harvard University.

Milkman R.D., 1960a, "The genetic basis of natural variation. I. Crossveins in *Drosophila melanogaster*", *Genetics*, 45, pp. 35–48.

Milkman R.D., 1960b, "The genetic basis of natural variation. II. Analysis of a polygenic system in *Drosophila melanogaster*", *Genetics*, 45, pp. 377–391.

Milkman R.D., 1961, "The genetic basis of natural variation. III. Developmental lability and evolutionary potential", *Genetics*, 46, pp. 25–38.

Milkman R.D., 1964, "The genetic basis of natural variation. V. Selection for crossveinless polygenes in new wild strains of *Drosophila melanogaster*", *Genetics*, 50, pp. 625–632.

Milkman R.D., 1970, "The genetic basis of natural variation. X. Recurrence of *cve* polygenes", *Genetics*, 65, pp. 289–303.

Mivart S.G., 1871, *On the Genesis of Species*, London, Macmillan.

Morgan C.L., 1896a, "On modification and variation", *Science*, 4, pp. 733–740.

Morgan C.L., 1896b, *Habit and Instinct*, London and New York, Edward Arnold.

Morgan T., 1934a, *Embryology and Genetics*, New York, Columbia University Press.

Morgan T., 1934b, "The relation of genetics to physiology and medicine", *Nobel Lecture*, www.nobelprize.org/uploads/2018/06/morgan-lecture.pdf

Müller U., Carpendale J.I.M., Smith L. (eds.), 2009, *The Cambridge Companion to Piaget*, Cambridge, Cambridge University Press.

Nagy Z., 1985, "In memoriam, Berenice Kindred, 1928–1985", *Immunogenetics*, 21, pp. 199–200.

Naumenko V.A., 1941, "Stabilization of specific mutations in the artificial selection of respective modifications", *Reports from the USSR Academy of Science*, 32, pp. 75–78 [in Russian].

Neel J.V., 1987, "Curt Stern, 1902–1981", in *Biographical Memoir, National Academy of Sciences*, Washington, pp. 443–473.

Nicholson D.J., 2014, "The return of the organism as a fundamental explanatory concept in biology", *Philosophy Compass*, 9, pp. 347–359.

Nicoglou A., 2015, "The evolution of phenotypic plasticity: Genealogy of a debate in genetics", *Studies in History and Philosophy of Biological and Biomedical Sciences*, 50, pp. 67–76.

Nicoglou A., 2018, "Waddington's epigenetics or the pictorial meetings of development and genetics", *History and Philosophy of the Life Sciences*, 40, pp. 1–25.

Odling-Smee F.J., 1988, "Niche constructing phenotypes", in H.C. Plotkin (ed.), *The Role of Behavior in Evolution*, Cambridge, MIT Press, pp. 73–132.

Odling-Smee F.J., 2010, "Niche inheritance", in M. Pigliucci, G.B. Müller (eds.), *Evolution – The Extended Synthesis*, Cambridge and London, The MIT Press, pp. 175–207.

Odling-Smee F.J., Laland K.N., Feldman M.W., 1996, "Niche construction", *The American Naturalist*, 147, pp. 641–648.

Osborn H.F., 1897a, "Organic selection", *Science*, 6, pp. 583–585.

Osborn H.F., 1897b, "The limits of organic selection", *The American Naturalist*, 31, pp. 944–951.

Palmer A.R., Strobeck C., 1986, "Fluctuating asymmetry: Measurement, analysis, patterns", *Annual Review of Ecology, Evolution, and Systematics*, 17, pp. 391–421.

Peirson B.R.E., 2015a, *Evolution under our feet. Anthony David Bradshaw (1926–2008) and the rise of ecological genetics*, PhD Dissertation, Arizona State University.

Peirson B.R.E., 2015b, "Plasticity, stability, and yield: The origins of Anthony Bradshaw's model of adaptive phenotypic plasticity", *Studies in History and Philosophy of Biological and Biomedical Sciences*, 50, pp. 51–66.

Pennisi E., 2018, "Buying Time", *Science*, 362, pp. 988–991.

Peterson E.L., 2011, "The excluded philosophy of Evo-Devo? Revisiting C.H. Waddington's failed attempt to embed Alfred North Whitehead's 'organicism' in evolutionary biology", *History and Philosophy of the Life Sciences*, 33, pp. 301–320.

Peterson E.L., 2016, *The Life Organic. The Theoretical Biology Club and the Roots of Epigenetics*, Pittsburgh, The University of Pittsburgh Press.

Petino Zappala M.A., 2024 (to be published), "A framework for the integration of development and evolution: The forgotten legacy of James Meadows Rendel", *Studies in History and Philosophy of Science*.

Piaget J., 1929a, "Les races lacustres de la *Limnaea stagnalis* L. Recherches sur les rapports de l'adaptation héréditaire avec le milieu", *Bulletin biologique de la France et de la Belgique*, 63, pp. 424–454.

Piaget J., 1929b, "L'adaptation de la *Limnaea stagnalis* aux milieux lacustres de la Suisse romande", *Revue suisse de zoologie*, 36, pp. 263–531.

Piaget J., 1965, "Notes sur des *Limnaea stagnalis* L. var. *lacustris* Stud. Elevées dans une mare du plateau vaudois", *Revue suisse de zoologie*, 72, pp. 769–787.

Piaget J., 1967, *Biologie et connaissance*, Paris, Gallimard.

Piaget J., 1974, *Adaptation vitale et psychologie de l'intelligence: sélection organique et phénocopie*, Paris, Hermann.

Piaget J., 1976, *Le comportement moteur de l'évolution*, Paris, Gallimard.

Piaget J., 1978, *Behavior and Evolution*, New York, Pantheon Books.

Pigliucci M., 2010, "Phenotypic plasticity", in M. Pigliucci, G.B. Müller (eds.), *Evolution – The Extended Synthesis*, Cambridge and London, The MIT Press, pp. 355–378.

Pigliucci M., Müller G.B. (eds.), 2010, *Evolution – The Extended Synthesis*, Cambridge and London, The MIT Press.

Pigliucci M., Murren C.J., Schlichting C.D., 2006, "Phenotypic plasticity and evolution by genetic assimilation", *Journal of Experimental Biology*, 209, pp. 2362–2367.

Pimpinelli S., Piacentini L., 2020, "Environmental change and the evolution of genomes: Transposable elements as translators of phenotypic plasticity into genotypic variability", *Functional Ecology*, 34, pp. 428–441.

Plutynski A., 2019, "Speciation post synthesis: 1960–2000", *Journal of the History of Biology*, 52, pp. 569–596.

Pocheville A., 2010, *La niche écologique: Concepts, modèles, applications*, PhD Dissertation, Ecole Normale Supérieure de Paris. https://tel.archives-ouvertes.fr/tel-00715471.

Price T.D., Qvarnström A., Irwin D.E., 2003, "The role of phenotypic plasticity in driving genetic evolution", *Proceedings of the Royal Society, London, Biological Sciences*, 270, pp. 1433–1440.

Prout T., 1962, "The effects of stabilizing selection on the time of development in *Drosophila melanogaster*", *Genetics Research*, 3, pp. 364–382.

Provine W.B., 1980, "Epilogue", in E. Mayr, W.B. Provine (eds.), *The Evolutionary Synthesis, Perspectives on the Unification of Biology*, Cambridge and London, Harvard University Press, pp. 399–411.

Provine W.B., 1992, "Progress in evolution and meaning in life", in C.K. Waters, A. van Helden (eds.), *Julian Huxley: Biologist and Statesman of Science*, Houston, Rice University Press, pp. 165–180.

Provine W.B., 1994, "The origin of Dobzhansky's *genetics and the origin of species*", in M.B. Adams (ed.), *The Evolution of Theodosius Dobzhansky*, Princeton, Princeton University Press, pp. 99–114.

Provine W.B., 2004, "Ernst Mayr: Genetics and Speciation", *Genetics*, 167, pp. 1041–1046.

Rabaud E., 1922, *L'adaptation et l'évolution*, Paris, Chiron.

Radick G., 2017, "Animal agency in the age of the modern synthesis: W.H. Thorpe's example", *British Journal for the History of Science*, 2, pp. 35–56.

Raju A., Xue B.K., Leibler S., 2023, "A theoretical perspective on Waddington's genetic assimilation experiments", *Proceedings of the National Academy of Sciences (PNAS)*, 120(51), e2309760120, doi: 10.1073/pnas.2309760120.

Rao V., Nanjundiah V., 2011, "J.B.S. Haldane, Ernst Mayr and the BeanBag genetics dispute", *Journal of the History of Biology*, 44, pp. 233–281.

Rendel J.M., 1959, "Canalization of the scute phenotype of *Drosophila*", *Evolution*, 13, pp. 425–439.

Rendel J.M., 1967, *Canalisation and Gene Control*, London, Logos Press & Academic Press.

Rendel J.M., Sheldon B.L., 1960, "Selection for canalization of the scute phenotype in *Drosophila melanogaster*", *Australian Journal of Biological Sciences*, 13, pp. 36–47.

Rice S.H., 1998, "The evolution of canalization and the breaking of von Baer's laws: Modeling the evolution of development with epistasis", *Evolution*, 52(3), pp. 647–656.

Richards R.J., 1987, *Darwin and the Emergence of Evolutionary Theories of Mind and Behavior*, Chicago and London, The University of Chicago Press.

Robertson A., 1956, "The effect of selection against extreme deviants based on deviation or on homozygosis", *Journal of Genetics*, 54, pp. 236–248.

Robertson A., 1977, "Conrad Hal Waddington", *Biographical Memoirs of Fellows of the Royal Society (London)*, 23, pp. 575–622.

Robinson B.W., Dukas R., 1999, "The influence of phenotypic modifications on evolution: The Baldwin effect and modern perspectives", *Oikos*, 85, pp. 582–589.

Roe A., Simpson G.G., 1958, "Introduction", in A. Roe, G.G. Simpson (eds.), *Behavior and Evolution*, New Haven, Yale University Press, pp. 1–3.

Rundell R.J., 2011, "Snails on an evolutionary tree: Gulick, speciation, and isolation", *American Malacological Bulletin*, 29, pp. 145–157.

Rünneburger E., Le Rouzic A., 2016, "Why and how genetic canalization evolves in gene regulatory networks", *BMC Evolutionary Biology*, 16, p. 239, doi: 10.1186/s12862-016-0801-2.

Sapp J., 1987, *Beyond the Gene. Cytoplasmic Inheritance and the Struggle for Authority in Genetics*, Oxford, Oxford University Press.

Scharloo W., 1964, "The effect of disruptive and stabilizing selection on the expression of a cubitus interruptus mutant in *Drosophila*", *Genetics*, 50, pp. 553–562.

Scharloo W., 1991, "Canalization: Genetic and developmental aspects", *Annual Review of Ecology and Systematics*, 22, pp. 65–93.

Scheiner S.M., 1993, "Genetics and evolution of phenotypic plasticity", *Annual Review of Ecology and Systematics*, 24, pp. 35–68.

Scheiner S.M., 2014, "The Baldwin effect: Neglected and misunderstood", *The American Naturalist*, 184, pp. ii–iii.

Schlichting C.D., 1986, "The evolution of phenotypic plasticity in plants", *Annual Review of Ecology and Systematics*, 17, pp. 667–693.

Schmalhausen I.I., 1941, "Stabilizing selection and its place among the factors of evolution", *Journal of General Biology*, 2, pp. 306–350 [in Russian].

Schmalhausen I.I., 1949 [1986], *Factors of Evolution, The Theory of Stabilizing Selection*, Chicago and London, The University of Chicago Press (English translation by Isadore Dordick).

Schmalhausen I.I., 1960, "Evolution and cybernetics", *Evolution*, 14, pp. 509–524.

Schmitt S., 2000, "L'œuvre de Richard Goldschmidt: Une tentative de synthèse de la génétique, de la biologie du développement et de la théorie de l'évolution autour du concept d'homéose", *Revue d'histoire des sciences*, 53, pp. 381–399.

Siegal M.L., Bergman A., 2002, "Waddington's canalization revisited: Developmental stability and evolution", *Proceedings of the National Academy of Science*, 99, pp. 10528–10532.

Simpson G.G., 1944, *Tempo and mode in Evolution*, New York, Columbia University Press.

Simpson G.G., 1949a, *"Factors of evolution, a review"*, *The Journal of Heredity*, 40, pp. 322–324.

Simpson G.G., 1949b, *The Meaning of Evolution*, New Haven, Yale University Press.

Simpson G.G., 1953, "The Baldwin effect", *Evolution*, 7, pp. 110–117.

Simpson G.G., 1958, "The study of evolution: Methods and present status of theory", in A. Roe, G.G. Simpson (eds.), *Behavior and Evolution*, New Haven, Yale University Press, pp. 7–26.

Slack J.M., 2002, "Conrad Hal Waddington: The last Renaissance biologist?", *Nature Reviews Genetics*, 3, pp. 889–895.

Smocovitis V.B., 1996, *Unifying Biology: The Evolutionary Synthesis and Evolutionary Biology*, Princeton, Princeton University Press.

Sober E., 1984, *The Nature of Selection*, Chicago, The Chicago University Press.

Spalding D.A., 1873 [1954], "Instinct, with original observations on young animals", *British Journal of Animal Behaviour*, 2, pp. 2–11.

Sterenly K., 2005, "Made by each other: Organisms and their environment", *Biology and Philosophy*, 20(1), pp. 21–36.

Stern C., 1958, "Selection for subthreshold differences and the origin of pseudoexogenous adaptations", *The American Naturalist*, 92, pp. 313–316.

Stern C., 1959, "Variation and hereditary transmission", *Proceedings of the American Philosophical Society*, 103, pp. 183–189.

Tahar M., 2022, "The history of the Begsonian interpretation of Charles Darwin's theory of evolution", *Bergsoniana*, 2, doi: 10.4000/bergsoniana.740.

Te Velde J.H., Gordens H., Scharloo W., 1988, "Genetic fixation of phenotypic response of an ultrastructural character in the anal papillae of *Drosophila melanogaster*", *Heredity*, 61, pp. 47–53.

Tebb G., Thoday J.M., 1954, "Stability in development and relational balance of X-chromosomes in *Drosophila melanogaster*", *Nature*, 174, pp. 1109–1110.

Thoday J.M., 1958, "Homeostasis in selection experiment", *Heredity*, 12, pp. 401–415.

Thoday J.M., 1959, "Effects of disruptive selection – I. Genetic flexibility", *Heredity*, 13, pp. 187–203.

Thomas F., Lefèvre T., Raymond M. (eds.), 2016, *Biologie Evolutive* (2nd edition), Louvain-la-Neuve, De Boeck.

Thorpe W.H., 1930, "Biological races in insects and allied groups", *Biological Reviews of the Cambridge Philosophical Society*, 5, pp. 177–212.

Thorpe W.H., 1938, "Further experiments on olfactory conditioning in a parasitic insect. The nature of the conditioning process", *Proceedings of the Royal Society*, B, 126, pp. 370–397.

Thorpe W.H., 1939, "Further studies on pre-imaginal olfactory conditioning in insects", *Proceedings of the Royal Society, B*, 127, pp. 424–433.

Thorpe W.H., 1940, "Ecology and the future of systematics", in J. Huxley (ed.), *The New Systematics*, Oxford, Oxford University Press, pp. 341–364.

Thorpe W.H., 1945a, "Animal learning and evolution", *Nature*, 156, p. 46.

Thorpe W.H., 1945b, "The evolutionary significance of habitat selection", *Journal of Animal Ecology*, 14, pp. 67–70.

Thorpe W.H., 1956, *Learning and Instincts in Animals*, London, Methuen and Co. Ltd.

Thorpe W.H., Jones F.G.W., 1937, "Olfactory conditioning in a parasitic insect and its relation to the problem of host selection", *Proceedings of the Royal Society, B*, 124, pp. 56–81.

Van Valen L., 1962, "A study of fluctuating asymmetry", *Evolution*, 16, pp. 125–142.

Waddington C.H., 1940a, "The genetic control of wing development in drosophila", *Journal of Genetics*, 41, pp. 75–139.

Waddington C.H., 1940b, *Organisers and Genes*, Cambridge, Cambridge University Press.

Waddington C.H., 1941, "Evolution of developmental systems", *Nature*, 147, pp. 108–110.

Waddington C.H., 1942, "Canalization of development and the inheritance of acquired characters", *Nature*, 3811, pp. 563–565.

Waddington C.H., 1943, "Polygenes and oligogenes", *Nature*, 151, p. 394.

Waddington C.H., 1952, "Selection of the genetic basis for an acquired character", *Nature*, 169, p. 278.

Waddington C.H., 1953a, "Genetic assimilation of an acquired character", *Evolution*, 7, pp. 118–126.

Waddington C.H., 1953b, "The 'Baldwin effect,' 'genetic assimilation,' and 'homeostasis'", *Evolution*, 7, pp. 386–387.

Waddington C.H., 1953c, "The evolution of adaptation", *Endeavour*, 12, pp. 134–139.

Waddington C.H., 1955, "The resistance to evolutionary change", *Nature*, 175, pp. 51–52.

Waddington C.H., 1956, "Genetic assimilation of the bithorax phenotype", *Evolution*, 10, pp. 1–13.

Waddington C.H., 1957a, *The Strategy of the Genes*, London, Allen and Unwin.

Waddington C.H., 1957b, "The genetic basis of the assimilated bithorax stock", *Journal of Genetics*, 55, pp. 241–245.

Waddington C.H., 1958a, "Inheritance of acquired characters", *Proceedings of the Linnean Society, London*, 169, pp. 54–61.

Waddington C.H., 1958b, "Theories of evolution", in S.A. Barnett (ed.), *A Century of Darwin*, London, Heinemann, pp. 1–18.

Waddington C.H., 1959a, "Evolutionary adaptation", *Perspectives in Biology and Medicine*, 2, pp. 379–401.

Waddington C.H., 1959b, "Canalization of development and genetic assimilation of acquired characters", *Nature*, 183, pp. 1654–1655.

Waddington C.H., 1960, "Experiments on canalizing selection", *Genetics Research*, 1, pp. 140–150.

Waddington C.H., 1961a, "Genetic assimilation", *Advances in Genetics*, 10, pp. 257–290.

Waddington C.H., 1961b, *The Nature of Life. The Main Problems and Trends of Thought in Modern Biology*, New York, Atheneum.

Waddington C.H., 1966a, *Principles of Development and Differentiation*, New York, The Macmillan Company.

Waddington C.H., 1966b, "Selection for developmental canalization", *Genetics Research*, 7, pp. 303–312.

Waddington C.H., 1968a, "Preface", in C.H. Waddington (ed.), *Towards a Theoretical Biology, 1. Prolegomena*, Edinburgh, Edinburgh University Press, no pagination (2 pages).

Waddington C.H., 1968b, "The basic ideas of biology", in C.H. Waddington (ed.), *Towards a Theoretical Biology, 1. Prolegomena*, Edinburgh, Edinburgh University Press, pp. 1–32.

Waddington C.H., 1968c, "Does evolution depend on random search?", in C.H. Waddington (ed.), *Towards a Theoretical Biology, 1. Prolegomena*, Edinburgh, Edinburgh University Press, pp. 111–119.

Waddington C.H., 1969a, "Sketch of the second Serbelloni symposium", in C.H. Waddington (ed.), *Towards a Theoretical Biology, 2. Sketches*, Edinburgh, Edinburgh University Press, pp. 1–9.

Waddington C.H., 1969b, "The practical consequences of metaphysical beliefs on a biologist's work. An autobiographical note", in C.H. Waddington (ed.), *Towards a Theoretical Biology, 2. Sketches*, Edinburgh, Edinburgh University Press, pp. 72–81.

Waddington C.H., 1969c, "Paradigm for an evolutionary process", in C.H. Waddington (ed.), *Towards a Theoretical Biology, 2. Sketches*, Edinburgh, Edinburgh University Press, pp. 106–128.

Waddington C.H., 1972a, "Form and information", in C.H. Waddington (ed.), *Towards a Theoretical Biology, 4. Essays*, Edinburgh, Edinburgh University Press, pp. 109–145.

Waddington C.H., 1972b, "Epilogue", in C.H. Waddington (ed.), *Towards a Theoretical Biology, 4. Essays*, Edinburgh, Edinburgh University Press, pp. 283–288.

Waddington C.H., 1974, "A catastrophe theory of evolution", *Annals of the New York Academy of Science*, 231, pp. 32–42.

Waddington C.H., 1975a, *The Evolution of an Evolutionist*, Ithaca, Cornell University Press.

Waddington C.H., 1975b, "Fifty years on", *Nature*, 258, pp. 20–21.

Waddington C.H., Robertson E., 1966, "Selection for developmental canalisation", *Genetics Research*, 7, pp. 303–312.

Waddington C.H., Woolf B., Perry M.M., 1954, "Environment selection by Drosophila mutants", *Evolution*, 8, pp. 89–96.

Wagner G.P., Booth G., Bagheri-Chaichian H., 1997, "A population genetic theory of canalization", *Evolution*, 51, pp. 329–347.

Wake D.B., 1986, "Foreword", in I.I. Schmalhausen (ed.), *Factors of Evolution, The Theory of Stabilizing Selection*, Chicago and London, The University of Chicago Press, pp. v–xii.

Walsh D., 2015, *Organisms, Agency, and Evolution*, Cambridge, Cambridge University Press.

Weber B.H., Depew D.J. (eds.), 2003, *Evolution and Learning, the Baldwin Effect Reconsidered*, Cambridge, The MIT Press.

Weinstein A., 1977, "How unknown was Mendel's paper?", *Journal of the History of Biology*, 10(2), pp. 341–364.

Weismann A., 1894, *The Effect of External Influences Upon Development, The Romanes Lecture*, Oxford, Clarendon Press.

Weismann A., 1895, *Neue Gedanken zur Vererbungsfrage*, Jena, Gustav Fischer.

West-Eberhard M.J., 1989, "Phenotypic plasticity and the origins of diversity", *Annual Review of Ecology and Systematics*, 20, pp. 249–278.

West-Eberhard M.J., 2003, *Developmental Plasticity and Evolution*, Oxford, Oxford University Press.

Whittington H.B., 1986, "George Gaylord Simpson, 16 June 1902–6 October 1984", *Biographical Memoirs of Fellows of the Royal Society*, pp. 527–539.

Wilkins A.S., 2002, *The Evolution of Developmental Pathways*, Sunderland, Sinauer.

Wilkins A.S., 2003, "Canalization and genetic assimilation", in B.K. Hall, W.M. Olson (eds.), *Keywords and Concepts in Evolutionary Developmental Biology*, Cambridge, Harvard University Press, pp. 23–30.

Wilkins A.S., 2008, "Waddington's unfinished critique of neo-Darwinian genetics: Then and now", *Biological Theory*, 3, pp. 224–232.

Williams G.C., 1966, *Adaptation and Natural Selection. A Critique of Some Current Evolutionary Thought*, Princeton, Princeton University Press.

Woltereck R., 1909, "Weitere esperimentelle Untersuchungen über Artveränderung, speziel über das Wesen quantativer Artunterschiede bei Daphniden", *Verhandlungen der deutschen zoologischen Gesellschaft*, 19, pp. 110–173.

Zakharov V.M., 1992, "Population phenogenetics: Analysis of developmental stability in natural populations", *Acta Zoologica Fennica*, 191, pp. 7–30.

Zakharov V.M., Shadrina E.G., Trofimov I.E., 2020, "Fluctuating asymmetry, developmental noise and developmental stability: Future prospects for the population developmental biology approach", *Symmetry*, 12, 1376, doi: 10.3390/sym12081376.

# Index

For Product Safety Concerns and Information please contact our EU
representative GPSR@taylorandfrancis.com
Taylor & Francis Verlag GmbH, Kaufingerstraße 24, 80331 München, Germany